Ecological Strategies of
Aquatic Insects

Ecological Strategies of Aquatic Insects

Charles W. Heckman
Retired Biologist
Formerly Research Biologist
Institute für Hydrobiology und Fischereiwissenschaft der Universität Hamburg

Max-Planck-Institut für Limnologie, Plön
Universidade Federal de Mato Grosso, Cuiabá

CRC Press
Taylor & Francis Group
Boca Raton London New York

CRC Press is an imprint of the
Taylor & Francis Group, an **informa** business

A SCIENCE PUBLISHERS BOOK

Cover illustrations: Cover photos provided by the author of the book, Charles W. Heckman.

CRC Press
Taylor & Francis Group
6000 Broken Sound Parkway NW, Suite 300
Boca Raton, FL 33487-2742

First issued in paperback 2021

Version Date: 20180402

ISBN-13: 978-0-367-78116-3 (pbk)
ISBN-13: 978-1-4987-1922-3 (hbk)

Library of Congress Cataloging-in-Publication Data
Names: Heckman, Charles W., 1941- author.
Title: Ecological strategies of aquatic insects / Charles W. Heckman, Max-Planck-Institut fèur Limnologie (retired), Plèon and Inst. fèur Hydrobiologie und Fischereiwissenschaft, Hamburg, Germany.
Description: Boca Raton, FL : CRC Press, [2018] | "A Science Publishers Book." | Includes bibliographical references and index.
Identifiers: LCCN 2018006390 | ISBN 9781498719223 (hardback)
Subjects: LCSH: Aquatic insects--Ecology.
Classification: LCC QL472 .H43 2018 | DDC 595.7/176--dc23
LC record available at https://lccn.loc.gov/2018006390

Visit the Taylor & Francis Web site at
http://www.taylorandfrancis.com

and the CRC Press Web site at
http://www.crcpress.com

*This book is dedicated
to my wife, Wai-Yuen Sylvia Heckman*

Preface

The arthropods that were once grouped within the class Insecta have recently been separated into several separate classes. Those with fully developed or vestigial wings, the vast majority of the species, are still classified in this way, while the completely wingless ones have been moved into new classes. All of these are grouped in a sub-phylum called Hexapoda. It is by far the largest taxon, both in terms of the number of species it encompasses and in the number of living organisms.

The typical hexapod lives on land or under the earth and breathes air throughout all stages of its life. Entomology, the scientific study of the group, is vastly complex, and it will provide interesting fields of study for the natural sciences for centuries to come. This book is meant to focus on the opportunities for needed research on a very small number of the relatively small minority of the hexapod species that live on, in, or under water. From among this group of aquatic insects and wetland collembolans, some of the most remarkable groups of species are selected for special consideration because they display adaptations, which could be the subjects of interesting and rewarding scientific studies.

For students of entomology and for entomologists seeking research areas in which to specialize, this book is meant to highlight aquatic species displaying some of the most unusual adaptations to extreme habitats and discuss the state of our knowledge about the morphological, physiological, biochemical, ecological, and behavioral modifications necessary to sustain them in the biological communities that they have been able to colonize. It is meant to point out the peculiarities of these aquatic species and indicate the greatest gaps in our knowledge about their ways of life.

For aspiring biologists with interest in pursuing careers in the fields of taxonomy, systematics, physiology, anatomy, ecology, zoogeography, behavioral studies, or related fields, the study of hexapods offers the best opportunity for exploring new fields of science and making new and surprising discoveries, as well as going deeper into the still poorly explained phenomena that have been observed and reported by earlier researchers.

First, as the largest group of described species by far, hexapods present the largest number of individual research subjects. Second, insects go through a series of metamorphoses, and each stage has its own distinct morphology, physiology, diet, behavior, and ecology. All of these and other features are distinctive, and there is still a great deal to learn and to find explanations for. There are undoubtedly many surprising facts to learn, because there are vast regions of the earth that can be considered *terra incognita* for entomologists.

It is hoped that this book will inspire scientists to turn their eyes and intellects toward the world of aquatic insects and begin unravelling some of the greatest remaining puzzles in the world of these many insects and collembolans.

Contents

Dedication   v

Preface   vii

1. Hexapods and Water: An Introduction to Aquatic Insects and Collembolans   1

2. General Classifications of Water Bodies Inhabited by Hexapods   28

3. Life on the Surface Tension   52

4. Herbivorous Insects on Aquatic Plants   67

5. Hexapods on the Roots and Rhizoids Beneath the Pseudoterrestrium   79

6. Active Submerged Hunters of Microorganisms and Small Prey in Lentic Water   93

7. Insects that Construct Underwater Shelters   106

8. Ambushers in Streams   119

9. Underwater Tunnel-Diggers, Aquatic Crickets, and Swimming Grasshoppers   128

10. Insects Inhabiting Rainwater   135

11. Hexapods in Mosses and Lichens   141

12. Insects of Phytotelmata   150

13. Insect Fauna Living in Insectivorous Plants   160

14. Aquatic Hexapods Active on Ice and Snow   167

15. Adults that do not Eat   179

16. Life in Water Without Oxygen   188

17. Insects on the High Seas   199

18. Marine Littoral Hexapods   208

19. Living in Hypersaline Water   222

20. Petroleum Flies   230

21. Larvae that Live in Sponges   235

22. Parasitoid Larvae in Aquatic Insects   239

23. Flies and other Insects that Eat Snails   248
24. Flies that Give Wedding Presents   253
25. Aquatic Insects that Kill People   257
26. Insects that use People for Bait   281
27. Flies in Hot Water   286
28. Ants and Water   291
29. What We Still Need to Discover   297

Literature   325
Index   337

1

Hexapods and Water
An Introduction to Aquatic Insects and Collembolans

Although most insects are not aquatic, almost every freshwater body in the world is the habitat of aquatic insects. Where conditions are not too extreme, at least for part of the year, a large variety of insects can be found living on, in, or beside and in direct contact with water. This chapter provides a cursory introduction to the colonization of typical inland water bodies by insects and serves as a prelude to the story of the successful adaptation of certain insects to specific conditions under which most other insects would quickly perish.

Most hexapods are terrestrial throughout all of their life stages, and most of these avoid contact with water. For many such species, just a short immersion in shallow water can be fatal. The location and morphology of the spiracles on the abdomens of most insects permit water to enter them easily and block the passage of gases until the insect has drowned. Wetting the wings of many insects prevents them from flying and almost always ends in death. One drop of rainwater landing on the wing as it unfolds just after certain insects have emerged from the final larval or pupal stage can prevent the insect from ever being able to fly and result in its nearly immediate death. Nevertheless, an impressive number of hexapod species can usually be found in, under, or on the surface of almost every stream, river, lake, or pond in the world, as long as it is not permanently frozen over. Bodies of freshwater, which are present throughout the year and support stable communities of aquatic and littoral plants, are usually especially rich in insects and collembolans belonging to diverse species.

It can be assumed that every hexapod inhabiting water during any life stage has morphological modifications, physiological adaptations, and behavioral patterns, which deviate from the basic body plan and way of life of terrestrial collembolans and insects. Their specific set of characteristics permit them to thrive in locations where the vast majority of other hexapods would not long survive.

This book concentrates on the ecology of those aquatic hexapods with unusual and unexpected habits. Some of these inhabit biotic communities in which few insect species of any kind could thrive. Before discussing their remarkable adaptations, however, we should look at the various ways the most familiar aquatic hexapods have adapted themselves to life in or on bodies of water without additional adverse factors for the insects to overcome than simply permanent or periodic contact with water.

The concept of the ecological niche is one that has been frequently discussed. A species can occupy a niche in a highly complex biotic community when its biological, ecological, and behavioral peculiarities are suited to take advantage of the physical, chemical, and biotic features of the biotope, utilize the available nutrient resources, and overcome or avoid adverse impacts, such as predation, extreme weather, and periods of severe cold, dryness, or pollution. In some cases, there is competition for a niche. This occurs frequently when additional species suitably adapted to fill the same niche are introduced from remote geographical locations. The chapters in this book describe common, as well as rare and unusual niches, which highly specialized aquatic insects have been able to fill. In many cases, the subtle adaptations that enable certain species to overcome factors which eliminate most other aquatic insects from the habitat remain unknown.

Three major groups of aquatic hexapods are (1) those adapted to aquatic life during all stages of development, (2) those inhabiting water as larvae and dispersing through the atmosphere as adults, and (3) those which live at the water's edge or in wetlands, entering and leaving the water for longer or shorter periods throughout their lives. The habits of the third group are often called semi-aquatic. This chapter provides a cursory survey of the general habits and adaptations common to the aquatic insects and collembolans most frequently encountered throughout the world. Subsequent chapters, however, focus on highly specialized, unusual, and remarkable species, which have adapted themselves to habitats in which few people would expect to find any hexapods at all or which display behavior that is worthy of special mention.

SECTION 1

Systematic Positions of Aquatic Hexapoda

Aquatic springtails and insects belong to a great variety of orders and families, indicating that certain species among fundamentally terrestrial groups of hexapods have been able to adapt to life in or on water again and again throughout geological history. A subphylum of Arthropoda, called Hexapoda, meaning six legs, is now being used when referring to members of the classes Insecta, Collembola, and at least three other taxa formerly classified as Apterygota, before this subclass was separated from Insecta and broken up into separate classes (Giribet et al., 2001). Several orders belonging to Hexapoda consist almost exclusively of species inhabiting water bodies or small assemblages of water during at least one of their life stages. The rest are individual species of predominantly terrestrial families, which have adapted to life in or on water for at least part of their lives.

a. The status of Collembola

For more than 200 years, springtails have been classified as members of the order Collembola, which until recently has been regarded as being within the class Insecta. Because they have neither wings nor vestiges of them, they were placed in the sub-class Apterygota along with other six legged arthropods that also lack all traces of wings. The number of known Collembola species far exceeds the numbers of species in each of the other taxa, which were formerly placed together with it in the sub-class Apterygota, namely Diplura, Protura, and Thysanura. No known species in any of these other higher taxa is aquatic, although a few live in detritus-rich soils, which always remain moist.

Where Collembola belongs within the Animal Kingdom has long been discussed, and it has been suggested that these species and other members of the Apterygota should belong to the class Crustacea rather than Insecta (Giribet et al., 2001). However, the similarities with Insecta are greater, both in quantity and importance.

New analytical results and various calculations have led to conclusions that Collembola should be placed in a group distinct from other orders long assigned to the Apterygota. To find a position for Collembola in the classification of Arthropoda, it has been suggested that springtails should be placed in a class by themselves. The springtails retain the name Collembola, but it is now being accepted as a taxon of equal rank with the class Insecta rather than an order of insects.

It has been remarked at various times that taxa higher than genera are classified artificially according to subjective criteria. Should the line separating springtails from all winged insects be above or below the level of class? This is still a matter of conjecture. Whether the present classification system will be accepted over the long term remains to be seen. Collembola species have morphological features that only this taxon possesses, and the preferred habitats of the majority are already well known. They live in biotic communities that almost always include insects, and they play important roles in many food webs.

This book therefore includes information about aquatic species of Collembola as well as aquatic insects with wings. However, strictly speaking, springtails are no longer considered to be insects by many specialists in arthropod systematics.

b. Collembola in aquatic habitats

The greatest obstacle to life in or near water for hexapods is the difficulty in respiration. Even those species that normally remain on the surface tension layer of the water can expect to be forced below the surface by sudden rain showers, wind storms, or the actions of large animals, and their survival in such cases depends upon their ability to cope with such emergencies. Hexapods associated with aquatic habitats during all life stages include a variety of species that breathe through spiracles and almost continuously remain on the surface tension layer of the water. Some of the most familiar of these include the tiny members of the Collembola, with a general common name in English: springtail. Aquatic springtails can live throughout their entire lives on the surface tension layer. Like their terrestrial relatives, some depend upon only spiracles for gas exchange with the atmosphere, although a few also possess a rudimentary tracheal system. Some are also able to exchange gases through the integument, either with the atmosphere or directly with the water (Wesenberg-Lund, 1943).

The body covering of most springtails is not heavily chitinized, and gas exchange between the ambient water and body of some species of these hexapods can apparently take place directly through it. Other species seem to have integuments that are impervious to water, allowing little or no water or gases to pass through it. The final developmental stages of few aquatic species measure much more than 5 mm in length, while most are significantly smaller. The surface tension layer of most lentic water bodies easily supports their tiny bodies. It also supports the force of the snapping furcula, also called the furca, which is the organ at the posterior end of the abdomen used to send the collembolan hurtling through the air when danger threatens. To the unaided eye of a casual observer, these tiny arthropods resemble fine bits of debris on the surface of the water, which suddenly disappear when something relatively large, like a human hand, approaches them. Collembolans are the most widespread and abundant members of what was formerly called the Apterygota, although they are no longer considered to be directly related to any of the other groups. The only common feature of all of these taxa was their lack of wings or vestiges thereof. This lack of wings was a feature that led entomologists of earlier eras to consider the members of the Apterygota to be collectively "more primitive" than insects with wings. However, most species of Collembola display unique features possessed by no insect or taxon formerly considered to be a subgroup of Apterygota, making it apparent that Collembola appeared as a separate and distinct taxon of Hexapoda very early in the geological history of the earth. The ventral tubus (Figs. 1.1 and 1.2), an organ possessed exclusively by species of Collembola, seems to be capable of giving them still unknown sensory information about the chemistry of the water upon which they live, and the furcula is a structure possessed only by some springtail species. It is capable of catapulting the tiny hexapod far away from a small, attacking predator in less time than it takes most winged insects to take off.

The furcula consists of three parts, which are easily distinguishable on the abdomens of species on which this organ is fully developed. The manubrium is the

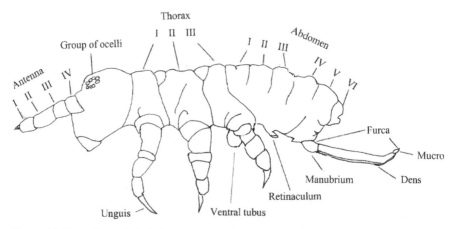

Figure 1.1 External morphological structures on the aquatic species of Collembola, Arthropleona, *Podura aquatica.* Redrawn and modified from Heckman (2001).

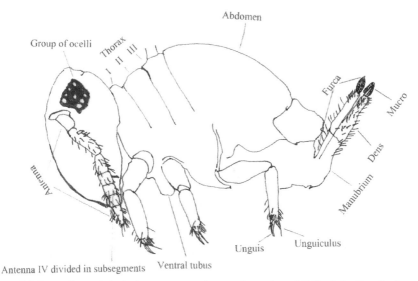

Figure 1.2 External morphological structures on the aquatic species of Collembola, Symphypleona, *Sminthurides ringueleti* Delamare-Deboutteville and Massoud, 1963. Redrawn and modified from Heckman (2001).

basal part to which a pair of middle sections, called dentes, are attached. At the end of each dens, a mucro is present. In species with fully developed furculae, the entire appendage is normally kept under great tension, but it is prevented from extending straight posteriad by a small structure called the retinaculum, which restrains the entire organ at a point usually near its apex to keep it in place ventral to the abdomen. When the retinaculum is released because the springtail perceives a danger, the furcula snaps rapidly in an arc of nearly 180°, catapulting the individual upward and away from its former position. This movement is so rapid that the collembolan seems to disappear, and it flies through the air a considerable distance, usually taking it out of the danger zone. This process accounts for the common name, springtail, for members of the class Collembola (Gisin, 1960).

Some species show various degrees of reduction of the furcula or its parts. In other species, the entire furcula is vestigial, and the springtail is unable to spring. Figures 1.1 and 1.2 illustrate the structure of the furcula of species in the sub-classes Arthropleona and Symphypleona, respectively. The furcula of some terrestrial species are vestigial, and the springtails cannot spring.

Modifications of the mucro often distinguish the aquatic species of Collembola. Those of terrestrial species are usually sharply pointed near the apex and have subapical claws or hooks. This gives the apex of each branch of the furcula a firm hold on the ground when it is released by the retinaculum, allowing the collembolan to spring many centimeters away in an attempt to escape danger. For aquatic species, a sharply pointed mucro or subapical claws would tend to puncture the surface tension layer of the water, causing the mucro to break through its substrate and not launch a jump that would take it sufficiently far from the danger. Instead, most aquatic species have broadened, paddle-shaped mucros, which make contact with a wider area of the surface tension layer. The results are easy to observe in the field and are impressive.

Much still needs to be learned about the habits of collembolans. In addition to the furcula, members of this hexapod class possess several other organs found in no group of insects. One of these is the ventral tubus or collophore, which is a short, usually cylindrical structure extending from the ventral surface of the first abdominal segment and covered by a membrane permeable to various substances dissolved in or mixed with the ambient water. The name collophore suggests that this organ glues the springtail to its substrate, but this does not truly describe what this organ does. The integuments of many collembolan species are thought to be completely impervious to water, which is repelled on contact. The membrane of the ventral tubus seems to be the only place on the body surface of most springtail species that maintains a direct contact with the water beneath it on which the animal is walking, allowing osmotic exchange of water and salts as well as a sensory detection by the individual of the water quality in the substrate on which it lives.

Although the rest of the body is impermeable to water, its surface is capable of gas exchange in various places, according to some reports. Only a few collembolans, species of the family Sminthuridae, have rudimentary elements of a tracheal system through which gas exchange can occur. Although a few species can live in very dry places, almost all of those species encountered on the surfaces of water bodies and many found among leaf litter on forest floors require a very humid microclimate, and they perish quickly from desiccation in the open air at room temperature. It would seem that there must be more areas of the integument permitting water to pass through and evaporate than are presently known. It does not seem likely that only evaporation through the ventral tubus could explain the deaths of the springtails as rapidly as it has been observed after exposure to dry air. More must be learned about the many variations in the morphology and physiology of the different species, as well as of the adults and immature stages of the individual species to understand the adaptations of springtails to their environments. One species, *Seira domestica* (Nicolet, 1842), lives in the dry air of heated houses far from any sources of water, indicating that it possesses several adaptations fundamentally different from those that permit other species to live on water bodies and marshes in direct contact with water.

With regard to the habitat requirements of aquatic and semi-aquatic hexapods, the temperature, humidity, air movements, and immediate proximity of ice and water in its liquid phase are of immediate importance. The meteorological conditions in close proximity to the body of the hexapod itself determine the survival of small individuals with relatively little protection of the body against drying or drowning. These conditions all contribute to the microclimate, which is a combination of meteorological factors in the immediate surroundings of the animal. This is often the most important consideration for the distribution of small hexapod species within large geographical areas and one explaining why aquatic and semi-aquatic species almost always must seek the proximity of a water body in order to survive.

One of the things distinguishing members of the Collembola from other hexapods is their vast areas of distribution. Some species occur on most continents. Although the number of species in this hexapod class is not nearly as great as that of several insect orders, Collembola is believed to account for a greater number of individual organisms present on the earth than any order of the class Insecta. While the systematic classification of species began in the middle of the 18th century, modern worldwide intercontinental shipping had already begun during the 15th century. At the present time, there is not enough evidence to accurately judge whether the distribution patterns of many species of springtails has existed since prehistoric times or whether it is the result of the transportation of species from one continent to another with the cargo carried by ships, in ballast water, or in cracks in the wood of the sailing vessels. Collembolans are simply too small and inconspicuous to be noticed, and prior to the advent of efforts to prevent the transportation of pathogenic microbes and their insect vectors, nobody would have cared whether these tiny creatures were expanding their ranges by stowing away on ships.

Most collembolans reportedly feed on small organic particles, including pollen and parts of fungi. Unfortunately, studies of the diets and differences therein have been undertaken on only a tiny percentage of the known species. Indications are that most or all of them feed on a wide variety of microorganisms, and the diet depends more on availability than choice.

The aquatic species of Collembola with the best adaptations to survive in their habitat has been found to be *Podura aquatica* Linnaeus, 1758. It is the only species in the family Poduridae which has been the subject of more than cursory scientific studies (Pichard, 1973). Because it occurs in northern Europe, where systematic studies of the species began, it has been studied more frequently than any of the many aquatic springtails known only from other parts of the world. It is widely distributed throughout the North Temperate climatic zone, and it can be encountered from the subtropics to subarctic water bodies and wetlands on three continents: Europe, northern Asia, and North America. Recent studies of its habits in northern Europe demonstrate that it spends the winter in large groups below the surface of the water in small water bodies or near the shore among emergent plants on larger ones. Before the water body freezes over, many individuals come together and form compact groups on the surface, which usually appear as grayish or bluish gray patches on the water. These usually move below the surface when the water starts to freeze, and they remain inactive on submerged plants until the ice thaws. Apparently, respiration in winter takes place by gas exchange directly with the water, and this is sufficient to keep the individuals alive during extended periods of inactivity. During the spring thaw, patches of these springtails appear on the surface of the water, and the individuals disperse and spread over the surface as the temperature rises. In autumn, they again form small groups in preparation for hibernation.

Not only the adults but also the eggs can spend time underwater. When the adults release their eggs, which do not shed water, they sink below the surface and remain there until hatching. The newly emerged first instars rise quickly to the surface. The water-repellent integument of the hatchlings keeps them dry during periods of rainfall as they rest on the surface tension layer (Pichard, 1973).

The various ecological niches occupied by the aquatic species of Collembola will be described in more detail in Chapter 3.

SECTION 2

Insecta

If the innovations in the systematics of arthropods are fully accepted, the class Insecta includes only the species that had been classified as species of the sub-class Pterygota in the past. All possess two pairs of wings, or vestiges thereof, as imagoes, a term used for insects in the final adult stage. One order of insects, Ephemeroptera, also possesses wings during a pre-imago stage. Shortly after the first winged instar appears, it sheds its exoskeleton again, and the second and final winged imago emerges.

No other order of insects is known to have two instars bearing wings. Several of them have only one pair of wings used for flying, with the other pair extensively modified to serve a different purpose. For example, the fore-wings, also called mesothoracic wings, of beetles are modified to serve as hard wing covers, called elytra, to protect the hind wings, which are used for flight. The hind wings, or metathoracic wings, of Diptera are greatly reduced in size and used for aerodynamic control purposes, permitting the flies to execute extremely rapid flight maneuvers to evade capture by predators. The flight is powered only by strong mesothoracic wings.

Most orders of insect and a large percentage of the extant families include at least a few aquatic or semi-aquatic species. However, only a relatively small percentage of insect families include exclusively or almost exclusively aquatic species, either as larvae or both as larvae and adults. There are very few major families of insect which include known species that develop from terrestrial larvae and then adopt an aquatic way of life as adults. One of the few is the family Dryopidae, in the order Coleoptera and sub-order Polyphaga, which includes a relatively small number of species widely distributed throughout the world. The larvae are terrestrial and develop in moist soils near streams, while the adults hide themselves in waterlogged mosses along the edges of fast flowing streams. More exceptions to this may be discovered in parts of the world in which long periods of desiccation are interspersed with periods of flooding, that is, if they exist at all. In a few cases, larvae may leave water bodies to spend a pupal stage on the shore or riverbank and then return to an aquatic or semi-aquatic habitat for mating and egg-laying.

SECTION 3

Taxonomic Overview of Insects and Springtails Considered to be Aquatic

An overview of the common aquatic hexapod species belonging to the major insect taxa is provided in Table 1.1.

Table 1.1 shows the most important higher taxa to which the majority of aquatic insect species belong. However, the species have distinct preferences for specific kinds of biotic communities in which they can find a suitable ecological niche to occupy. Assigning the major groups of aquatic insects to specific categories of aquatic habitats must be done before the role of each can be determined. Some examples of this work have led to such elementary conclusions like the one that species in the order Plecoptera must be sought in or near fast flowing streams, while most beetles in the

Table 1.1 A brief outline showing the representation of some major taxa of Hexapoda among the biotic communities in different kinds of habitat.

Class Collembola – hexapods lacking wings; no wing vestiges are present.

Collembola: tiny wingless hexapods living on the surface tension layer, along the seashore, in water-soaked mosses on the ground, in epiphytic lichens and mosses, and in bromaliads. Some species in the family Poduridae hibernate in water under ice, and their eggs hatch underwater.

Arthropleona – Elongate collembolans; aquatic species have well developed furcae, but terrestrial species sometimes have only vestiges of a furca; even rudimentary tracheae are absent.

Symphypleona – Short-bodied collembolans, sometimes almost spherical; rudimentary tracheae may be present.

Class Insecta – winged hexapods and those with a secondary loss of wings.

Hemimetabola – Insects without a distinct pupal stage.

Ephemeroptera – mayflies, typically with fully aquatic larvae and short-lived adults, some in streams, others in lentic water. They are the only insects with two instars bearing functional sets of wings.

Odonata – dragonflies.

Suborder Zygoptera – damselflies, all larvae aquatic, most in streams, ponds, and lakes; a few in bromeliads. Adults usually remain near water bodies.

Suborder Anisoptera – dragonflies, all larvae aquatic in ponds, lakes, and streams, with larvae among water plants, on the bottom, or buried in sediments. Adults long-lived, strong fliers, and often seen far from water.

Plecoptera – stoneflies – all known larvae but two are aquatic, and almost all develop in fast-flowing water of streams.

Orthoptera – Aquatic and semi-aquatic grasshoppers, crickets, and wetas; sublittoral mole crickets. Very few aquatic species are known.

Hemiptera – Heteroptera – Two major taxa and one minor group of aquatic heteropterans are well known throughout most of the world.

Nepomorpha – all but two families encompass only species that live underwater throughout all life stages. The major families in this group include Belostomatidae, Nepidae, Naucoridae, Notonectidae, Pleidae, Hydrotrephidae, Aphelocheiridae, Potamocoridae, and Corixidae. Only adults of species in the families Gelastocoridae and Ochteridae leave the water to walk or hop around on stones or gravel at the water's edge. Their movements are amphibious, and they are seldom far from the water.

Gerromorpha – Species in this group live on the surface tension layer of the water and avoid going beneath the surface. The major families in this group include Gerridae, Veliidae, Hermatobatidae, Mesoveliidae, Hebridae, Hydrometridae, Paraphrynoveliidae, and Macroveliidae. A few species in the family Gerridae can live on the surface of the open ocean.

Leptopodomorpha – Species known as shore bugs live along the shores of water bodies and water courses. Like the nepomorph species in the families Gelastocoridae and Ochteridae, they are noted for jumping. The semi-aquatic species belong to the family Saldidae.

Holometabola – these have distinct larvae, pupae, and imagoes, that is, adult stages.

Neuroptera – a heterogeneous group of ancient insects with disputed systematics.

Suborder Megaloptera – treated by some as an order and by others as a sub-order; all known species have aquatic larvae and adults that usually remain near water. The two major families are Sialidae and Corydalidae.

Suborder Plannipennia – treated by some as an order called Neuroptera by those who consider Megaloptera an independent order; few have aquatic or semi-aquatic larval stages, except for those in the families Sisyridae, which include some with larvae inhabiting freshwater sponges and possibly also colonies of ectoprocts, or in moist littoral habitats, and Osmylidae, with larvae sometimes associated with aquatic habitats.

Subgroup Rhaphidia, treated as an order or a suborder; the larvae are aquatic.

Table 1.1 contd....

... *Table 1.1 contd.*

Trichoptera – Larvae are typically aquatic and build either fixed shelters out of stones or sand grains or portable shelters from stones, sand grains, pieces of aquatic plants, or other detritus. Members of some families do not build shelters. Most adults remain near water bodies.

Lepidoptera – The great majority of species live on terrestrial plants, but the larvae of a few species of moth from several families build submerged shelters from aquatic plants, and some live underwater.

Coleoptera – Several families are fully aquatic as larvae and adults, and many individual species from overwhelmingly terrestrial families have adopted an aquatic life style during at least one of their life stages.

Suborder Adephaga – Most families are fully adapted to life in or on water during all life stages, but the family in this suborder encompassing the greatest number of species, Carabidae, includes, with few exceptions, terrestrial species. Among these few exceptions are a few which are adapted to flooded wetlands or the littoral zones of ponds, lakes, and streams. Most of them would be classified as semi-aquatic. Other families in this suborder consist almost entirely of aquatic insects. By far the largest of these is Dytiscidae. Smaller but with large numbers of species found throughout the world are Noteridae, Laccophilidae, Haliplidae, and Gyrinidae. Noteridae and Laccophilidae are sometimes classified as subfamilies of Dytiscidae, known as Noterinae and Laccophilinae, respectively.

Suborder Polyphaga – One large family and several small ones are fully adapted to aquatic habitats during all life stages. The rest of the suborder consists mainly of terrestrial species, but they include a very small percentage of aquatic and semi-aquatic ones, specialized mainly in feeding on aquatic plants, including fungi. Hydrophilidae encompasses the largest number of aquatic species of all families of Polyphaga, including mainly those best adapted to life in water during both the larval and adult stages. Small families encompassing mainly aquatic or semi-aquatic species include Dryopidae and Hydraenidae.

Diptera – This order encompasses a large number of species with aquatic larvae, but very few are aquatic as adults.

Suborder Nematocera – The adults of these insects usually have relatively thin antennae consisting of many more than three segments. All larvae in several families are aquatic, while other families include only a few species with aquatic larvae. Adults of some aquatic species live on blood and can spread serious diseases to both animals and human beings.

Brachycera – Several families are aquatic as larvae, and some of these remain near water as adults. Several species are noted for developing in strongly polluted, hypersaline, or oxygen-poor waters. Others seek the blood of vertebrates as adults. Most families include species with larvae that develop in terrestrial, aquatic, or muddy habitats. A few are predominantly aquatic as larvae, such as species of Tabanidae, which feed on the blood of various mammals as adults; as well as Sciomyzidae, most of which seem to be parasitoids that develop as larvae in aquatic snails or bivalves. Most other families include species occupying a whole spectrum of habitats, from streams and ponds to dry, terrestrial soils.

Cyclorrhapha – The species in this suborder develop into greatly skilled fliers, which are fast and maneuverable in flight. The larvae develop in a great variety of substrates, including decaying animals, dry or liquid manures, organically polluted muds, and shallow accumulations of highly polluted water.

Hymenoptera – This order includes a few poorly known families, which apparently encompass thousands of species, most of which have not been described. These species develop as parasitoids, which deposit their eggs on or in selected species of insect so that the larva can bore into the body of the host and feed on its non-vital organs until it is almost ready to pupate. The larva then feeds on the vital organs, killing the host, and sometimes its pupa attaches itself to the dead carcass of the host. Most of these aquatic hymenopteran parasitoids belong to the superfamilies Ichneumenoidea or Chalcidoidea. The systematics of this taxon has been subjected to recent revisions as more knowledge about them has been gained, and the family with the most aquatic species seems to be Braconidae, a family belonging to Ichneumenoidea.

family Hydrophilidae can only be found in the lentic or slowly moving water of lakes, ponds, canals, and water-filled ditches, especially those containing many submerged plants. In this case, the selective factor that makes the habitat suitable for the insect is whether or not the water flows, and if so, how fast.

However, most factors cannot be analyzed on the basis of a positive or negative determination. For example, larvae in the dipteran family, Simuliidae, require lotic water in which to develop. However, the individual species show considerable differences in habitat preference. Some species have been found in the relatively slow-flowing water of the Elbe Estuary, while others require the rapid flow of water in mountain streams. Therefore, as a water course flows downstream from the mountains to much slower, small lowland rivers in Europe, changes in the simuliid fauna produce continual changes in the insect communities. The presence of certain species can be taken as indications that the rate of water flow is approximately within a certain range, although this range may also be affected somewhat by the temperature and oxygen concentration of the water. Any given species that develops along the course of the stream will first appear at the point upstream at which condition first become suitable and disappear from the community again where conditions downstream become unsuitable for that species again. Theoretically, it would be replaced in the community by the larvae of other species of *Simulium* which can better tolerate water flowing more slowly.

In addition to water flow, geographical factors play a fundamental role in the distribution of the aquatic insect species. Unlike water flow, however, the adaptations to climate and weather are most frequently unrelated to the families and orders of the species present. Therefore, the dipteran family Culicidae is distributed in all kinds of lentic habitats from the equator to the Arctic tundra, but they are almost never encountered in lotic habitats anywhere. Conversely, members of the family Ceratopogonidae develop as larvae in many parts of the world, but almost always in relatively fast-flowing streams.

The climatic zones present in most parts of the world change in almost predictable ways from the equator to the poles. They also change in almost the same ways from sea level to the tops of high mountains. Thus, the biotope occurring on the sides of a high mountain from its base on the equator to its peak would change from that of a dense tropical rainforest to that of a permanent glacier, just as the biotope at sea level transitions from rainforest to glacier when moving from the equator to one of the poles. Typically, the rainforest gradually changes with elevation to resemble a temperate forest and then an evergreen forest of the taiga. The evergreen conifers cover an area reaching to a distinct tree line at a relatively high elevation, above which trees do not grow. Tundra above permafrost develops at a still higher level above sea level, and the mountain is topped by a glacier.

Moving northward at sea level from the equator to one of the poles, the same general changes are encountered. Usually, an arid zone is present in the subtropics, followed by a typical subtropical forest or grassland, depending upon local rainfall patterns. Indistinct borders separate the forests subject to seasonal temperatures above freezing and those experiencing winter periods of frost. Still farther from the tropics, conifers replace deciduous trees as the dominant forest plants. Then comes tundra with permafrost below it, followed by permanent ice cover.

The differences between the changes occurring due to terrain elevation or latitude include local differences in insolation depending upon the side of the mountain on which the changes are acting and other topographical features on the mountainsides between the base and the top. At sea level, a much greater variety of factors influence the climate between the equator and the poles. Some of these include conditions influencing local climatic features, such as wind patterns and resulting rainfall, the locations of warm and cold ocean currents, the presence and heights of nearby mountain ranges, and the presence of lakes and rivers.

The survival of an individual insect species depends upon its successful adaptation to specific local conditions. Those species which possess the necessary adaptations for survival and are already present in the same zoogeographical region as any given habitat are the ones most likely to be successful in occupying a vacant niche in that habitat. If no insects with suitable adaptations are native to the region at the time the niche begins to exist, there is a good possibility that members of a suitable insect species would stray out of its own region and colonize the new habitat. Other species native to more distant regions have been deliberately or accidently introduced by man to water bodies as far away as other continents and thrived there. Many of these introduced species have become problems by spreading dangerous diseases or causing damage to local flora or fauna. It must therefore be deemed important to determine the ranges of the insect fauna as it exists now so that future change can be recognized as such and introductions of dangerous species avoided.

A small percentage of the species seek the aquatic larvae of certain insects as hosts. Needless to say, before a parasitoid that develops on or in a specific host can settle in a new biotic community, at least one of its host species must be present. Some of the parasitoids are even able to deposit their eggs directly on hosts, which are insect larvae of other orders that live as tunnel borers inside the stems and leaves of aquatic plants. It is therefore not surprising that insects belonging to the order Hymenoptera live in close association with other species upon which they are completely dependent for the successful development of each of their successive generations.

Obviously, we are finding that the interrelationships between insect species and between insects and other forms of life are surprisingly complex, and that the kinds of symbiotic relationships of hymenopteran species, especially ants, are so intricate that they can hardly be conceived of by those familiar only with classical principles of biology. Most of the species with extremely complex relationships with other species in their biotic communities inhabit the tropics, and only a few of their possible relationships have actually been documented.

Among the simpler of the behavioral patterns employed by ants which build underground nests on dry floodplains of the Pantanal of Mato Grosso, Brazil, are those of small ants which much cope with periods of flooding by water at depths of well over 1 m lasting for about half of a year followed by periods of complete desiccation lasting from two to several months. During the seasonal flooding of their underground nests, they leave their burrows and form a constantly moving mass of living ants, in which each individual ant spends some of its time at the top to breath and some if its time at the bottom to support other ants while they take their turn in contact with the atmosphere. As the plants grow, they move onto their emergent parts to make more permanent nests concealed in the emergent vegetation. Months later, after the flood waters have receded or evaporated, they return to life in underground nests.

Other species of ants are plant gardeners, one of which plants large bromeliads. The species they plant will not grow unless their seeds are planted and tended by ants. The young plants seem to need the care of ants to thrive, and after they do, they begin to catch rainwater in their leaf axils and provide aquatic habitats for species for insects with the ability to develop as aquatic larvae in the water trapped in such plants. This will be further discussed in Chapter 12.

SECTION 4

Ecological Groups

a. Ecological roles of predominantly aquatic hexapod taxa

Several orders of winged insects, which had been placed in the subclass Pterygota before recent revisions separated the Apterygota from the class Insecta, consist entirely of aquatic species, with extremely rare exceptions. These orders include Odonata, Ephemeroptera, Plecoptera, Mecoptera, and Trichoptera. Megaloptera would also have been included in this group if it had not been reunited with Neuroptera, which includes both aquatic and terrestrial species. Almost all of the insects in these orders or suborders develop as larvae in water, exchanging gases directly with the surrounding medium through the integument, gills, or gill-like body surfaces.

This chapter identifies and defines some of the broad categories of the aquatic hexapods that are most commonly encountered. By far the most aquatic insects can be classified ecologically in one or more of these categories, and they can be considered to be those to which the hexapods have become best adapted. In subsequent chapters, the ecological categories of the habitats will be more narrowly defined, and then those that are apparently most hostile to insects and collembolans will be described in order to show what adaptations the few species that have been able to survive in them have led to their success.

b. Insects that are aquatic as larvae but leave the water to mate and spawn

There are also major families of insects that consist exclusively, or almost so, of species with aquatic habits, at least during one developmental period of their lives. One large ecological grouping encompasses all of those families characterized by larval development under water and adult activity in terrestrial habitats until mating and deposition of the eggs in or near water. Another encompasses those taxa which remain in water after metamorphosis to adults and do not leave, except, in some cases, during short mating flights or flights to seek out new aquatic habitats. A very large ecological group encompassing parasitoids that develop in other insects apparently consists of species of Hymenoptera, most of which develop as larvae in terrestrial insects. A minority of the large number of poorly known species in this group are aquatic and have adaptations unlike those of any other insects.

Some examples of the first group, which leave the water as adults, include the families Sialidae and Corydalidae of the group Megaloptera, now included by most as part of the Neuroptera. Also included in the Neuroptera are the families Sisyridae and Osmylidae, which are all aquatic or semiaquatic in habits as larvae, as far as has been determined (Table 1.1).

In general, most aquatic Diptera also display life cycles including aquatic larvae and often pupae followed by terrestrial adult stages. Diptera is not the order of insects encompassing the most species, but it is one arguably with the greatest variety of ecotypes and adaptations to unusual habitats, including aquatic and semi-aquatic ones.

c. Species inhabiting water as imagoes

The second group of families includes members of the largest insect order, Coleoptera, the beetles, as well as the Hemiptera, which encompasses the true bugs. Most species in both Coleoptera and Hemiptera are terrestrial during all stages, and most exceptions belong to families known to consist almost exclusively of aquatic species. Coleoptera includes two large suborders, Adephaga and Polyphaga, and both include predominantly aquatic families. Hemiptera also includes two large suborders, Heteroptera and Homoptera. Heteroptera includes several large families of aquatic species, while Homoptera encompasses almost exclusively terrestrial species, although there are a few species of aphid which live on floating or emergent aquatic plants. Personal observations of a few aphid species living on the emergent parts of aquatic plants in the Pantanal of Mato Grosso, a vast wetland near the geographic center of South America, showed that they were not harmed if shaken from leaves of floating plants onto the underlying water (Heckman, 1998). They were then adept at climbing or flying back onto their host plants. Such contact with water would have been fatal to most terrestrial aphid species, which appear frail and unable to move freely on the surface of water.

Members of the Hemiptera are hemimetabolic, that is, all larval stages usually resemble small, wingless adults. There is no pupal stage. The larval stages were formerly called nymphs, while those of aquatic species were referred to as naiads. Much of the entomological literature from earlier years employs the terms "nymph" and "naiad" for the immature instars of all hemimetabolic insects, which are now recognized as larvae. As already mentioned, species of the hemimetabolic family Ephemeroptera, produce a winged subimago stage. The mayfly subimago bears wings but quickly sheds the integument together with the pre-imaginal wings, often within about a day, and then makes a short mating flight prior to depositing its eggs in or near water.

The terms larva, pupa, and adult, also called imago, are used for holometabolic insects, regardless of whether they are terrestrial or aquatic. This kind of development is described as "complete metamorphosis", especially in older publications.

Among the suborder Adephaga of the order Coleoptera, the largest family of aquatic species by far is Dytiscidae, commonly called predacious diving beetles. Most are streamlined and most can swim faster than species in other families. Many authors have included Noterinae as a subfamily of Dytiscidae, but others treat it as an independent family: Noteridae. Two smaller families of Adephaga are also almost entirely aquatic, although not as well modified for swimming as the dytiscids. These families, Haliplidae and Gyrinidae, are widely distributed around the world.

Members of the Gyrinidae can swim rapidly, but they constantly change direction. They belong to a distinctive ecotype with the surface tension layer as the preferred habitat for their activities as imagoes. Some gyrinids spend a considerable amount of time on land (Wesenberg-Lund, 1943). However, they are most familiar to observers as the shiny beetles that swim chaotically in swarms, constantly changing direction. This swimming accounts for their common name, whirligig beetle. The ventral surface and legs remain below the surface, while the dorsal side remains above the surface in contact with the atmosphere. The compound eyes are divided exactly along the plane of the water surface, permitting the beetles to simultaneously observe what is going on in the air above them and in the water below. Like most aquatic species in the suborder Adephaga, they carry a plastron of air adhering to a dense layer of short, hair-like setae on their ventral surfaces, including parts of the legs. This often appears silvery while the insect is swimming on the surface.

As mentioned above, there is one large terrestrial family in the suborder Adephaga: the Carabidae. A very small percentage of carabids live in habitats that could be considered aquatic or marginally aquatic. Several aquatic or semi-aquatic carabid species belong to rare or unique ecotypes. Some of these will be discussed in subsequent chapters. They appear somewhat like insects that inhabit marshy areas or the floors of forests, but several of them live in the spray from small waterfalls, in mosses growing at the edge of fast-flowing streams, or in flowing water at the edge of streams. A few such beetles have been placed in their own families, and their taxonomic positions have not yet been agreed upon.

Among the suborder Polyphaga, the largest family of aquatic beetles is Hydrophilidae, called water scavenger beetles. Many are almost as agile in the water as dytiscids, but their movement appears clumsier because they typically move their legs alternately instead of simultaneously while swimming, as dytiscids do. Some resemble dytiscids in size and form, but they can immediately be distinguished by the shape of their antennae and palps. Dytiscids have long, filiform antennae consisting of many segments, which are much longer than the palps, while hydrophilids have short, clubbed antennae with antennal flagella consisting of a few short basal segments followed by a few greatly broadened ones forming the terminal club. In contrast to aquatic Adephaga species, their palps are relatively long.

In addition to Hydrophilidae, Polyphaga includes a variety of small aquatic families, many of which consist of beetles that do not swim at all but rather crawl over submerged substrates or move inverted on the underside of the surface tension layer. These families include Elmidae, Helodidae, Dryopidae, Limnichidae, Psephenidae, Hydroscaphidae, and Heteroceridae. In addition, a few aquatic species belong to overwhelmingly terrestrial families, such as Curculionidae, Chrysomelidae, and Lampyridae, but they live in or on water, either as larvae, as adults, or both. The ecological roles of these beetles will be discussed in subsequent chapters, where appropriate.

In summary, species in the large aquatic beetle families, Dytiscidae and Hydrophilidae, with few exceptions, live as larvae and imagoes within the water column, swimming to the surface to renew the air supply carried with them. Species belonging to several smaller beetle families also remain in the water column and come to the surface for air periodically. Those families in the suborder Adephaga encountered in both lentic and lotic water bodies include the following: Dytiscidae, which typically swim rapidly in the water column between the surface and the sediment; Haliplidae, which swims in the water column, usually among submerged vegetation, and Gyrinidae, which typically swim directly on the surface of the water, although these beetles may leave the water at times and spend their pupal stage along the shore. Aquatic species of other families are also believed to spend their pupal stages in moist places out of water. Families in the suborder Polyphaga with similar habits include Hydrophilidae, Helodidae, and individual species belonging to other families. Aquatic species in additional families of Polyphaga creep over submerged substrates or move inverted beneath the surface of the water carrying large air bubbles for respiration with them. Many of these beetles lack the ability to swim at all because the air supply prevents all but the apices of the legs from coming into contact with ambient water.

Aquatic Hemiptera species are grouped in families displaying more uniform behavior patterns than members of other insect orders. They are predominantly if not entirely predatory. Their piercing and sucking mouthparts are poorly suited for feeding on plants or detritus, although feeding on plants has sometimes been reported, especially for members of the family Corixidae. Such feeding has never been demonstrated by careful studies to distinguish between piercing and sucking water plants and pipetting microorganisms living on or among such plants. Most larval stages are also more similar to adults in feeding and habitat choice than insects belonging to other orders. The free swimming larvae in the families Corixidae, Notonectidae, and Naucoridae closely resemble the adults. They can be distinguished by their smaller size and lack of wings. However, the adults of some species can also be apterous or brachypterous, that is, they lack wings entirely or have wings too short for flying during the imago stage. In addition to the species noted for swimming in the water column, Nepomorpha also encompasses species excellently adapted for life on both sides of the surface tension layer.

The upper side of the surface tension layer is the habitat preferred by members of the Gerromorpha, which encompasses the families Hydrometridae, Gerridae, Veliidae, Hebridae, Mesoveliidae, Macroveliidae, and two small families: Hermatobatidae and Paraphrynoveliidae. They can be found there during both larval and imago stages, and the imagoes of many species include some with fully developed wings and others that are apterous or brachyperous, like some species of Nepomorpha. All species in these families seem to be predators, feeding mainly on other insects, including many terrestrial species that fall into the water. Because their mouthparts are adapted to piercing their prey and injecting digestive enzymes that liquefy the soft tissues, species of Hemiptera must take their time devouring their prey to allow digestion to take place before the digested material can be consumed through the narrow, sucking mouthparts. Thus, one insect may steal dead prey from another, and dead insects may be found and consumed. Many and perhaps all members of the Gerromorpha seem to contain noxious chemicals, which discourage predators from attacking them.

The second habitat for members of the Hemiptera is the water column. Members of the families Corixidae, Notonectidae, and Naucoridae are strong swimmers and actively seek their prey. They rest at the surface, among benthic detritus, and in masses of aquatic plants. These families are included in the Nepomorpha. All of these families are apparently predatory, although Corixidae has been reported to pierce and suck the juices of aquatic plants, including filamentous algae. However, many such reports are doubtful because several species observed carefully seem to be pipetting microorganisms for food. Many corixids share common habitats, and the foods of each may be distinctive, minimizing competition for food.

Members of the family Pleidae belong to the Nepomorpha and live on the underside of the surface tension layer. The ventral margin of the body is linear, and the insect swims inverted to pass directly along the flat surface of the standing water, usually not far from beds of water plants. The insects are relatively small and seek any terrestrial insects that fall into the water or other aquatic arthropods small enough to overpower. They seem to possess the chemical defenses that protect most aquatic species of Hemiptera.

A third group of insects occupies benthic habitats in water bodies as both larvae and adults or live among the masses of dense aquatic vegetation. They differ from the insects inhabiting the open water by not usually swimming but rather creeping over surfaces or through masses of submerged plants. These include members of the following families of Hemiptera, suborder Heteroptera: Nepidae, Belostomatidae, Naucoridae, and Ranatridae, accompanied by families of Coleoptera: such as Amphizoidae, which is represented in Europe by one species, which lives in benthic mud. The largest group of insects that belongs to the benthos or lives concealed in masses of plants, however, is the Diptera. Some of the larvae inhabit clean and oxygen-rich water. Among those that must survive in water with relatively low concentrations of oxygen, many of the species make use of snorkels or breathing tubes, some of which are remarkably long. Larvae of most species of Diptera are smaller and usually more abundant in water bodies they share with species of Hemiptera. In most such water bodies, they probably account for the most abundant prey of the water bugs.

Among the members of the Nepomorpha that seldom swim, some typically remain on the bottom of shallow water bodies while others remain motionless among submerged plants and await prey, which fail to detect their presence and approach them. Once within range, the organism suitable as prey is grabbed with a quick snap of the powerful fore-legs. The nepomorph benthic predators belong to three families. Those of the family Belostomatidae become relatively large, and a few are among the largest aquatic insects. They are considered to be gourmet treats by people in tropical Asia.

Members of the family Nepidae are generally smaller but have similar habits. Species in the family Ranatridae are very slender and appear to be sticks or other parts of dead plants as they rest motionless awaiting the approach of small animals to capture. These insects breathe through long tubes extending from the apex of the abdomen. Many are large enough to feed on small fishes and amphibians.

The final group of aquatic Hemiptera classified as Nepomorpha are those adapted to life at the edge of the water. They can live in or out of the water and hunt actively along the shore. The species belong to the families Gelastocoridae and Ochteridae. Gelastocorids are called toad bugs, and their movement at the water's edge resembles that of tiny toads undergoing metamorphosis and leaving the water. Ochterids resemble small, terrestrial bugs that often walk along the shore seeking prey.

For most members of the Nepomorpha, respiration takes place directly with the atmosphere. Those that live along the shores are in direct contact with the air most of the time. As long as the pleids are in contact with the surface tension layer, they also have continual access to the air. Corixids and Notonectids maintain a plastron of air, part of which may be stored between the body and the inner surfaces of the wings. The plastron is refreshed when the insect comes to the surface of the water, although it can be partially relieved of carbon dioxide by the surrounding water.

The largest of the Nepomorpha are able to remain on the sediment in shallow water and breathe through long tubes that serve as snorkels. The immature stages of these insects must actively swim only during the earliest larval stages, after which they behave in ways similar to the adults. Most of the species in the families Nepidae and Gelastocoridae display parental care. Females glue their eggs onto the backs of males, which protect their eggs and presumably remain with the first instar larvae, protecting them from would-be predators, many of which could serve as prey for the protective male adult.

Species with remarkably long breathing tubes include many in the families Belostomatidae and Ranatridae and dipterans in the family Syrphidae. Some larvae in the syrphid subfamily Eristalinae are known as rat-tailed larvae because their long, narrow breathing tube is sometimes longer than the body of the insect. Other insects making use of breathing tubes, although usually shorter, include members of the Tipulidae and related families in the superfamily Tipuloidea, as well as larvae of syrphid species with short breathing tubes, which are inserted into the aerenchyme tissue of emergent aquatic plants in waters polluted by large amounts of organic material. Such insects are capable of remaining entirely submerged beneath completely anaerobic water bodies, often with appreciable quantities of substances in the water and sediment that would be toxic to other insect species. These will be discussed in detail in subsequent chapters.

d. Adult insects on the surface of fast-flowing streams

Several families of Coleoptera include species that can move on the surface tension layer of lotic water. These beetles are typically smaller than 1 cm long and often escape notice. Interestingly, several such species are known from the surface tension layers of fast-moving streams, where they congregate in shallow, somewhat sheltered water near the banks. Some of these species belong to small families mentioned in the next part of this book.

Beetles of the family Gyrinidae typically swim almost as a part of the surface tension layer in lotic waters, as do the species of that family which prefer lentic water. The dorsal part of the adult body extends above the surface of the water and the ventral part remains below. The compound eyes have two fields of vision simultaneously, as do the species living is lentic water. The facets of the compound eyes are divided by a commissure along the horizontal plane, which is kept at the level of the surface tension layer as the insect swims. It is safe to say that the nervous system separates the fields of vision so that the beetles can react appropriately to the approach of a large predator from any direction. When approached too closely from above the water by a human observer, the beetles typically dive beneath the surface. When approached by a predator from the water beneath them, members of at least some of the species quickly take to the air and fly to a new location.

The adults form swarms swimming without any sense of direction on the surface of the water, where their short swimming legs produce the typically rapid changes of direction which characterize these insects and account for their common name: whirligig beetles. Some of the larvae develop in moist terrestrial soils, but others seem to develop below the surface of the water in very shallow places and breathe through external gills. As already mentioned, their pupae rest along the shores.

Like other aquatic imagoes belonging to the major families of aquatic beetles, whirligig beetles maintain a layer of atmospheric air along body surfaces coated by fairly short, closely spaced, water repellant setae (Heckman, 1983). This layer holds a plastron of air and appears silvery while the ventral side of the beetle is submerged. Additional air can be stored beneath the elytra. The extent of the plastron on the submerged surfaces of the beetle and the amount of air stored beneath the elytra vary from species to species. The dorsal surface of the elytra may be shiny or matt from sculpturing on the head, pronotum, or elytra, and this feature is sometimes useful for distinguishing species.

At least some species of whirligig beetles are capable of taking off quickly from the surface of the water if danger approaches from below. The body is well designed for escaping the water body with almost no water trapped on or in the body parts. When removed from the water, all water instantly separates from the plastron, and the opening of the elytra to release the metathoracic wings instantly liberates all water that might have accumulated along the edges of the wings. Apparently, most whirligig beetles have little trouble flying from one water body to another. Capturing these insects in large cloth nets in South American wetlands proved difficult because as soon as the net was moved in deep water beneath a group of these beetles and raised rapidly toward the surface to capture the whole swarm, the beetles began to open their wings and take off quickly, leaving very few behind to capture.

Although the gyrinids in the South American wetlands can obviously fly very well, they are seldom found beneath outdoor lamps where migratory beetles belonging to other families accumulate during the mating season at the end of the rainless period. This indicates that they migrate during daylight hours, migrate seldom or not at all, or do not navigate using fixed lights as reference points.

Adults in other beetle families live on the underside of the surface tension layer. Swimming with the ventral side up allows them to keep in nearly continuous contact with the atmosphere while hunting for food on the underside of the surface tension layer. Adults in the families Elmidae and Dryopidae typically live along streams. The habits of many are not known. Adults of species in the family Helodidae in North America are terrestrial, but their offspring develop as larvae in water.

e. Insects with aquatic larvae and adults that fly

Several orders of Insecta are typically aquatic during the larval stage but leave the water as adults to fly, mate, and deposit their eggs. These insects almost all show a typical alternation between aquatic and terrestrial phases, but the duration and behavior of each of the two phases varies greatly from species to species.

The habitat choice of the adults for depositing eggs determines the kind of habitat in which the larvae must develop. Many deposit their eggs in lentic or slowly flowing lotic habitats, while others seek out sites along fast-flowing streams. Generally, a species can survive in only one of these habitats. Those water bodies with little or no water flow are suitable only for species that can develop in relatively warm water containing moderate to low oxygen concentrations, while only species with the ability to hold firmly to the substrate or find shelters in the sediment can survive in fast-flowing streams.

Insects must also choose habitats with adequate food supplies. Some species are able to hunt for food and survive for long periods of time after they have emerged from the water. Others cannot eat at all after they emerge as adults, so they must be able to find enough food as larvae in order to sustain themselves while mating and producing eggs. The larvae of Ephemeroptera and the large family of Diptera, Chironomidae, do not eat at all after they have emerged from their final aquatic instar (Chapter 14). In contrast, adults in the orders Odonata, Plecoptera, Neuroptera suborder Megaloptera, and Trichoptera all live as adults for relatively long periods of time, that is, at least a week but usually several weeks, during which most of them catch enough food to continue spawning as long as the weather conditions permit. If the adult insects cannot eat, egg production and spawning are limited to relatively short periods of time, and the spawning period is often synchronized with an astronomical or meteorological event, such as a phase of the moon or calm weather.

Other factors that play a role in the choice of spawning sites include the speed of the current, the amount of specific food items in the water, the presence of certain kinds of water plant, and any number of other factors that the spawning females can detect. Entomologists have only just begun to perform detailed studies to determine how each of the many thousand aquatic insect species look for suitable places to spawn. The female insect depositing the eggs must limit her search for a spawning site to those water bodies in which the larvae of the species are able to survive. A careless mistake will almost certainly assure that none of her offspring will survive.

How spawning choices are made by each species along a stream can be inferred from the locations at which the adults of the species emerge from that stream. However, this may only hold true if little or no drift occurs during the development of the larvae. Where drift occurs due to strong currents, the adults must return to the spawning sites at which they were left to prevent a slow movement of the spawning site downstream from year to year. This gradually takes each year class of larvae farther and farther away from the optimal conditions for the species.

The large order, Diptera, presents special problems for determining the locations of the optimal spawning sites for the individual species. Even for economically important families, such as Culicidae and Ceratopogonidae, the knowledge of how and where each species spawns is still far from perfectly known. As we will see in later chapters, some of the species that transmit bacteria and viruses to human beings and domestic animals do not always develop in natural water bodies. They may also develop in water trapped on old tires and other kinds of refuse, and a few can also develop in water trapped in epiphytic plants.

It is already possible to provide general outlines concerning the kinds of spawning sites that species of aquatic Diptera choose for spawning. For example, species of

Psychodidae generally develop in eutrophic or hypertrophic assemblages of water. Many larvae in the family Culicidae develop in standing water. Larvae of the family Dixidae develop in standing water containing submerged aquatic plants, while those in Chaoboridae choose large lakes with clear water. Those of Ceratopogonidae inhabit fast-flowing water of natural streams, and those of Simuliidae develop in fast to slow-moving water courses, with some species preferring large rivers or the freshwater sections of estuaries. Still others develop in the small water courses formed above the permafrost as its surface melts during arctic summer. However, the ranges in the parameters used by the insects of each family when choosing their spawning sites are enormous, and exceptions can be found in almost every factor which is considered by the females when selecting places to deposit their eggs.

For insect species to survive, it is necessary for the females to choose sites for depositing their eggs that maximize the chances of their larvae to survive after hatching. There are many strategies for doing this that are already known, but the search for suitable places and the actual deposition of eggs remains undescribed for the majority of aquatic insect species. Once the eggs are deposited, the larvae are completely on their own to find food and avoid predators, but the actions of the parents in placing the eggs in advantageous locations goes a long way to determine whether they will survive to produce a new generation.

Some of the strategies used by aquatic insects to improve the chances of their progeny include the following:

1. Deposition of eggs on plants overhanging a water body so that they are not devoured by aquatic omnivores before they hatch. Usually, when the eggs hatch, the larvae drop into the water.

2. Spawning simultaneously with many conspecific individuals to assure that the maximum number of larvae hatch at the same time. In this way, predators can catch and eat as many larvae as they can for a short period of time, after which their surviving prey have already found hiding places, making subsequent predation difficult. This permits predators only one good chance to feed on the larvae of a species present in great abundance for only a short period of time. If the larvae continued to hatch over a longer period of time, members of the species would be subject to predation for a long period of time, during which the predators would be able to digest their meals and continue to eat repeatedly, assuring that much greater percentages of the larval population would perish prior to metamorphosis. Aquatic insects employing this strategy of reproduction include most species in the order Ephemeroptera as well as many species of such families of Diptera as Chironomidae, Chaoboridae, Culicidae, and Empididae.

3. Other species employ the strategy of laying eggs very gradually over a long period of time, but carefully selecting the sites at which one or a few eggs are deposited at a time. Most species of Odonata employ this strategy, which is best used by long-lived species that are strong flyers. The eggs can be distributed over a large area, and some of the sites will almost certainly be conducive to the successful development of the larvae. Laying eggs in this way reduces the population density of the species and the competition for food among the larvae. Dragonflies, especially the large species belonging to Anisoptera, frequently

employ this strategy, which also tends to spread the larvae of the species into new water bodies.

4. A few insects employ parental protection of the eggs and early larval stages, as already described. The females of species in the families Nepidae and Belostomatidae paste their eggs on the dorsal surfaces of the males. These predators are large and fearsome, and they can easily ward off most other predators in a small water body, including fishes. After the larvae hatch, the first instars usually remain near the parents and continue to enjoy a certain amount of protection.

5. The most careful method of spawning is employed by the parasitoid species of the Hymenoptera in the superfamily Ichneumenoidea. These species carefully seek an egg or a larva of a suitable host and deposit one or a few eggs on it. The larva of the parasitoid invades the body of the host, where it consumes organs that are not vital as they grow. When it is ready for its own metamorphosis, the parasitoid consumes vital organs, killing the host, and then undergoes metamorphosis to its own pupal stage.

f. Aquatic insects of flooded meadows, marshes, wetlands, bogs, and small aquatic habitats

The groups of species inhabiting diverse small wetlands are the most difficult to categorize and describe. Many are known only from the remarks noted by the taxonomists who described them. Others have been observed and collected only once or twice, and it is not yet known whether or not their collecting site is typical for their preferred habitat. Many of the species in subsequent chapters fall into this category.

In many families, there are species which develop in a wide variety of habitats, most of which are terrestrial. A few species, however, develop in fully aquatic but small bodies of water. This makes it difficult to define their typical mode of existence. A problem in working with such species is the lack of reliable information concerning everything about them other than the morphological description of their imago. One such family is Tipulidae, the species of which develop as larvae in a wide range of soils, from dry to inundated. In most parts of the world, the habitats of the larvae remain unknown. It is therefore not possible to classify them as aquatic or terrestrial, or perhaps somewhere in between.

The closely related family, Limoniidae, which has sometimes been treated as a subfamily of Tipulidae, seems to encompass a greater percentage of species with aquatic larvae, but fewer of its larvae have been described at all. Because the species in these families in Europe are best known, discussion about the families will be confined mainly to that continent.

Other families of Diptera can similarly be arranged along a spectrum from fully aquatic to fully terrestrial. These include families such as Ephydridae, Empididae, and Spheroceridae, in contrast with Tabanidae, and Sciomyzidae, most of which develop from fully aquatic larvae. The members of large families of Coleoptera also fall into the category of species with larval development falling across a large spectrum of ecological characteristics. Out of many species in such large beetle families as Carabidae, Staphylinidae, and Curculionidae, only a tiny percentage have been categorized as aquatic. Some of the smaller beetle families, such as Lampyridae, are represented by larger percentages of species with aquatic larvae, but still encompass a majority of fully terrestrial species. Similarly, out of hundreds of species of small moths in the family Crambidae, only one in northern Europe, *Cataclysta lemnata* (Linnaeus, 1758), is known to be fully aquatic during its larval stage. Its larvae live in water inside of a tube it constructs, usually using whole *Lemna minor,* which is a floating plant, very simple in its form.

The tube containing the larva remains at the surface of the water, camouflaged by whole plants, which are incorporated into the tube when it was constructed using silk secreted by the caterpillar. A supply of air held in the tube keeps it just below the surface tension layer. *Lemna,* called duckweed or water lens, forms dense masses of plants, which can completely cover the surface of eutrophic water bodies, obscuring the water below. The tube of a larval *Cataclysta lemnata* blends in with this green carpet of tiny plants so well that it can only be detected when the larva opens one end of its tube to move around. The pupae rest in the tubes until metamorphosis. Thereafter, adult moths can be seen flying just above the beds of *Lemna* seeking suitable places to deposit their eggs.

When investigating the most unusual and inhospitable kinds of aquatic habitat, insects belonging to unexpected families are often encountered. Among these habitats are mountain streams with strong currents, the spatter zone of rocks under waterfalls and near the rapids along mountain streams, waterlogged masses of mosses and lichens, tundra with permafrost, water trapped in plants, bogs and marshes, highly polluted water bodies, and hypersaline water, just to name a few.

Many of the insects with remarkable adaptations remain undiscovered or have been described and named without any information being recorded about their habits or their preferred habitats. Research is often hindered by the remoteness of their habitats and the limited ranges of the insects. For example, an interesting group of insects are members of a community inhabiting bromeliads in tropical rainforests. The adaptations of many such species are poorly known because of the difficulties in reaching their aquatic habitats high in the trees of tropical rainforests.

The aquatic species found in unusual habitats are often uniquely adapted to fill the available niches, and their aquatic habits are not shared by closely related members of their families. In some cases, the species are so poorly known that their larvae have not yet been identified sufficiently to know whether they should be classified as terrestrial or aquatic.

The European fauna has been studied systematically since the mid-18th century. The insect fauna there is the best known. North American species have been studied since the early 19th century, but the land mass is much greater, and there are far more gaps in the knowledge of the native insects. Some regions of the earth, such as the entire Neotropical Region and Southeast Asia, are settled by far more native species than most of the rest of the world, but information about them is still fragmentary. For that reason, information about the biological niches occupied by individual insect species is taken mainly from literature on the European species. Unfortunately, there are more unusual niches available in tropical regions than in Europe, and some of the most unusual habitats exist only outside the regions that have been well studied.

g. Main species aggregations of aquatic insects

As already mentioned, typical biotic communities among which aquatic insects live are considered to be those including the largest numbers of insect species. A paucity of species often indicates that the habitat causes stress to aquatic insects in some way. Those habitats in which relatively few insects can survive are those in which conditions are most hostile to insects, in general. We will therefore make only brief mention of the biotic communities in which most aquatic insects can be found in subsequent chapters so that emphasis can be given to non-typical water bodies and the species that have been able to develop special adaptations to survive in them. It is research concerning the way certain insect species have overcome the stresses of unusual habitats and thrived where other insects cannot live that offer some of the best research opportunities for entomologists at the present time.

This chapter has provided a cursory summary of information on the typical habitats in which the greatest numbers of aquatic hexapods are found. They have to be considered relatively unstressed. There are clearly zones in the aquatic habitats, each with its typical species of Hexapoda. To survive, these collembolans and insects depend upon the sources of oxygen available for respiration, the kinds of food available for them in or near the water, places of concealment from predators, and the permanence of water in the habitats in a part of the world where the seasonal temperatures are amenable to the well-being of the insects.

1. Life on the Surface Tension Layer

Although the surface tension layer is strong enough to support relatively large insects, those without special morphological adaptations will fall through and drown. To those species that are equipped to move across the water on the surface, the habitat provides them with food and a refuge from those predators that are not able to live on the surface. The insects on the surface need no additional modifications for respiration and can breathe by direct gas exchange with the atmosphere, just as terrestrial insects do. However, most have water-repellant surfaces on the body surfaces to prevent them from being forced under water by rain or large animals. The surface tension layer is usually augmented as a habitat by a nearby shore and floating aggregations of small water plants that cover part of the surface. The small plants provide resting places for the hexapods, where they are not visible to predators in the water below them.

The biotic community living on the upper side of the surface tension layer covering water bodies is sometimes called epineuston, and the animals inhabiting it have been called superaquatic fauna (Rapoport and Sanchez, 1963). This fauna includes most aquatic species of the class Collembola; insect species in the order Hemiptera belonging to the families Hydrometridae, Gerridae, Veliidae, and Hebridae; and individual species belonging to a great many families of Diptera, including those which rest briefly on the surface after metamorphosis to adults, those which frequently land on the surface tension layer to find food, mate, or rest; and those that hunt for prey on that layer. A few of the species can skate rapidly across the surface of swiftly flowing water. Veliids in the genus *Rhagovelia* are particularly noted for this. It should be remembered that life on the surface depends on the physical properties of water. When attempts to reduce evaporation from California reservoirs led to the use of heavy alcohols to coat the surfaces of the water, gerrids and other insects on the surface fell through the surface and drowned.

Typical for this kind of habitat is the opportunity it gives to the small hexapods which live on it to safely rest, move, and remain on the surface tension of the water. All other characteristics of the members of this community are similar to those of terrestrial species, including respiration directly with the atmosphere, feeding on food items that also rest on the surface tension layer or emergent plants, and reproduction very close to the shore, where the shallowness water excludes most aquatic predators from moving below the groups of insects and collembolans to feed on them.

2. Life in the Water Column

A far greater number of species of Hexapoda are found in the water column of both lentic and lotic water bodies than are present on the surface tension layer. Generally, the number of species present decreases with the speed of the current, and the structure of the community changes, as well. In addition, the presence of appreciable amounts of dissolved oxygen in the water at all times is also conducive to a rich and varied structure of the biotic community.

The water column extends from the underside of the surface tension layer to the bottom of the water body and even somewhat below, as long as the sediment remains loose enough to permit ambient water to penetrate into the interstices of the material at the bottom. Depending upon the physical features of the water body and the local climate, distinctive niches are provided for a great variety of insects. Geography, climate, speed of the current, size, depth, and trophic level all contribute to both the number of species that can survive in the habitat and the number of individual insects that will be present seasonally.

The order Plecoptera is represented predominantly by species with larvae that inhabit lotic waters. Relatively few are associated with relatively slow-flowing streams or large lakes. Most of the larvae are adapted to living under large rocks, among smaller stones, and in shallow waters near the edges of the streams.

Lotic water courses are also preferred by most members of the suborder Zygoptera in the families Calopterygidae, Protoneuridae, and Perilestidae, at least those with larvae that have been investigated. The leaf-like gills of the larvae hinder them from seeking sheltered waters beneath rocks on the sediment of streams. The typical larva in one of these damselfly families inhabiting flowing streams is most frequently encountered in masses of submerged plants, which flourish only in parts of streams with a somewhat attenuated water flow.

Other species of insect inhabit the lentic but oxygen-rich water of clear lakes. The larvae of species in the dipteran family Chaoboridae are among the larvae that remain at middle depths of the water column of such lakes. Basically, the requirements of an insect to survive in the open water column include any of several methods to obtain oxygen and an effective method to find food and avoid predators. Methods of respiration adopted by these aquatic insects are most frequently gas exchange with the water through gills or, for smaller hexapods, through the integument; transportation of atmospheric air held to a plastron by a coating of hydrophobic setae; storage of atmospheric air within other structures of the body; and respiration through a snorkel-like breathing tube, which may be long enough to reach the surface of the water body in shallow water or short and strongly constructed to penetrate the lower stems or roots of aquatic plants and obtain air from the aerenchyme tissue that carries atmospheric air from the emergent parts of the plants into the lower parts rooted in sediments at the bottom of a water body, which are temporarily or permanently anaerobic.

3. Life in the Benthic Community

Survival at the bottom of ponds, lakes, and streams presents problems not encountered by insects that live at the surface or swim in the water column. The difficulty of survival increases with the depth of the water, with deep habitats being poorly suited for species that require contact with the atmosphere for respiration. Meeting the needs of the insect for oxygen is also made more difficult by periodic shortages of the gas in dissolved form due to the rapid breakdown of organic detritus in the water column by microorganisms.

Shallower benthic habitats are colonized by insects which have the ability to reach the surface, either by periodic short trips to exchange the air or by use of an organ functioning as a snorkel. Some large species in the hemipteran family Belostomatidae possess snorkels longer than their bodies, as do some larvae of syrphid flies in the subfamily Eristalinae. This permits them to remain submerged indefinitely, concealed in benthic detritus, but not at depths greater than the lengths of the breathing tubes.

An even better adaptation is that of larvae in a few syrphid species possessing short spikes rather than a long, flexible tube like a snorkel. Where emergent aquatic plants are present, the larvae drive their spike-like snorkels into the lower stems or roots of emergent plants and penetrate the aerenchyme tissue, which conducts air to the roots, permitting the plants to survive in anaerobic water. As mentioned above, this permits the larvae to remain in deep water and survive in fully anaerobic habitats for indefinite periods of time, constantly supplied with atmospheric oxygen through the stem and root system of plants adapted to this kind of habitat.

4. Special Adaptations for Survival in Aquatic Environments

This chapter was meant to provide a basic review of some of the most common kinds of aquatic habitats to which species of Hexapoda are adapted. Most of the familiar aquatic species of Collembola and Insecta are encountered in such habitats. However, the world has countless water bodies, both temporary and permanent, and there are no two which are exactly alike. Subsequent chapters will continue with enumerations and descriptions of different kinds of water body and discuss various hexapods that live in or on them. In general, those insects that have modifications permitting them to adapt to the most extreme aquatic habitats are the ones about which the least is known. For someone contemplating a career in entomological research, it is these insect species that offer the best opportunities for making interesting and valuable discoveries.

The species that are encountered in many kinds of aquatic habitat require other morphological, physiological, biochemical, and behavioral modifications than those that permit respiration, resistance to extreme temperatures, rapid water flow, and chemical peculiarities of the water. These are to be described in subsequent chapters of this book. It is important to remember that there are great gaps in the knowledge of how these modifications from the basic hexapod body plan enable individual species to thrive in unusual habitats in which almost all of their related aquatic species cannot survive. The great majority of such modifications remain either unknown or poorly understood, and detailed studies of the species that possess them provide a vast field for future research.

It is the purpose of this book not simply to describe the ways of life adopted by collembolans and insects to thrive in bodies of water with unusual characteristics but also to discuss those phenomena that are not yet fully understood. Therefore, aspects of the modifications of hexapods that have not been fully elucidated will be emphasized, and additional studies will be proposed to increase the understanding of some of the more perplexing observations that have been made since aquatic organisms have first been studied by scientists.

Obviously, the material in subsequent chapters will be taken mainly from published reports from those parts of the world in which most of the research has been performed. It is well to remember that there are certainly many more new and astounding discoveries still to be made in those parts of the world in which not very much research has yet been undertaken. In the last chapter, recommendations will be made for research projects still needed to find out how specific insect species are able to survive in waters where most other aquatic insects are absent.

2

General Classifications of Water Bodies Inhabited by Hexapods

The great many of aquatic hexapod species are encountered in and around bodies of freshwater classified in general terms as ponds, lakes, streams, rivers, impoundments, canals, drainage ditches, and freshwater sections of estuaries. The aquatic habitat extends into the areas of shallow water at the edge of these water bodies and into waterlogged patches of soil that may or may not seasonally dry up. Depending on their physical and chemical characteristics, such flooded or marshy places are called marshes, swamps, bogs, raised bogs, moors, flooded meadows, and a variety of local names. Especially in tropical regions, such marshy places are often used for agriculture. Those rich in nutrients are used to raise such familiar crops as rice, taro, lotus, and water chestnut.

Bogs and raised bogs, also called moors in northern Europe, are usually present in the temperate zones. They are notably poor in inorganic nutrients and are therefore well suited for the cultivation of such plants as cranberries and blueberries. They are also sources of peat, which is locally used in many parts of the world as a fuel. It is thought to be the first step in a multimillion year process leading to the natural formation of coal. The peat is formed by the annual deposition of detritus produced by mosses and other small plants capable of surviving in nutrient-poor soils. Undisturbed, the dead plant material forms layers that are arranged sequentially, and the age of each of its visible layers can often be attributed to a specific year, at least as far back in the past to the latest ice age.

All of these habitats are among the most important environments for aquatic hexapods. One comprehensive work can only describe all of the habitat characteristics in a general way and mention examples of the best known species from each general ecotype. Each of the general kinds of environment is suitable for so many aquatic species world-wide that only general descriptions of the fauna can be provided. Therefore, only species that are best known are mentioned, and these must be from geographical areas in which a sufficient number of research studies have been completed to provide accurate information. Insecta alone includes more species by far than any other class of animals or plants, and even the small percentage of these species that are known to be aquatic number in the tens of thousands distributed throughout the world. To this number must be added the many species that remain to be discovered, described, and named. Estimates of the number of such undiscovered species differ considerably, and it is not yet possible to offer even a reasonable guess.

Most detailed species lists have been compiled for water bodies located in tropical, subtropical, and temperate zones. Polar and sub-polar regions, as well as high mountain lakes, require fundamentally different kinds of survival strategies due to the presence of ice during much of the year. The species inhabiting such water bodies are treated in separate chapters on insects with exceptional adaptations. They will be discussed in much more detail in subsequent chapters. This chapter will focus on the environments themselves, especially those in which the most species of aquatic hexapods have been found.

However, a few hexapods have ventured into habitats that are more hostile than the inland water bodies filled with relatively pure freshwater. Most are required to have all of the adaptations to water possessed by other aquatic species, but aside from these, they must have additional modifications of their morphology, physiology, biochemistry, and behavior permitting them to venture where thousands of other aquatic species would fail to survive. Subsequent chapters in this book will examine how such adventuresome species succeed where almost all other hexapods fail to survive. We now know that there are insects that live in the pelagic region of the ocean, in shallow marine habitats and along the shores of the seas, in intertidal zones, in hypersaline inland waters, in highly polluted waters lacking all oxygen most of the time, in hot springs, in waters from slowly melting permafrost, on ice, in small assemblages of water trapped in plants or hollow rocks, in the perpetual darkness of caves, inside of other insects which they gradually consume, in the presence of chemicals produced specifically to kill insects, and in many other kinds of extreme aquatic habitats.

This chapter will focus on the most common kinds of habitat in which hexapod species of many kinds thrive. Subsequent chapters will describe some of the most extreme cases of adaptation to more hostile kinds of aquatic and marine habitats and focus on what we still need to learn about how they can survive where most aquatic hexapods cannot.

SECTION 1

Classifications of Habitats According to Water Movement

a. Lentic habitats

Lentic habitats, such as ponds, small lakes, large lakes, and impoundments are characterized by little or no water flow during periods of relatively calm weather (Fig. 2.1). Vertical water movements usually occur in parts of the temperate zones where the water freezes at times during the colder seasons. Water reaches its maximum density at about 4°C, so as the temperature of the water approaches that temperature, it sinks to the bottom of the water body, causing a strong vertical circulation, which brings the water at the bottom to the surface as the water at the surface sinks to the bottom. When the water temperature descends below 4°C, the coldest water rises

Figure 2.1 A standing water body in the Bung Borapet, a protected freshwater floodplain in Nakorn Sawan Province, Thailand, at the end of the rainy season during an exceptionally dry year. The large emergent leaves in the fore-ground are those of the lotus, *Nelumbo nucifera* (Gaertner).

above the warmer water. Therefore, the coldest water winds up at the surface, where it freezes first. Other liquids would start to freeze at the bottom rather than the surface, but water is unique in freezing first on the surface. For living freshwater organisms, this is a very important factor for their survival during cold winters. It provides for a layer of water in the liquid state below the ice cover. The temperature at the bottom of a water body covered by ice is usually 4°C, at which many aquatic organisms can live and even remain somewhat active.

The characteristics of ponds and lakes conducive to supporting aquatic life are determined mainly by size, depth, content of organic and inorganic material, as well as dissolved oxygen, water temperature, pH, and the concentrations of essential mineral nutrients. These are influenced by the daily weather and can be characterized according to the climatic zone in which they are located. It should always be remembered that each of the characteristics displays a four dimensional pattern, which is distinctive for each water body and for the respective seasons of the year. Taking dissolved oxygen as an example, the values determined in the middle of a large, deep lake will differ from the values in shallow zones near the shores, which are usually filled with aquatic plants. In the middle of the lake, recording the oxygen concentrations every 10 cm from the surface to the sediment at the bottom will usually produce a distinctive curve. During periods of sunshine in summer, the highest concentrations will generally be at or near the surface, and they will decrease continually toward the bottom. In deep lakes, the oxygen concentration may reach 0 mg per liter at or just above the benthic sediment. The samples taken shortly after noon near the bottom of a deep lake may remain at or near mg per liter, but as the samples taken at intervals of 10 cm from the bottom to the tip are collected and analyzed, increases in the amounts of dissolved

O_2 are to be expected with each sample. The maximum values for dissolved oxygen can be expected at or just below the surface of the lake in the early afternoon, but all of the values will show marked differences dependent upon all chemical parameters determining the water quality in the lake. The concentrations of oxygen themselves will result from a combination of the biotic activities of photosynthesis by green plants and decomposition of organic substances in the water by saprobic microorganisms.

The curves for oxygen concentration with the depth of the water will therefore be influenced primarily by the amounts of organic detritus and inorganic nutrients present and the influences of insolation and temperature on the biological activity. In oligotrophic lakes, in which there is little biotic activity because of the lack of nutrients of all kinds available to the organisms, the water is usually clear, and concentrations of dissolved oxygen are always relatively high all the way to the bottom, even during the night. Light may penetrate all the way to the bottom during the day, even in deep water, and small quantities of oxygen may be released by plants attached to objects resting on the sediment.

Mesotrophic water bodies are characterized by moderate amounts of nutrients. This category is subdivided into α-mesotrophic and β-mesotrophic, with the alpha category being richer in nutrients than the beta category. Such water bodies support entire biotic communities encompassing a whole array of organisms from algae and bacteria to birds and small mammals. These water bodies typically have their maximum dissolved oxygen concentration at the surface, and it decreases toward the sediment. During the summer, as photosynthesis and water temperature both increase during the day, the concentration of oxygen will rise to exceed 100% of saturation. As the oxygen concentration continues to rise, tiny bubbles of oxygen start to form in the water by the early afternoon. Not only does the continued insolation stimulate photosynthesis by autotrophic micro-organisms, thereby increasing the absolute concentration of O_2, it also raises the water temperature, which lowers the solubility of oxygen in the water. Both of these processes increase the relative percentage of oxygen saturation in the water.

Water bodies with amounts of nutrients greater than those found in mesotrophic water bodies are classified as eutrophic. They are generally very rich in individual organisms, which frequently obscure the water. In the tropics, during periods of maximum photosynthesis in eutrophic water bodies, stirring the water gently often produces a foam of tiny oxygen bubbles coming to the surface. At such times, the relative concentration of the oxygen in the water was recorded to be more than 500% of saturation in seasonally eutrophic water bodies of the Pantanal, a large wetland near the geographic center of South America. However, the analysis was inaccurate because while it was being conducted, gaseous oxygen was constantly bubbling out of the solution and interfering with the analysis. This showed that oxygen was being produced faster than it could leave the solution and bubble into the atmosphere, but it is no longer possible to determine any exact value for the amount dissolved in the water (Heckman, 1998).

This phenomenon of oxygen effervescence may continue to be evident until the sunlight becomes less intense in the late afternoon, and the relative concentration of oxygen decreases to 100% or lower. Depending upon the depth of the water body, the relative concentration of oxygen near the bottom may remain only slightly above 0%

while the upper water layer is supersaturated. A wide variety of weather conditions can disrupt the pattern of photosynthesis and supersaturation of oxygen, regardless of the depth of the water body. Heavy rainfall can create turbulence that hastens the release of oxygen into the atmosphere, or it can cause more atmospheric oxygen to dissolve in nearly anaerobic water. Cloud cover can reduce the light necessary for photosynthesis or slow the warming of the water during the day, permitting it to hold more oxygen before reaching saturation. In large, open lakes, temperature changes produce vertical circulation, as described above. Strong winds produce horizontal circulation, which is sometimes strong enough to move both living organisms and detritus to one side of the lake, relocating the part of the lake in which both photosynthesis and decomposition of organic materials occur at maximum rates.

While photosynthesis requires light, the organic detritus present in eutrophic water bodies is broken down for 24 hours each day. This process sometimes consumes most or all of the oxygen produced during the day, reducing the dissolved oxygen concentration to about 0% of saturation in the early morning hours before sunrise. This extreme variation in the oxygen concentration was personally observed in shallow waters on drying floodplains in the Pantanal of Mato Grosso near the end of the rainless season. Although activity can also be extreme in the temperate zones, such large variations in water chemistry were never observed anywhere in either temperate zone, if for no other reason than the lower intensity of the sunlight.

During the course of 24 hours, the water in a eutrophic lake in the Temperate Zone, that is, one that has a rich supply of inorganic nutrients to support photosynthesis, may show an increase from 0 mg of oxygen per liter to 100% of saturation at its ambient water temperature in a vertical profile from the bottom of the lake to the surface. Where the lake is heavily fertilized with all inorganic minerals necessary for plant growth, the amount of oxygen in the surface water can be expected to reach a minimum shortly before dawn. As the amount of light from the rising sun increases, the oxygen concentration begins to increase, slightly at first, but then relatively rapidly, depending upon the richness of the water in plant nutrients. A maximum relative concentration of O_2 is typically reached during the early afternoon. The exact time will depend upon a combination of factors. While the sunlight supports photosynthesis with energy, thereby increasing the absolute amount of oxygen dissolved in the water, it also raises the water temperature, thereby lowering the amount of oxygen needed to saturate the water. Therefore, it takes less oxygen to bring the dissolved oxygen to saturation, and the excess amounts of oxygen are usually released from the water as gas when the concentration exceeds 100%. On the other hand, slightly supersaturated water can exist without releasing any oxygen, especially when the water is not agitated enough to cause the gas to form small bubbles. Meanwhile, the increasing temperature brings about an increase in the rate of decomposition of organic detritus by saprobic microorganisms, increasing their utilization of the dissolved oxygen that is being produced.

From the difference between the diurnal and nocturnal concentrations of oxygen in lake water, a good deal can be inferred about the amount of plant growth a water body can support. Where plant nutrients are in excess, there is a large difference between the daytime and nighttime concentrations of O_2. Several factors influence this difference. In shallow waters containing much dead plant material, bacteria, fungi, and aquatic animals break down the organic detritus, thereby rapidly exhausting the oxygen supply. This can cause the dissolved oxygen concentration to decrease to 0% after the sun sets and photosynthesis stops until sunrise. By morning, large amounts of plant nutrients have been released, and the sunlight available after sunrise permits photosynthetic species to produce large amounts of new organic material, thereby releasing O_2 as a waste product. By sundown, the saprophytic species breaking down the detritus no longer can take advantage of large supplies of excess oxygen in the water, and they begin to rapidly exhaust the oxygen, which can only be replenished slowly by direct absorption from the atmosphere.

In lakes containing very pure water and nearly free from nutrients necessary to support photosynthesis or plant growth, the diurnal curves showing the changes in the oxygen concentration over a 24 hour period show that there is almost no change at all during the course of one full day. This is typical for oligotrophic lakes. The scarcity of photosynthetic organisms in such lakes is indicated by the great clarity of the water. Oligotrophic and eutrophic water bodies therefore contrast strongly in the diurnal changes which occur in them over periods of 24 hours. Curves plotted to show the changes that occur in the amount of dissolved oxygen in the water as a function of the time of day can give insights into the biotic activity in the water, while the results of single water analyses must be viewed as meaningless artifacts dependent only on the time of day the water is sampled and analyzed. That shows how important it is to use dynamic factors as criteria for studying aquatic habitats because conditions can change rapidly during the course of a single day. When the oxygen concentration in the water changes from 0% to more than 100% of saturation, or even reaches a dubious value of more than 500% in the tropics, it is obvious that the value recorded will indicate only a momentary condition. It also must be remembered that the analysis of dissolved oxygen must be made immediately after sampling in order to be accurate. In spite of how obvious these observations are, researchers have sometimes used single determinations of the oxygen concentration to characterize a water body, as the following example shows.

In a eutrophic pond and in an oligotrophic lake at 10 a.m., identical O_2 concentrations of 6.0 mg l^{-1} may be recorded, but that certainly provides no clue to what is happening in the water. What would be significant for determining the activity in these aquatic habitats is a record of the changes in the concentration during the course of a 24 hour day. If only two analyses instead of one were conducted at each of the two water bodies, a comparison of the results would begin to show the great differences between eutrophic and oligotrophic waters. For example, if the results of the analysis at 10 a.m. could be compared with those of a second analysis made at 11 a.m. on the same day, the results might well show that the O_2 concentration in the eutrophic pond had increased to 9.5 mg l^{-1}, while the concentration in the oligotrophic lake had remained unchanged at 6.0 mg l^{-1}. Clearly, the more analyses that have been made, the more correct the conclusions drawn about the conditions in the water bodies.

If analyses were made once per hour for 24 hours, curves could be plotted showing little change in the oligotrophic lake and great changes in the eutrophic pond. In the oligotrophic lake, very small increases in the oxygen concentration might appear during the early afternoon, and decreases of equal magnitude would occur during the night. Under most circumstances, the curves plotted would intersect twice during the day; once during the morning and again during the night.

Along with the oxygen concentration, measured either as an absolute value in $mg \cdot l^{-1}$ or as a percentage of saturation, in percent, other physical and chemical parameters show extreme diurnal changes, as well, if the waters are eutrophic. The pH values of the water tend to rise as the daylight supplies the energy for photosynthesis. After sundown, the breakdown of organic detritus by saprobic organisms lowers the pH during the night. In oligotrophic lakes, the pH usually changes little during the course of a day, but there are sometimes more significant changes with the depth from which the water sample is taken. Isolated pH values tell little or nothing about mesotrophic or eutrophic water bodies, except perhaps a little general information about the geology in the place at which the water body is located. However, diurnal curves of the pH are very useful to determine what chemical reactions are occurring in most water bodies that are not frozen over. Generally, oligotrophic water bodies will show little change in any of the parameters related to photosynthesis during the course of 24 hours. Mesotrophic water bodies will provide curves with distinct differences during both day and night. The more eutrophic a water body becomes, the more pronounced the diurnal and nocturnal differences will be. Photosynthesis increases the pH value, while the decomposition of organic material lowers it. However, the magnitude of the change in pH during the course of a day also depends somewhat on the buffering of the water in the lake. An oligotrophic lake containing few dissolved chemicals will have less of a buffering capacity less than one in which many inorganic ions are present which are not involved in photosynthesis. Thus, there can be noticeable differences in the diurnal curves of two oligotrophic lakes that have nothing to do with photosynthetic activity. In eutrophic lakes, the presence of inorganic ions released by saprobic activity may noticeably buffer the reactions that result in pH changes.

With the advent of automated electronic data recorders connected to sensors that continually record physical and chemical conditions in water bodies, many difficulties in recording abundant data have been eliminated. This will be extremely valuable for determining the preferred habitats of countless aquatic insect species in all parts of the world.

A lentic water body cannot be regarded as a single kind of habitat for insects. The larger the pond or lake is, the more different individual habitats for insects are likely to be found in it. Examples of such distinct kinds of aquatic habitat include the surface layer of the water, the open water, masses of algae, beds of aquatic tracheophytes, the epibenthos, aquatic sediments, the littoral zone, shallow-water sublittoral zone, deep water benthos, and the bodies of living hosts for parasitoids. An aquatic species may be limited to a specific kind of habitat during one of its life stages, or it may remain in fairly homogeneous lentic habitats throughout its entire life cycle.

In most cases, water movements other than a one-way flow from a higher to a lower elevation are not considered when classifying a water body as lentic or lotic. Large lentic water bodies, such as lakes and impoundments, often display considerable vertical movements during periods of diurnal and seasonal temperature changes. Unlike most liquids, water is at its maximum density as a liquid at a temperature 4°C, which is significantly above its freezing point. That means that colder water will sink below warmer water until it reaches a temperature of 4°C. During autumn in the temperate zones, water cooled at the surface will sink below the warmer water until it reaches

4°C at the bottom. This can produce fairly strong vertical currents and brings the water at the bottom of the pond or lake to the surface, as already explained.

If the water continues to cool, it will remain at the surface and eventually freeze there to form an ice layer on the top, which is less dense than the liquid below. For this reason, the ice and water remain stratified according to temperature during the winter, and a layer of ice on the top provides some insulation against further cooling. In spring, when the ice melts and the water on the surface warms above 4°C faster than the water at the bottom, it remains stratified, and there are usually only weak vertical currents if there are any at all. In the autumn, there is typically a turnover within the formerly stratified water. This brings the cooling surface water down to the bottom of the water body and the deep water to the surface, more rapidly when the water at the top cools faster.

A second cause of water currents in lentic water bodies is wind, which acts most strongly on large lakes and impoundments. Wind is usually a factor to be reckoned with seasonally according to the local climate or during storms. Storms often bring heavy rainfall, and the rainwater is usually considerably cooler than the water of the lake into which it is falling, especially in tropical regions. As a result, periodic, vertical circulation may occur as a result of extreme weather events.

b. Lotic water courses

Many aquatic insects are adapted only to life in lotic, or flowing water. In most such habitats, but not in all streams, aquatic insects enjoy the advantages of a rich supply of oxygen dissolved in the water and the availability of a food supply that is continually carried to them by the moving water. Streams in higher mountainous regions tend to be considerably colder than lowland water courses, and their water typically flows rapidly (Fig. 2.2). Cold temperature and rapid water flow are stressful

Figure 2.2 A fast flowing mountain stream near the entrance to Sequoia National Park, California. Although the current is rapid in all parts of the stream, slower flow is observed behind rocks and near the shores, allowing insect larvae to find suitable sites to anchor themselves.

for aquatic insects, eliminating many of the insect species that are familiar in lowland streams and rivers. This means that those insect species with suitable adaptations to cope with the low temperatures and rapid current have the advantage of not having to compete for food with many other insect species. While cooler temperatures tend to slow the metabolism of invertebrates, it also increases the solubility of oxygen in the water, making it relatively easy for the insects adapted to cold water to meet their respiratory demands. It also reduces their rate of metabolism, permitting more of the nutrients consumed by the insects to be stored up for their reproductive activities after metamorphosis to the imago stage.

Conditions in some streams change gradually along a gradient, while those in others show irregular changes as they pass through mountain ranges or geologically irregular regions. In general, those with alternating rapids and slower moving zones are generally more interesting ecologically and richer in fauna. Many important factors vary strongly from steam to stream, making most generalizations inaccurate. It is important to realize that even streams located near each other in the same mountainous regions may have great differences in dissolved oxygen concentrations, water hardness, pH, concentrations of heavy metal ions, anthropogenic pollution with organic wastes, fine suspended matter, exposure to sunlight each day, annual inputs of leaf litter, size and composition of particulate matter in the sediment, ice cover during winter, and desiccation during irregular or seasonal periods of dryness, to name only a few of the most obvious factors that can be conducive to some species and hostile to others.

In fact, the enormous number of variable factors in streams during the course of each year and the possible combinations of such factors have led to speculation that every stream has unique characteristics, making it different in important ways from every other water course in the world. Added to the physical, chemical, and biological characteristics prevailing at the present time are the events in geological history that have determined which species have been able to colonize a stream since it first came into existence and which have been successful in surviving in it up until the present time.

Observing the geographical ranges of the aquatic insect species reinforces the impression that lotic water courses are almost all unique habitats, although it also reveals many similarities of streams caused by their proximity to each other. This especially applies to streams in the same catchment areas. Ecological studies also reveal a zonation of the species within single streams, which are obviously related to the special physiological requirements of each life stage of each species.

Both temporal and spatial distribution of the insect species also determine their distribution within an individual stream. Many aquatic insect species have seasonal life cycles which cause it to be completely absent from certain streams for weeks or months at a time. Many species in the orders Odonata, Plecoptera, Neuroptera, and Diptera inhabit streams as larvae and, in some cases pupae, for most of the year every year, but during several weeks or months each year, all of their life stages are completely absent from the habitat. The period of absence of most European insects varies from less than a week to roughly two months, depending upon the life-span of the adults and the amount of time the eggs remain dormant before hatching.

What was said about the importance of collecting as many physical and chemical data as possible about lentic habitats applies to lotic habitats, as well. However, even more understanding of the biological processes influencing the water in streams and rivers is required for planning and conducting monitoring projects for lotic waters. The main difference to keep in mind is that lotic water moves, and any particular moving water mass can be influenced by the biota at any location along the water course for a matter of a few seconds to the better part of a minute as it flows. This produces a great many possible differences in the water quality, supply of individual nutrients, and biota with which each aquatic species interacts. A stream flowing through a large forest, one flowing past agricultural land, and another flowing though several towns can be expected to develop completely distinct conditions for the species of aquatic insects inhabiting each.

An example of what can result from ignoring individual factors in planning a survey was a project to record physical and chemical parameters along the course of a rather slow flowing, lowland stream in northern Europe. The person making on-site analyses and recording the data went to the source of the stream at sunrise on a spring day and began making a set of analyses. He then drove downstream to a series of sampling sites along the course of the stream and recorded such data as temperature, pH, oxygen concentration, and several others parameters. Shortly before sunset, he arrived at the mouth of the stream and made his last set of analyses.

The results he presented included a map of the stream and the values he had recorded at his sampling sites located along its course. He showed that the oxygen concentration increased from the source of the stream to its mid-length and then decreased as it flowed toward its mouth. What he had failed to consider is the fact that he might have been sampling the same moving mass of water throughout the day. The water was flowing slowly enough for him to have caught up with it using his car at every sampling site and sampled it repeatedly throughout the day. However, even if a person had gone to a pond or lake on the same day and taken samples at one spot at the same intervals during daylight hours, he would have arrived at the same conclusions from the data. Obviously, at any one spot, the oxygen concentration would continue to increase from sunrise to early afternoon, and then it would have decreased gradually until sunset. The problem with the presentation of the results was the correlation of the oxygen concentration with the geographical location along the stream instead of with the time of day.

c. *Where the water flows in both directions*

Research involving different habitats makes it clear to the researcher that whenever general categories are established, exceptional places are discovered that do not fit into any of the categories. One such water course comes to mind that cannot be described as either lentic or lotic. This large freshwater habitat changes the direction of its flow twice each year, and there are two interim periods during which it does not flow at all. There are also smaller water courses that fit this description in different parts of the world, but these are not as well-known and do not have such well-defined seasonal patterns of flow.

There are two reasons why the water changes direction periodically in water courses: tides and patterns of seasonal rainfall. At the mouths of many rivers, the flow changes with the tides. These regions, called estuaries, are defined according to how far inland the tides can be detected. The freshwater sections of estuaries are limited to the parts farthest inland influenced by the tides. The sites at which the salinity begins to increase due to the influx of seawater with the flood tide are the upper limits of the brackish water section of the estuary. This section continues past the mouth of the estuary into the sea, which become less saline during the ebb tide. How far seaward this dilution of the seawater is detectable depends upon the seasonal outflow of freshwater from the river.

Other rivers in which water flows in both directions occur where there is a pronounced season cycle in rainfall. The largest and best known such water course is the Tonle Sap in Cambodia. It involves only freshwater carrying a variable amount of silt and suspended matter according to the season. Its middle section forms a great lake that covers a large but highly variable area of Cambodia. Its drainage area encompasses much of the country. However, during the Southwest Monsoon, which occurs between April and October, enormous amounts of rainfall are drained through the Tonle Sap into the Mekong River where its delta is beginning to form. The water cannot flow seaward fast enough to promptly dispose of the amounts of rain falling on the country, so the central lake begins to grow until it reaches is maximal size, which may vary somewhat from year to year.

Relatively little rain falls on Cambodia between mid-October and early April, and the water in the lake drains continually into the Mekong until it reaches the same water level in that river. Thereafter, evaporation of water from the lake continues to reduce its water level, causing a flow from the Mekong into the Tonle Sap. By the end of the dry season, the water in the lake has become so low, that the flow from the Mekong into the lake becomes fairly rapid. The onset of the rainy season begins the annual rise in the water level in the Tonle Sap, and the flow slows, stops, and then reverses itself again. This great lake is extremely productive in fishes, and it supports a highly diverse fauna. It borders on great areas of cultivated rice fields and provides a habitat for much terrestrial wildlife. Relatively little is known about its aquatic hexapods, although many of its species can be assumed to have been described based on specimens from neighboring countries.

The estuaries of the world are distinctive, and there are few that are very similar to any other. Rivers with a seasonal, two-way flow are extremely rare, and one the size of the Tonle Sap is unique. These habitats merit special ecological studies, and their insect fauna should be more thoroughly investigated.

SECTION 2

Classification of Water Bodies According to Available Nutrients

Nutrient content is a multifaceted concept. In addition to carbon sources, plant growth is promoted by several essential nutrients, such as nitrogen in the form of ammonium, nitrite, or nitrate, and phosphorus, present as phosphates or related compounds. These are incorporated in algae and the tissues of tracheophytes, which are eaten by animals or broken down by bacteria and fungi. In the temperate zones, large amounts of nutrients in a water body may be manifested by abundant growth of green plants and high diurnal oxygen concentrations in spring, large amounts of vegetation in or beside the water body during summer, much decomposing material and low oxygen concentrations during autumn, and low oxygen concentrations and sometimes the presence of hydrogen sulfide under ice cover during especially cold winters. Water bodies containing low concentrations of nutrients are characterized by less particulate material in the water, leaving it clear throughout the year. The tiny particles that make mesotrophic and eutrophic water appear turbid usually consist of detritus from the previous year accompanied by microorganisms consuming and digesting the detritus or engaged in photosynthesis.

It must be remembered that such communities of photosynthetic microorganisms as well as larger vascular plants require several different kinds of nutrients to successfully engage in primary production. The rate and amount of photosynthesis will be affected most by the single nutrient that is in the shortest supply. In some places, for example, a water body may contain a considerable amount of dissolved nitrates but only a low concentration of phosphate. The species that require much phosphate might be displaced by those which require only small amounts, but when the phosphate concentration sinks to almost zero, plant growth will almost stop. In natural water bodies, however, there is almost always a breakdown of detritus continuing at a slow rate and consumption of some of the primary producers by animals, both of which release phosphates into the water. As a result, there will continue to be some photosynthesis occurring, but at a relatively slow rate. As a result, the oxygen concentration changes little during 24 hour periods. If certain water bodies are lacking in most nutrients, photosynthesis will remain slow from season to season. There will be little or no growth of vascular plants during the course of a year. As mentioned earlier, such aquatic habitats are classified as oligotrophic. Familiar examples of such habitats include large, pristine lakes, such as Lake Superior, one of the Great Lakes of North America, and Lake Baikal in Siberia. A different kind of oligotrophic water body is exemplified by raised bogs, which usually contain much humic matter but few nutrients, limiting the primary production to a few mosses, which die and produce a new layer of humus each year. Only a few species of vascular plant are capable of surviving in the soil of such moors, and a few of these capture and digest insects to obtain the nitrates and phosphates needed for their growth. The same conditions used to exist in lowland moors, but over the years, many of these have been drained or so contaminated with nutrient-rich pollutants that there have been fundamental changes in the flora and fauna.

In tropical and subtropical regions, seasonal changes somewhat reminiscent of the temperate zones are often observed where seasonal differences in rainfall are considerable and occur regularly. Obviously, growth of aquatic plants is more rapid during rainy seasons, while their decomposition occurs during the rainless months as the water evaporates or runs downstream out of the wetland, leaving dry detritus, which will decay quickly with the return of the rains. The trophic level of the water appears to depend upon the season, as well. While the area of a water body increases rapidly during periods of heavy rainfall, the nutrients also become more diluted. As desiccation reduces the amount of surface water covering the area, the nutrients become more concentrated, and the water becomes green and turbid due to the accelerated growth of microscopic algae.

The difference between high and low nutrient concentrations in the tropics becomes quickly evident when diurnal curves of the oxygen concentration are plotted. By the end of the dry season, many formerly flooded areas have dried up completely and are covered by short-lived terrestrial plants. In lands under the influence of the Monsoon, many of the areas flooded during the rainy season are already covered by the stubble from harvested rice fields. During the freshet, many of these lands are flooded again by the first heavy rainstorms. This begins a dynamic process that repeats itself with great regularity each year in Southeast Asia. Other wetlands influenced by smaller and less predictable weather systems than the Monsoon do not begin at about the same time each year, and some years they might not occur at all.

Immediately after the first flooding, the nutrients in the wetlands are still not freely available in the water. While the amounts of available nutrients are still low, the oxygen concentration changes little during a 24 hour period. Once under water, the nutrients begin to be released through the action of microbial saprophytes or an influx of organic substances from external sources, and the oxygen concentration begins to decline, especially during the day when the water temperature is high. In spite of the rapid decomposition of the organic detritus during this season, the oxygen concentration seldom decreases to a concentration near 0% because frequent heavy rainstorms refresh the oxygen supply. As the flood water deepens, the nutrients released by the saprobic community promote the growth of primary producers, including algae and aquatic macrophytes. During the day, they release increasing quantities of oxygen into the water. In places where rice and other aquatic plants are cultivated, further development of a natural biotic community is interrupted by the transplanting of the rice plants into the fields. However, together with the rice, it is possible for a rich community of fishes, amphibians, and insects to develop together with a large number of microorganisms, which develop along with the rice (Heckman, 1979).

In the frigid zones, inorganic nutrients are seldom limiting factors to the maintenance of a food web. Solar radiation may be available 24 hours per day at times, but it is weak, and subfreezing temperatures limit biological activity to a small number of species that can remain active or at least survive below the freezing point of water for much of the year. In most Arctic water bodies, more chemical nutrients are present than the plants can utilize during their short growing season. Trees and other woody plants that might slowly store up phosphates and nitrates in their wood are absent from tundra habitats, permitting the lichens, mosses, and other plants that are predominant in such places to utilize only a small proportion of the plant nutrients that are available during their short growing seasons.

Biologists classify water bodies according to their trophic level in general terms because so many different factors are involved. The causes and effects of nutrient input into water bodies are simply too complex to be quantified in any simplistic way. Attempts at finding precise, objective index values out of the broadest spectrum of different factors have never been successful, and it is doubtful that any single value can be calculated that amounts to anything but an artifact. However, a set of terms to describe the trophic state of a water body are in general use. They are descriptive, and exactly how they should be applied to lentic or lotic water bodies depends upon accepted local usage (Table 2.1).

Table 2.1 Names and characteristics of the trophic levels of water bodies, as used in Central Europe.

Category	Oligotrophic	β-Mesotrophic	α-Mesotrophic	Eutrophic	Hypertrophic
Nutrient content	Low	Medium low	Medium high	High	Polluted

Characteristics

Oligotrophic: the water is always clear, but it may be tinged with yellowish brown; at least one mineral nutrient essential for plant growth is lacking or almost lacking in the water during all seasons; oxygen concentration remains near saturation to considerably less in raised bogs, and there is little difference between the day and night concentrations; vegetation in the water and along the shore is sparse; resident fish populations are small; local species characteristic of oligotrophic habitats are present.

β-Mesotrophic: the water appears slightly to moderately turbid during the warmer seasons; all essential mineral nutrients are detectable during the cooler seasons; the oxygen concentration increases moderately during the day and decreases during the night; growth of algae in the water is evident, and vascular plants are present submerged in the water, on the surface, or along the shore during the warmer seasons; moderate numbers of fishes, amphibians, and water birds are usually present; some of the species present are characteristic of β-mesotrophic water bodies.

α-Mesotrophic: the water appears moderately to considerably turbid during all seasons; all essential mineral nutrients are detectable throughout the year; the oxygen concentration increases to saturation during the day in the warmer seasons and decreases considerably during the night, and there is a great difference between the concentration at the surface and that just above the sediment during the afternoon; growth of algae in the water is obvious seasonally, but vascular plants may displace them at times; fishes, amphibians, and water birds may be abundant at times; some of the species present are characteristic of α-mesotrophic water bodies.

Eutrophic: the water appears turbid because of the presence of microorganisms and detritus, limiting visibility to only several cm deep throughout most of the year; free mineral plant nutrients are always present because saprophytes release them from detritus about as fast as they are re-incorporated into living organisms; the oxygen concentration surpasses saturation on every sunny day during the warm seasons and approaches zero during the night at the bottom of the water body and often near the surface, as well; both algae and vascular plants grow abundantly during the warm seasons, and some algae also grows during the winter; vertebrates are present but some species avoid eutrophic waters; many species characteristic of eutrophic waters are present.

Hypertrophic, also called hypereutrophic: This state differs radically from other sets of trophic conditions in that it brings about fundamental changes in the water chemistry, such as a nearly perpetual oxygen concentration at or near zero, resulting in photosynthesis by green and purple sulfur bacteria and the production of considerable amounts of hydrogen sulfide. Only a few, specialized insect species can thrive in hypertrophic water bodies. Notable among these are the larvae of species in the order Diptera, such as those in the sub-family Eristalinae of the family Syrphidae. Many of these water bodies regularly receive organic pollution.

The biotic communities in water bodies under each set of trophic conditions undergo seasonal successions of species typical for the location. The seasonal changes that are repeated each year are produced, in part, by the life cycles of the insect species present, which is, in turn influenced by the trophic level of the water body. The combinations of the species, each with its own seasonal schedule for its life cycle, provides an annual time table of expected events in one or a group of similar aquatic habitats. At certain locations around the world, agricultural activities play a significant role in the seasonal succession of species, both by the enrichment of the water with nutrients and by the schedule of planting and harvesting.

Inland bodies of freshwater typically undergo a process of eutrophication that does not follow a repeating cycle from year to year. It is rather progressive and ends up converting a lake, pond, or artificial canal into dry land. Typically, a water body produced in mountainous regions by geological events, such as earthquakes or volcanic eruptions, or else by human activities, will start out as an oligotrophic water body. Nutrient chemicals enter slowly, and both plants and animals establish populations in the water. As the water bodies progress to the eutrophic state, sediments produced fill it up. Eventually, only a marsh remains, and when the water dries up, the lake has been converted to dry land. For this process to completely convert a moderately large lake to dry land can require 10,000 years or much more if the lake is particularly deep. Studies of small drainage ditches in a fruit growing district along the Elbe Estuary near Hamburg, Germany, revealed that about 12 years were required for them to reach the hypertrophic state after they were dug out and cleaned of sediments (Caspers and Heckman, 1981, 1982). After about 20 years without maintenance, these ditches would be completely gone from the orchards through the natural process of filling with detritus, including leaves and fallen fruit from the trees. After many of the ditches were no longer maintained, it was found that harvests were reduced due to spring frosts which froze the flowers on the trees. The water in the ditches had previously kept the air warm enough to prevent the blossoms from freezing during short nighttime frosts.

SECTION 3

Classification of Freshwater Bodies According to Climatic Conditions

The main climatic zones of the earth are classified primarily according to the seasonal amounts of solar radiation reaching the earth's surface, the temperature during the individual seasons, and the flora and fauna present. The borders between the typical regions of the frigid zones follow the lines of latitude only approximately. These are modified by a great variety of factors that fundamentally change the local conditions. Among the most important factors is the proximity of the location to the ocean and whether the ocean currents near the shore are warm or cold.

Other factors influence habitats that are not close enough to the sea to be heated or cooled appreciably by the water masses near the shore. Where the terrain is mountainous, the changes encountered while ascending to higher elevations mimic the changes encountered while moving north or south from the equator. For example, a mountain on the equator will display typical equatorial weather conditions at its base near sea level. As a person ascends the mountain, the standard environmental lapse rate in temperature is about 6.5°C per 1000 m. In dry air, the lapse rate is about 9.8°C per 1000 m. That means that if the temperature at sea level is 26°C, the temperature at 4000 m will be about 0°C if the air contains much water vapor. Still higher, and at night, freezing temperatures will prevail, accounting for the presence of permanent glaciers atop mountains located on or near the equator. Between the tropical water bodies at the base of the mountain and the glacier at the top, there is a series of intermediate aquatic habitats, mainly streams encountered flowing downward from the glaciers, through tundra with permafrost, past taiga in which evergreen conifers prevail, to an ever denser equatorial forest beginning well before the stream reaches the bottom.

This general plan is nowhere nearly as simple as it is presented here. The situation is complicated by many factors related to the individual shape and location of each mountain. Mountains located in deserts often lack forests, and many mountain ranges receive rainfall mainly on the windward side. In some locations, such as the part of Asia under the influence of the Monsoon winds, this means that for approximately half of the year, one side of mountain ranges will receive plenty of precipitation, while the other side will remain almost rainless. During the other half of the year, the situation will be reversed, and the formerly wet side of the mountains will receive little or no rain, while the formerly dry side receives all of the rainfall.

The effect of the Monsoon on the local climates of Southeast Asia is illustrated by the following example, explaining why the rainy season in Vietnam occurs during the Northeast Monsoon, while the rainy season in Thailand, Laos, and much of Cambodia occurs during the Southwest Monsoon. The Monsoon winds result from the rapid heating of the Gobi Desert and adjacent parts of Central Asia during the spring and summer and its rapid cooling during autumn and winter. When it is warmer than the surrounding land, a stationary low pressure system is formed. In the Northern Hemisphere, winds blow in a counterclockwise direction around low pressure systems, producing a huge system of winds that blow across most of eastern Asia and the adjacent oceans. The winds during the Southwest Monsoon pass over the Indian Ocean, saturating the warm air with water evaporated from the ocean. They then pass over parts of Myanmar, Thailand, Laos, and Cambodia, which are experiencing a hot season with little wind in April. The arrival of the southwest winds, bringing large amounts of water vapor with them, start to produce intense thunderstorms, bringing much water to the lowlands. When the winds reach the Annamite Mountains, separating coastal Vietnam from the rest of continental Southeast Asia, and the Luang Prabang Range and other mountains to the north, separating China from the Southeast Asian countries, the rest of the water held by the air becomes rain that falls on the southwestern faces of the mountains. When the air descends over the mountains, it has already expended its water, and there is little or no rainfall in Vietnam north of the Mekong Delta during the Southwest Monsoon.

By mid-October, the Gobi Desert has cooled down to a temperature equal to that over much of the adjacent parts of Asia, and as it cools more, a high pressure system is formed. The wind reverses itself and starts to circle clockwise around the high pressure system. This brings the air circling from northern Asia across the Pacific Ocean, where it cools and saturates itself with water. It is warmed by the ocean, making it warmer than the air over the mountains along coastal Vietnam. It is cooled as it rises over the colder air along the Vietnamese coast, producing a fine rain that persists for long periods of time, that is, the kind of rainfall typical of warm fronts. Not very far inland are the mountains, which cause the air to rise as the wind is directed upward and loses much of the rest of the water it is carrying. After it passes the mountain range, it descends and has no more humidity to lose. Therefore, Thailand, Laos, and much of Cambodia and Myanmar remain rainless for most of the period from late October through mid-April.

The Monsoon winds also have important effects on much of the rest of Asia, but for continental Southeast Asia, the tropical wet and dry climate is immensely important for agricultural production, the conditions in the local water bodies, and the species of aquatic insects that inhabit them.

All of these observations indicate that climate can be determined only by long-term studies of the daily weather, and many factors contribute to making it want it is. It has no fully predictable effect on water bodies that can be determined from consideration of any single factor. The nature of each individual water body in any climatic zone can be understood only when the climate is considered together with many other factors, such as topography, geography, geology, hydrology, wind patterns, characteristics of the adjacent soils, existing biota, and catastrophic events.

When dealing with small organisms, such as insects, the success of the species depends more on the microclimate than on the general climate for the region (Table 2.2).

Table 2.2 A list of common kinds of water bodies in climatic regions providing distinctive habitats for aquatic insects due to the microclimate they provide.

1. Permanent ice cover with temperatures near or below zero throughout the year produces microclimates hostile to all but a few species. Such zones are located in the Arctic and at high elevations on mountains throughout the world.

2. Permafrost in locations with short summers, during which only the surface of the ice melts, are characterized by extremely cold winters; short, hot summers; and enormous populations of a few aquatic insects. Zones located in treeless Arctic lowlands and rarely in high mountains located in many parts of the world are characterized by microhabitats that seldom vary much from the typical hot and cold seasonal pattern.

3. Water bodies in the taiga, with cold temperate climates and often deep permafrost below the tree roots. Temperate Zone forests like these are encountered over vast areas of Asia, Europe, and North America; relatively small areas in South America, Australia, and New Zealand, as well as on high mountains just at or slightly below the tree line.

4. Temperate forests with summers long enough for the ice beneath the tree roots to completely melt. Seasonal rainfall is sufficient to maintain permanent water bodies. As the local climate warms, evergreen trees are replaced more and more by deciduous forest. The microclimate changes follow the changes during four typical seasons, and cultivated land replaces much of the pristine deciduous forest in some places.

5. Semi-arid, subtropical microclimates in regions with many seasonal water bodies temporarily filled by irregular rainfall; the land includes desert only rarely receiving rainfall, in which permanent rivers and lakes are encountered carrying water from other geographical areas or receiving upwelling ground water.

6. Subtropical microclimates in regions receiving regular rainfall, usually which evaporated from adjacent regions of the ocean. Many islands of the northern and southern subtropical zones have such climates, which often show characteristic differences between the windward and leeward sides of the islands. The winds typically blow from one direction during most or all of the year. The temperature never drops below the freezing point of water.

7. Typical seasonal microclimates characterize water bodies and wetlands influenced by Monsoonal wind patterns, which bring regular, seasonal rainfall over a large area of Asia. The air circulation over the vast area alternates between a counterclockwise flow during the northern summer and a clockwise flow during winter.

8. Other cyclonic and anticyclonic air patterns over large, subtropical land masses affected by other seasonal cyclonic or anticyclonic air movements, such as those of Africa, South America, and Australia, produce similar effects as the Monsoon but occur in an irregular annual pattern. Because of their considerably smaller size, their annual patterns of local wind and rainfall are less predictable than those of the Monsoon.

9. Equatorial climates, usually characterized by little seasonal activity, strong insolation, and short, daily rainstorms. Wherever chains of high mountains cross the equatorial region, their climate is

Table 2.2 contd....

...Table 2.2 contd.

transformed to one or a series of the other kinds as the elevation of the terrain increases, and the microclimates produced vary, often with few traces of predictable seasonal variations. Exceptional patterns involving large seasonal changes in the flooding along major rivers are observed in places where the rivers carry much water drained from outside of the equatorial watershed, such as in the vast drainage area of the Amazon River.

10. Arid and semi-arid regions are not notably rich in aquatic insects, but there are many kinds of small habitats with microclimates adequate to support populations of such species, including those influenced by springs, permanent spring-fed water holes, small waterfalls on mountain sides, and on mountainsides covered by cloud forests fed mainly by dew from passing clouds. Aquatic habitats in such places are often inhabited by few species because their insect fauna is well isolated from other suitable habitats by large arid areas. However, studies often yield important and interesting results because some of the isolated aquatic habitats support a relict fauna, which has not been well studied. Some species present may not be found anywhere else in the world.

SECTION 4

Classification of Water Bodies According to Zoogeographical Regions

The major zoogeographical regions in which the lentic and lotic habitats of the familiar aquatic insects are situated have each been subjected to considerably different amounts of scientific study. From what is known about them already, each region seems to provide its native aquatic insect species with very similar ecological niches to occupy. What is different about the structure of their biotic communities seems to be due mainly to the distribution of the water bodies within their respective continental areas, the topography at their locations, their geological history, and the features of their climates. What is emerging as a very important factor in determining which insect species are presently predominant at each location within every region is the human impact. Human beings can and have excluded species from their former habitats and introduced new species from other biogeographical regions. Among the human activities with the greatest impact on both lentic and lotic water bodies are construction projects that greatly alter or eliminate them, intentional or unintentional introductions of new species from other regions, changes made to the local terrestrial vegetation, and the use of toxic chemicals.

This chapter employs those large and well-established names of the biogeographical regions that have been in use for many years. There have been various modifications of the zoogeographical concepts over the years and subdivisions of the major regions. Some of the introductions of names for smaller regions and subdivisions of the larger ones are found in such works as Proches and Ramdhani (2012). As knowledge of the fauna and flora increases, so will the number of subdivisions of the major zoogeographical and phytogeographical regions. For the purposes of this book, discussions will be limited to the largest and oldest of the regions, that is, those with major barriers between them, such as oceans and large deserts.

a. The Palearctic Region

The zoogeographical region surrounding the Arctic Ocean but also including North Atlantic and North Pacific parts of Eurasia and North America are grouped into what is called the Holarctic Region. This massive region is regarded as consisting of two easily recognizable subdivisions, which can be distinguished more by the history of the biological studies of their fauna and flora as by their physical, chemical, and biological features. The classical sciences of phytogeography and zoogeography originated and blossomed into important divisions of the biological sciences during the great age of exploration, which started with the commerce by Portuguese sailors between Portugal and India, Southeast Asia, and Japan, and the discovery of the Americas by Columbus. It was intensively pursued by the scientists of the European countries engaged in world exploration during the 18th century and had reached its zenith by the end of the 19th century.

The museums of Europe, with their collections of specimens from all parts of the world, were centers of classification and description of species, for which reasonably well maintained collecting data were compiled. For this reason, the best studied zoogeographical region has been the Palearctic. Many scientists without the means to undertake long sea journeys could perfect the knowledge of their local flora and fauna, including the aquatic insects that lived in the ponds or streams not far from their own doorsteps. The countries in which they lived had local clubs and organizations for both amateur and professional biologists, and many of these began publishing reports and journals as early as the 18th century. By the mid-19th century, many journals still published today had already been founded.

The European countries were rich enough to construct large museums to house collections of insects, both from their own lands in Europe and from the colonies and trading partners throughout the world. Their collections grew until it was possible to detect the obvious differences in the collections from various parts of the world in their own collection cases without the need for the scientists to travel anywhere. When people were educated in Europe, they learned the methods of collection and preservation of specimens, which spread the techniques being developed throughout all parts of the world.

For these reasons, zoogeography had its roots in the Palearctic Region. It has been studied there longer than anywhere else, and many of its major institutions are located in European countries. Today, the Palearctic Region is considered to be the parts of Europe, Africa north of the Sahara Desert, and western Asia located mainly in the North Temperate Zone, which have a recognizable and relatively well-known fauna. In the case of aquatic insects, it is generally bounded by two oceans and great expanses of arid lands, in which few freshwater habitats exist.

b. The Nearctic Region

The other division of the Holarctic Region is called the Nearctic. It is the part in the Northern Hemisphere located in the New World. Scientific studies of the flora and fauna of the Nearctic began in earnest at the beginning of the 19th century. It was officially initiated after Napoleon sold the vast Louisiana Territory to American President Thomas Jefferson in 1804. Jefferson immediately organized an expedition led by Meriwether Lewis and William Clark to survey the territory. Their duties included surveying the flora and fauna inhabiting the regions through which they travelled.

North American species of aquatic insects had found their way into European collections more or less as curiosities during the 18th century, but the new United States began to make systematic collections, which were later to be housed in large museums donated to the United States and its individual states and cities by various benefactors, as well as museums located at prominent universities. Their work of maintaining collections of the Nearctic biota began in earnest nearly a century after the research on the Palearctic species had begun.

Geographically, the Nearctic and Palearctic Regions are separated by the Atlantic and Pacific Oceans, although the northeastern parts of Asia are sometimes considered separately to be in a region of their own. Extremely low temperatures in Siberia certainly produce a barrier to the free movement of aquatic insects across the vast region and then across the Bering Straits. There is also a barrier to most aquatic insects to the south of the Nearctic Region, where a semi-arid belt in the Southwest of North America separates the North Temperate Zone from the Neotropical Region. This barrier is not as formidable for aquatic insects as the Sahara and Arabian Deserts, and aquatic insects have had an easier time travelling back and forth between the Nearctic and Neotropical Regions via rivers and mountains in the Southwestern United States and Mexico or by island hopping in the Caribbean.

Similarly, aquatic insects adapted to conditions in the watershed of the Arctic Ocean do not face serious barriers to spreading throughout the northern parts of Siberia, Alaska, and Canada. Because the areas in the Arctic Watershed are vast and have not been intensively studied, it is still not known precisely how similar the structure of the insect fauna inhabiting the northern parts of Asia and North America actually is.

In northern Europe, however, the Gulf Stream warms the seas, making the conditions in the inland bodies of fresh water more similar to those considerably farther south along the East Coast of North America. This would tend to facilitate the trans-Atlantic movement of insect species via ships between Europe to North America, where conditions in the freshwater bodies are similar.

c. The Oriental Region

The regions of East and Southeast Asia are separated by formidable land barriers from the Holarctic Region. Part of the region also has climates distinctive from those in the Temperate Zones. For those reasons, a separate Oriental Region is recognized. To the north, the differences are somewhat less extreme. The water bodies and streams in the region of Japan, Siberia, and Alaska have more similar climates than those of the lands between Mongolia and Chinese Turkestan (Xinjiang Province), and the area encompassing northern Malaysia, Thailand, Laos, Cambodia, and Vietnam, even though all of these countries, as well as parts of Burma and India, are under the influence of one large climatic system: the Monsoon.

Studies on the fauna in the Oriental Region began shortly after those in the Palearctic, but they progressed much more slowly. That is because the number of animal species inhabiting the countries of Southeast Asia, including aquatic insects, far exceeded the numbers inhabiting areas of similar size in Europe and western Asia, and, of course, because there were much fewer institutes devoted to investigations of the biological resources.

The Meiji Restoration in Japan started a rapid modernization of Japanese society. One of the changes involved the introduction of modern scientific methods in the country. During the last decades of the 19th century, Japanese scientists began to study the country's flora and fauna, and by the early years of the 20th century, Japanese taxonomists began to describe new plant and animal species that they discovered in Japan and neighboring Asian countries. The interest in biology spread throughout society, and even the Showa Emperor published papers in scientific journals on marine biology, in which he developed a great interest.

At the present time, interest in the flora and fauna of the Oriental Region has been increasing rapidly. High quality publications on insects have appeared in Thailand, written by both Thai and foreign authors, and a few technical publications of high quality have been published on individual insect families in China. The work is continuing, and publications are now appearing on a regular basis. However, there is still a great deal to be done to catalog the Asian insect fauna.

It might seem that the boundaries of the Oriental Region would be vague considering that large areas of Central Asia differ considerably in climate from tropical Southeast Asia. However, the Himalaya Mountain Chain is so high that it provides an effective barrier between Monsoonal Southeast, South, and East Asia, and the lands to the West. The frigid forests of Siberia also provide a barrier between the eastern Palearctic and the Oriental Regions, except for insects well adapted to such climatic zones. Even though southern Malaysia and Indonesia lie more in an equatorial climatic zone than one under the influence of the Monsoon, they are bordered to the east and south by one of the most classical boundaries between two zoogeographical regions of the earth. Although it looks unimpressive on a map, the Wallace Line runs between the islands of Indonesia inhabited by animals of the Oriental Region and those inhabited by the distinctive fauna of Australia and New Guinea. This is one of the boundaries, the discovery of which demonstrated the importance of pursuing the science of Zoogeography during the 19th century.

The Oriental Region is still not as well defined or delimited as the zones in the Holarctic Region. Its aquatic insect fauna warrants more systematic and taxonomic studies, which would certainly provide many useful results.

d. The Ethiopian or Afrotropical Region

The region encompassing tropical Africa south of the Sahara and southern Asia is even more isolated than the other regions discussed, with the exception of the Australian Region. The vast Sahara and Arabian Deserts are almost free of freshwater sources at the surface except for a few oases containing relict species. To the north and east of India, the Himalayan Mountain Range and its foothills in Burma separate the Palearctic and Paleotropical Regions, and the Pamirs serve to further isolate the Oriental Region from the Ethiopian and Paleotropical Regions.

The Paleotropical Region is less influenced by predictable climates than the other regions, especially with regard to rainfall patterns. Droughts lasting several years are not uncommon in parts of Central and Southwest Africa, but most of the Indian Subcontinent is better supplied with water, thanks to the precipitation that falls on the Himalayas each year and feeds the large Indian river systems. The isolation of the

Paleotropical and Ethiopian Regions is not complete, however. There is a connection with the Oriental Region along the coast of the Indian Ocean through Myanmar, and Thailand to the rest of Southeast Asia. After more research enhances the knowledge of the insect fauna of Asia, a more complete zoogeographical mapping of the region will be possible.

e. Neotropical Region

Apparently, the richest fauna of aquatic insects is found in the Neotropics. It is also thought by almost all entomologists that the largest number of insect species of all kinds inhabit Central and South America and nearby islands. It is difficult to document this with numbers because available information is not yet sufficient to even make an educated guess concerning the number of insect species in South America. Estimates of the number tend to vary greatly, and collections made in the Amazonian rainforests nearly always yield specimens belonging to undescribed species. South America has a large percentage of the available fresh water in the world. The Amazon is the world's largest river system. In addition, the Rio Paraguai drains a second vast system, which drains the parts of the continent south of the Amazon Basin into the Rio de la Plata. The middle sections of the river and its tributaries form the largest pristine wetland in the world during the rainy season. Parts of it are located in three countries. It is called the Pantanal. In the western part of the continent, a large number of small streams drain the steep western slopes of the Andes Mountains and provide habitats for a great many insects that live in fast-flowing streams. Many of these are species considered to be related to insects inhabiting the lands surrounding Antarctica rather than to typical Neotropical species. Others are thought to have migrated from the Nearctic Region through Central America and along the Andes Range. These cold water species indicate that streams of Patagonia and southern Chile represent a zoogeographical region distinct from the rest of South America and akin with the circumantarctic regions.

The Guiana Shield in northeastern South America is probably one of the least explored regions of the earth, and distinctive aquatic insects can be expected to occur there, adding to the enormously large diversity of the South American hexapod fauna.

f. The Australian Region

Although they started relatively late, the scientists studying the Australian Region encompassing Australia, New Zealand, New Guinea, and many small island south and east of the Wallace Line organized their programs well and are now using the computer to make very accurate surveys of the fauna in their sub-regions. The Wallace Line distinctly separates the Southeast Asian and Australian Regions. The zoogeographical homogeneity of most of the islands of Indonesia and The Philippines and continental Southeast Asia was already recognized by the researchers who took part in exploratory voyages around the world as early as the 18th century. However, after crossing the Wallace Line in the direction toward New Guinea, New Zealand, and Australia, named after the scientist who discovered and defined it during the early 19th century, a distinctly different fauna and flora was encountered. This was initially recognized from studies of the birds and mammals, but the distinctness of the two regions it separates can be seen from studies of most major taxa, as well.

g. Marginal regions for insects

As we shall see, the main kinds of ecological niches occupied by aquatic insects are present and often abundant in all parts of the world. Those best known are located in countries in which the most research has been done. The oldest tradition of scientific research is in Europe. Ecological studies involving insects in North America lagged behind until much of the country was settled, so taxonomic studies of the hexapods began about 60 years later, when scientists were sent to accompany exploratory expeditions, beginning with that of Lewis and Clark. However, far fewer researchers began to investigate a much larger area, and progress has therefore been markedly slower. Studies in South America and Africa are just beginning to reveal the richness of the insect fauna. It appears that South America will be found to have by far the richest fauna of aquatic hexapods, with vast numbers of species still to be described. Due to the nature of the tropical habitats located there, a relatively large proportion of the insects will be found to be aquatic.

Tropical Asia will also prove to be extremely rich in insects, but there are few pristine habitats left there because agriculture has eliminated or fundamentally changed the conditions in the water bodies. Except for Japan, which has a long tradition of biological research, there is still a large amount of work to be done to provide basic surveys of the Asian insect species, especially the aquatic ones. Recent surveys of the Chinese insects seem thorough, but most orders have not yet been treated. Australia is making much progress cataloging its insects, but that continent is one of the least hospitable to aquatic insects because of the great extent of its deserts, and it is apparently populated by fewer species. Because of the lack of surface water bodies, Antarctica is apparently the only continent with a substantially poorer insect fauna. Studies on the entomology of the many islands in the Pacific and Indian Oceans have been sporadic, and detailed studies of more than one or two endemic aquatic species on a single island have been few.

Based on the history of research focusing on aquatic insects around the world, most of our information about their natural history and ecology will continue to be based on studies of European species. An increasing number of recent publications have been augmenting the knowledge of individual species from all parts of the world, which will hopefully provide the basic information needed to understand how aquatic insects have been able to adapt to more extreme conditions. Some information needs to be added to the general knowledge about a few of the groups that are poorly represented in Europe, such as the insects with larvae that develop in bromeliads, aquatic ants, and aquatic insects that transmit serious tropical diseases, which have been largely eradicated in Europe many years ago.

In addition to the large continental areas in which nearly all aquatic insect species have already been described, there are still isolated islands in the Pacific and Indian Oceans, as well as several archipelagos in the Philippines and Indonesia with faunas of aquatic insects including species that are certainly not present anywhere else in the world and have not yet been described and named. There are also poorly explored regions of the Arctic in which aquatic and semi-aquatic insects abound. Other regions of the world that probably still have considerable numbers of undescribed species of aquatic insect will certainly be found in Siberia, northern Canada, and the high mountain ranges of Central Asia. Entomologists will certainly not lack opportunities for research on the ecology of the aquatic species during the coming century.

Most of this book does not focus on studies conducted on the common species, genera, and families that are encountered in the most familiar kinds of water body. It is rather concerned with species that have unusual adaptations possessed by few other species, those that have been neglected or ignored because of the technical difficulties in studying them, and those which are important for human health and well-being. The last chapter in the book concerns fields of study that have been neglected but would probably provide a great many valuable insights into processes that are still unknown or poorly understood. Such processes are those explaining why a few insect species are thriving where many others would fear to tread.

3

Life on the Surface Tension

The most readily visible aquatic hexapods are those that live on the surface tension layer of the water. The habitat is formed by a layer of water molecules arranged in a two-dimensional, horizontal plane according to the geometric pattern produced by the orientation of the one negative and two positive charges on each water molecule. The two hydrogen ions bound to the oxygen atom are separated from each other at an angle of about 105°, giving each water molecule a net positive charge at one end and a negative charge at the other. This arrangement of the magnetic charges brings the water molecules to arrange themselves into a regular network strongly bound together on the horizontal plane at the water's surface. Beneath the surface, a three dimensional arrangement of the charges on the molecules permits a random and constantly changing movement of the molecules, and no strong, two dimensional layer can be formed. These principles have long been known, but the details are still being tested both theoretically and experimentally by many physicists (Chaplin, 2009).

A second location at which a strong surface tension layer forms is at the interface between the plastron of air along the body surfaces of certain aquatic hexapods and the ambient water. The strong magnetic forces binding the water molecules prevent the loss of air through the tension layer at times the surrounding water becomes turbulent due to storms or attacks by large predators that enter the water. The many species of insect that carry a plastron of air with them below the surface display a great many morphological modifications to secure the air on a limited surface area of the body, while leaving the rest in contact with the ambient water. How the border lines between these two kinds of surfaces anchor the surface tension layer securely to the body in each of the many species employing this method of respiration under water has not been thoroughly investigated. However, it seems clear that there are hydrophilic surfaces to which the margins of the plastron of air anchor themselves. These surround a hydrophobic layer covered with fine, dense setae remaining covered by the layer of air. The air is held firmly in place by the surface tension of the layer of water in a two dimensional arrangement of the water molecules, which is similar to the arrangement on the surfaces of freshwater bodies.

Research on the surface tension layer alone has demonstrated that it is strong enough to support insects with suitable morphological adaptations to take advantage of this platform for moving rapidly across water bodies in search of prey or decomposing carcasses of insects and to find partners for mating. Needless to say, this gives the hexapods all of the advantages of respiration by direct exchange of gases with the atmosphere without elaborate modifications of the external respiratory structures to prevent the arthropods from drowning.

Because the water molecules below the surface orient themselves randomly in three dimensions, permitting random movements of the positions of the two positive charges, one each on the two hydrogen ions at one pole of the molecule, and the two negative charges on the one oxygen ion at the opposite pole, a firm lattice of regularly arranged intermolecular bonds cannot be formed. Thus, below the surface, there are no strong electrostatic arrangements of molecules, and the molecules are capable of only a minimal resistance to continual rearrangement due to movements of the water. Hence, an organism swimming below the surface can move in any direction without undue effort. It can be seen that each water molecule interacts with other molecules attracting or repelling it in every direction.

At the surface of the water, on the other hand, the water molecules are arranged in a two dimensional pattern with one positive hydrogen ion on each of two molecules orienting themselves to bond with a single negative oxygen ion on another. The other hydrogen ions on the two molecules immediately form their own bonds, each with single negative oxygen ion on two different molecules. These, in turn, also each bond with one of the hydrogen ions on two more water molecules, forming a very regular and stable two-dimensional pattern on the surface of the water. This results in a very regular orientation of the water molecules at the surface, with the electrostatic charges holding them in place within a homogeneous lattice with considerably more strength than any of the randomly arranged bonds below the surface (Fig. 3.1). The positive and negative charges on the molecules below the surface tension layer continue to attract or repel the molecules forming the surface tension layer randomly, according to the orientation of the individual molecules. The integrity of the layer remains strong because the forces between the water molecules are strong, while the forces between each individual water molecule and any of the molecules of the gases forming the overlying atmosphere are relatively weak (Chaplin, 2009).

Calculations of the surface tension on the water has yielded a value of 72.8 dynes cm^{-1} at 20°C, which is much higher than that of other non-metallic liquids. The units of the values, dynes cm^{-1}, are equivalent to and sometimes reported as millinewtons m^{-1}. The surface tension decreases with temperature. Liquid water at 0°C has a tension of 75.6 dynes cm^{-1}, while at 50°C, it is 67.9 dynes cm^{-1}.

At 20°C, the surface tension of most organic liquids is roughly 1/3 or less than that of water at the same temperature. A comparison of the surface tension of various liquid organic substances with that of water, all recorded at 20°C, is shown in Table 3.1. Not shown are mixtures of liquid organic substances with water, which are higher than those for the organic substances alone. All values shown were observed for the surface tension of the liquid at the interface with unmodified atmospheric air.

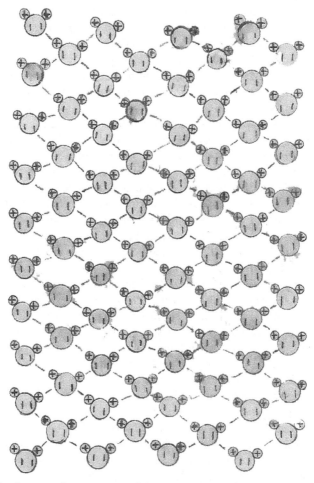

Figure 3.1 The diagrammatic arrangement of the water molecules forming the surface tension layer explains the reason for its great strength. Because the electrical charges force the molecules to arrange themselves in a two-dimensional, horizontal plane only at the surface, a very cohesive tension layer strong enough to support relatively heavy items is formed. Each molecule consists of an oxygen ion with a total negative charge of 2- and two hydrogen ions, each with a positive charge (+) separated by about 105° from each other. Each positive charge associates itself with one of the negative charges on a different oxygen ion, producing a very stable two dimensional pattern. Below the surface, the water molecules arrange themselves in a continually changing, random, three dimensional fluid pattern.

As the table shows, the surface tension of water is the highest and at least three times stronger than those of every other substance shown, except for glycerol. Mixing water with some of the liquids shown in the table increases the strength of the surface tension. Metallic mercury is one of the only substances with a greater surface tension value than water. Its surface tension is 487.0 dynes cm^{-1} at 15°C. A 55% sucrose solution has a surface tension only slightly less than that of pure water. Adding NaCl to pure water in order to make a 6.0 M aqueous solution at 20°C raises the surface tension from 78.2 to 82.6 dynes cm^{-1} (Dean, 1973). This indicates that the surface tension of

Table 3.1 The surface tension in dynes cm⁻¹ of selected undiluted liquids, all at a temperature of 20°C (Dean, 1973).

Liquid	Surface tension (dynes·cm⁻¹)
Acetic acid	27.6
Acetone	23.7
Diethyl ether	17.0
Ethanol	22.3
Glycerol	63.0
n-Hexane	18.4
Isopropanol	21.7
Methanol	22.6
n-Octane	21.8
Water	72.8

seawater is not lower than that of freshwater. The sea can potentially support hexapods just as bodies of freshwater do, but turbulence must also be taken into account when determining whether aquatic hexapods can survive on the surface of the ocean.

It is easy to observe that the surface tension layer on lentic water bodies is strong enough to support relatively heavy objects. Tap water in a bowl easily supports a needle or a flat safety razor blade. Once below the surface, any such metal object rapidly sinks to the bottom of the container. The presence of both detritus and living organisms on the surface tension layers of ponds can be seen at a glance. A variety of floating plants take advantage of the surface tension to form almost continuous layers on ponds and lakes. Species of the water fern, *Azolla* spp., and the common water lenses, often called duck weed, *Lemna* spp., and *Spirodela polyrrhiza* (Linnaeus) Schleiden, can be observed in most parts of the world. Such plants are buoyant, and their dorsal surfaces repel water. Therefore, the upper surfaces of the carpets of these tiny plants covering the water remain dry, even during heavy rainstorms. Such plants may support intact beads of water on their hydrophobic surfaces, but they shed these quickly after the rains cease without their surfaces becoming wet. After such plants are forced below the surface of the water, they rapidly float upwards and reorient themselves to form a layer of floating plants with the water-repellant upper surfaces facing upward, and the roots, downward.

Other varieties of adaptation to live on the surface include those of *Wolffia arrhiza* (Linnaeus) Horkel ex Wimmer, which is smaller than most species of *Lemna* and lacks roots on the ventral surface, and *Lemna turionifera* Landolt, a species that forms resting propagules, called turions, which are formed shortly before temporary water bodies dry up. Turions are also produced by the relatively large member of the family, *Spirodela polyrrhiza*, but not by any other known member of the genus *Lemna*.

Other plants frequently observed floating on the surface tension layer include various species of water fern, such as species of *Azolla*, a genus encompassing different species filling roughly the same ecological niche in each major biogeographical region. The plants rest directly on the surface of the water, with the fronds rising only slightly above the level of the surface tension layer. Most other water ferns have more complex fronds rising well above the surface, as well as dense masses of rhizoids below.

The masses of floating plants are particularly important for the tiny hexapods that live on the surface tension layer, such as the aquatic species of Collembola. The small aquatic plants not only provide resting places for individual collembolans on the surface, they also provide concealment from the eyes of the predators below, which would otherwise have an uninhibited view of their potential prey.

Similarly, larger water plants serve similar purposes for larger animals that live on the surface of water bodies. The leaves of water lilies, such as the native North African *Nymphaea caerulea* Savigny, and European *Nymphaea alba* Linnaeus, are also supported in position by the surface tension layer strongly enough to permit animals as large as frogs to rest on them. Buoyancy alone cannot account for the weights that flat plant leaves can support while they are resting on the surface of a pond. Like pins and razor blades, they are heavier than the water that supports them and would sink if it were not for the great strength of bonds formed between the water molecules on the surface.

An extreme example of this ability to support flat objects is exemplified by the large floating leaves of the South American species, *Victoria amazonica* (Poeppig) J.C. Sowerby, which can support itself and a small child, although there are physical factors contributing to its support other than the surface tension of the water. The edges of the enormous leaves are folded upwards, giving the leaves buoyancy like a small, shallow draft boat. The leaves are protected on the underside by a meshwork or large spines, which keep aquatic animals from biting or tearing holes in the leaves from beneath.

Plants with floating leaves are worth mentioning because many small animals that walk on the surface tension layers utilize the leaves of small floating plants to rest and to support them during heavy rainstorms. The leaves of *V. amazonica*, however, are generally freer of resting aquatic animals than the much smaller leaves of other plants due to the raised rim around their entire circumference. This forms a barrier sufficient to prevent the large leaves from being overloaded with resting aquatic animals. The commonest insects seen on such leaves are usually strong flyers, such as members of the orders Odonata, Trichoptera, Neuroptera, and Diptera.

One fundamental question must still be asked. Are the hexapods that live on the surface tension layer of the water aquatic, or should they be called semi-aquatic or even terrestrial? The community that lives on the surface tension of the water is often referred to as epineuston. Some species with the capability of moving across the surface tension layer are not aquatic because they spend most of their time hiding along the shore. Personal observations of spiders in the families Pisauridae and Lycosidae have been made in Laos and Germany, respectively. They may be called semi-aquatic depending upon whether or not the aquatic habitats are the places in which they can most frequently be found. A different species in the family Lycosidae, *Pirata piraticus* (Clerck, 1757), was observed to be truly aquatic, something which has also been frequently reported by other observers. Because a species of insect remains in constant contact with the atmosphere certainly does not mean that it is not aquatic, if it lives most or all of its life on the surface of a water body, finds its food on or in the water, and seldom, if ever, travels along the shore or farther inland. Those which remain near the water and venture out on the surface tension only occasionally could probably be correctly referred to as semi-aquatic.

SECTION 1

Nature of the Surface of Lentic Water

The surface tension layer forms quickly when freshwater comes to rest. The magnetic polarization of the water molecules immediately brings them into the typical two-dimensional orientation at the horizontal interface between the water and the atmosphere. It provides a platform for aquatic insects that is much stronger and more stable than the water below it (Heckman, 1983).

The surface of small standing water bodies also supports small pieces of detritus, pollen grains, and molecules of lipoproteins, which, in turn, provide nutrients for bacteria, fungi, and protozoa. Rapoport and Sánchez (1963) reported that some of these microorganisms may be harmful to the invertebrates living on the surface. However, this combination of detritus and microorganisms can make the surface tension layer even stronger by spreading the weight of the insects and other hexapods over a greater area of the tension layer. On larger ponds and lakes, however, much of the water surface is typically open to the wind, which moves the entire epineuston, including the larger living organisms, to the downwind part of the lake, reducing the area that the species of epineuston can settle upon.

Interestingly, the biotic community of the epineuston is quite similar throughout the world. While the species of hexapod involved are usually those confined to the zoogeographical region in which the habitat is located, there are various ecological niches that are usually occupied by hexapods belonging to closely related species. However, some of the species of Collembola frequently observed in the epineuston are indistinguishable from and probably identical with the species of Collembola found on other continents and widely separated parts of the world. Which of these species spread throughout the world in earlier geological periods and which have been transported unintentionally throughout the world by ships during the past few centuries must still be determined.

Large predators which come to prey on the insects resting on the relatively strong surface tension layer are too heavy and not suitably shaped to distribute their weight on the surface and fall through, allowing their potential prey to escape by moving to locations above relatively deep water. A few large spiders seem to be exceptions to this.

Many aquatic species of Heteroptera, a suborder of Hemiptera, have an additional defense against almost all predators. They produce strong smelling secretions that burn and discolor human skin and deter most predators from attacking them. This was confirmed by personal observations during the early rainy season in the Pantanal, a large wetland in Brazil, where mosquitos searching for blood were swatted during the early rainy season and fell on the surface of the very clear water. They were seized almost immediately by nearby *Hydrometra argentina* Berg, 1879, insects in the family Hydrometridae called water measurers. These insects are small, slender heteropterans, which move relatively slowly across the surface tension layer. They are common in the Pantanal and also found to the south and northward through Amazonia (Moreira et al., 2009). Many small fishes coveted these mosquitoes and swam quickly to the surface to steal them from the water measurers. When doing so, they were very careful to seize the prey without touching the heteropterans, which were small enough for the fishes to consume, as well. This demonstrated very clearly how chary the fishes were to come into contact with the noxious chemicals that the insects were capable of discharging.

Other surface-dwelling hexapods from the epineuston community employ different strategies to avoid predators. For example, members of the order Collembola, the springtails, keep an organ known as a furca continually under tension by a restraining structure called the retinaculum. As described in Chapter 1, when they observe potential predators approaching, the retinaculum releases the furca, which snaps open, catapulting the insect for a distance many times greater than its body length. To a human observer, the movement of the insects is so rapid that they seem to simply disappear from the surface of the water.

The most prominent hexapods that move across the surface tension layer on lentic water bodies as adults include the following:

Collembola – several species of both sub-classes, Arthropleona and Symphypleona, form dense populations on the surfaces of small water bodies and shallow areas of larger ones. The identity of these species depends upon the continent and climatic zone in which the water body is located.

Hemiptera, Suborder Heteroptera – most species in the families Hydrometridae, Veliidae, Macroveliidae, Mesoveliidae, Hebridae, and Gerridae are commonly observed on the surface of water bodies of all sizes throughout the world.

Coleoptera – As far as can be determined, all species in the family Gyrinidae live in the surface tension layer of the water as adults, at least during the breeding season of their species. Their bodies are divided into a dorsal part kept above the surface and a ventral part below it. Their compound eyes are divided exactly along this division so that a dorsal part can see what is going on above the water level, and their ventral part can see what is happening within the water column. These insects are strong swimmers and some species prefer to live in lentic water bodies, while others can live in fast flowing streams, as discussed further below. The species *Gyrinus substriatus* Stephens, 1828, was part of the biotic community in drainage ditches in agricultural regions along the Elbe Estuary near Hamburg.

Other insects from small families and a few from large families predominantly of terrestrial species live primarily on floating plants but make frequent trips across the surface tension layer of the water. For example, *Tanysphaerus lemnae* (Fabricius, 1792), is a tiny European weevil usually found on the surface of species of *Lemna* spp. It slowly but easily crosses open areas of water between floating clusters of the tiny duckweed plants on ponds and flooded ditches. It belongs to Curculionidae, the weevils, the family of beetles encompassing the largest number of species of all insect families. Among these species, only a tiny minority can survive in or on aquatic habitats.

Diptera – the imagoes of many species in the suborder Brachycera belonging to the families Ephydridae, Empididae, Dolichopodidae, Sphaeroceridae, and several others are often present in massive numbers on the surface of water bodies. Apparently, the majority of species in these families are terrestrial and develop as larvae in moist soils. In addition, many species in the suborder Nematocera have aquatic larvae, and the adults of some may rest on the surface tension layer when depositing their eggs.

In order to reduce the evaporation of water from California reservoirs, several of them were coated with heavy organic alcohols, which float on the surface and reduce contact of the water with the atmosphere. Such treatments were found to cause considerable mortality of insects and probably collembolans that normally rest on the surface of the water. The alcohol rises to the surface of the water body but does not form a strong surface tension layer as water does, so larger insects, such as heteropterans in the family Gerridae and Veliidae, fall through the alcohols into the water and drown, even if the alcohols do not kill them on contact.

SECTION 2

The Surface Tension on Lotic Water Bodies

Although it might seem that the surface tension layer would be more stable on lentic water bodies than on lotic ones, there are species well suited to live on the surfaces of rapidly flowing streams, as well. Among these are heteropterans in the family Veliidae, genus *Rhagovelia*, and adult flies in the families Ephydridae and Empididae, among others. They are often accompanied by beetles in the family Gyrinidae, which have unusual adaptations for living imbedded in the surface layer of the water, so that part of the body remains above the surface while the rest is below. Like the gyrinids that swim imbedded in the surface tension on lentic waters, their compound eyes are divided along the horizontal plane so that they have two separate fields of vision, one above the surface and the other below.

Most species that are suited for and prefer life on the surface tension layer of lotic water are noted for moving much more rapidly across the surface than most species of related insects living on ponds and lakes. Some can also easily take to flight if it is necessary for them to remain near a specific location along the course of the stream. In spite of the rapid flow of the water in the stream, these insects are usually seen moving erratically back and forth on the surface in somewhat sheltered locations near the banks. When disturbed, they descend below the surface and allow themselves to be carried away with the current to a place out of danger downstream. When the disturbance ceases, the insects move quickly back upstream to their favorite location.

Adult gyrinids form small swarms on the surface while swimming rapidly but constantly changing direction. Although they are easily seen on the surface, they would be difficult for any predator to catch. If any vertebrate or large insect were to try to catch one, either from above or below, it is doubtful that they would succeed. As soon as gyrinids are approached from above, they dive below the surface and do not return until the danger seems to have passed. If attacked from below, they take to flight. This is almost the exact method of escaping predators used by the species that live in lentic water. Their plastron on the body surface can hold enough air to allow them to stay underwater for a long time until danger passes. Even entomologists do not find them easy to capture, except by sweeping the water quickly with a very large net.

Collembolans also utilize the surface tension of the water in streams, but they are almost always observed in places in the immediate vicinity of the shore. It appears that the water just below the surface is moving rapidly downstream while the surface tension layer remains intact and anchored to emergent objects near the water's edge, unless the surface is broken by turbulence in the vicinity of rocks or other emergent objects. A well-established surface tension layer therefore seems to be electrostatically bound to fixed objects along the shore, restraining a thin film of water anchored in one place just above the fast flowing water below it. If insects moving on this surface tension layer feel that the layer is being disrupted and they are swept out onto the open water, they can usually be seen springing back toward the shore.

SECTION 3

Collembola Species that can be Observed Living on the Surface Tension Layer

A review of adaptations that permit life upon the surface tension demonstrates how this habitat for epineuston provides hexapods with one of the easiest modes of adaptation for terrestrial species to utilize an aquatic habitat. The only adaptation that is commonly seen when a terrestrial collembolan and an aquatic one are congeneric is in the shape of the mucro. Those of terrestrial species are armed with curved, pointed spines on the side opposing the substrate, while those of the aquatic species are flattened and sometimes paddle shaped with the flat side facing downward against the sediment, at least when they are being launched into the air.

a. Collembola – Arthropleona

Poduridae

Podura aquatica (Linnaeus, 1758)

Probably the springtail species best adapted to life in an aquatic environment is *Podura aquatica*. Its habits have been well studied in Europe, where it spends the winter in groups beneath the surface of the water. When the water bodies freeze over, these collembolans remain beneath the ice. After the water warms sufficiently in the spring, the hibernating individuals appear on the surface tension layer of the water, usually in groups.

Like other collembolans that live on the surface tension layer of water bodies, *Podura aquatica* has a long furca with a paddle-shaped mucro on the end. Terrestrial species usually have a claw-like dens, often with subapical teeth, so that they can gain tractions on hard, uneven surfaces when they spring. The mucros of aquatic species, on the other hand, are usually flat with a broad surface, which takes advantage of the cohesiveness of the water molecules in the surface tension layer to prevent the furca from sinking into the water column when the insect springs.

The diet of *Podura aquatica* has not been fully investigated. It is generally reported to feed on small particulate items, such as pollen, algae, and other microorganisms. The food items that provide it with the most important nutrients and the items most preferred by these collembolans have not been unequivocally identified.

The genus *Podura* was long thought to include only *Podura aquatica*. However, several other species from other parts of Europe were also described after Linnaeus named the species. All of these have become junior synonyms of *P. aquatica*. Because these tiny hexapods have the ability to survive on and under water, they are probably easily transferred from one part of the world to another as tiny stowaways in ships. It is therefore likely that the species will eventually be successfully transported from Europe to other continents outside of the Holarctic Region. Once reported from a new part of the world, a specimen of *Podura* will not be easy to recognize as either an undescribed species that was previously overlooked or a newly introduced *Podura aquatica* from Europe.

During the warmer seasons, the close proximity of these insects to open water prevents their dehydration. Many terrestrial collembolans cannot survive very long if the relative humidity of the air decreases very much below 100%. Just above the surface tension layer of a pond or stream, the typical humidity of the microclimate can be assumed to be close to 100%, even where the atmosphere is much drier a short distance from the water. At the same time, the hexapods are continually in contact with the atmosphere, except at times when they are temporarily forced below the water by heavy rainstorms or the activities of large predators.

The water also provides a rich food supply for collembolans that live among the epineuston. All pollen, fungus spores, tiny animals, and particulate detritus that falls on the water become a potential food supply for collembolans living on the surface in sheltered places near the shores of streams. There are also many species of bacteria, fungi, algae, lichens, mosses, protozoa, and tiny invertebrates that break down the detritus that falls into the streams, utilize the minerals released for primary production, or feed on the primary producers. All of these potential foods remain available as food for the springtails. As the streams empty into lakes or rivers, all of the potential food materials that they carry with them flow into these standing water bodies.

Finally, species of Collembola can obviously be preyed upon by a great many larger arthropods in or near water. The springtail species that live primarily on the surfaces of water bodies have a distinct advantage over their terrestrial relatives by keeping well out of reach of the general predators that would feed on them if they were not living on a surface tension layer through which most predators would fall if they attempted to reach their prey hiding in the epineuston.

Most collembolans in the temperate zones seem to spend the winter below moist soils, but few studies have sought to discover the interspecific differences and seasonal life histories of the individual species. *Podura aquatica* seems to differ from most other aquatic springtails by spending the winter in large groups beneath the surface of the water or beneath the ice layer covering the water body.

Like many collembolan species, *Podura aquatica* seems to have a range including several continents, in which suitable climatic conditions occur. The preferred conditions seem to be relatively mild temperate zones with winters during which temperatures below freezing occur.

Isotomidae

Probably the most numerically abundant aquatic springtail on the shallow water bodies of northern Germany is the isotomid, *Isotomurus palustris* (Müller, 1776), which is apparently most closely related phylogenetically to a group of terrestrial and semi-aquatic species, some of which are also occasionally encountered on or near small, shallow water bodies or wetlands. This species sometimes becomes numerically abundant where large amounts of decaying vegetation are present on or in the water.

In contrast to *Podura aquatica,* groups of closely associated individuals belonging to *Isotomurus palustris* do not appear only on the surface of the water. Typically, they spread out evenly over large marshy areas where foods are abundant. Waving a hand across such areas typically produces a shower of individuals flying through the air in all directions.

Other species in this family are also found occasionally on the surface of water bodies, but they are not nearly as abundant as *Isotomurus palustris* on aquatic habitats. Apparently, they are more characteristic of moist soils of terrestrial habitats. One of these that was observed regularly on the surface of the water of drainage ditches in northern Germany was *Isotoma viridis* Bourlet, 1839. Specimens were sometimes captured where sections of the marsh adjoin shallow water bodies.

Chang (1966) compared the behavior of two aquatic isostomids often found on the surface of lentic water bodies in North America: *Isotomurus palustris* and *Folsomia fimetaria* (Linnaeus, 1758). Reproduction was more successful when the springtails were kept at 8° to 10°C than it was at 25° to 28°C. He reported that *Isotomurus palustris* produces ovoviviparous young, brooding one egg at a time. It can also release the unhatched eggs if disturbed, and they can develop on the surface tension layer.

The epicuticles of both *Isotomurus palustris* and *Folsomia fimetaria* shed water to keep the springtails dry, which Chang (1966) attributed to a coating of a wax-like substance coating the body. Water can apparently be absorbed only through the ventral tubus and the mouth. Respiration occurs through tiny tracheal openings, although an internal tracheal system has apparently not developed. *Folsomia fimetarius* seems to prefer remaining in darkness for longer periods of time than *Isotomurus palustris*. Their food was reported to be microscopic invertebrates, including annelids, rotifers, nematodes, and protozoans.

b. Collembola – Symphypleona

Sminthuridae

While members of the Arthropleona are typically elongate, species of Symphypleona appear as tiny spheres on the surface of the water. Characteristic species among the community of aquatic springtails in northern Germany were *Sminthurides aquaticus* (Bourlet, 1843) and *Sminthurides malmgreni* (Tullberg, 1876). Both were abundant at times during the warmer periods of the year.

The aquatic species encountered are almost always found on the surface of the water or on small plants floating on the surface. They are seldom observed on the shore of the water body. Little is known about their activity pattern. It is sometimes reported that they feed on pollen that lands on the surface of the water.

SECTION 4

Insecta

Hemiptera - Heteroptera

Aquatic members of the Heteroptera are typically insects that live throughout all life stages on or in the water. They seldom fly, although many species can be observed on mating flights during certain seasons of the year. Although there are many terrestrial species that feed on plants, most aquatic species seem to be predatory. However, some species of the aquatic family Corixidae may consume microorganisms, but none of them live on the surface tension layer.

Heteropterans are frequently described as having piercing and sucking mouthparts, used by the aquatic species to inject poisons and digestive juices into their prey and after a period of time, suck out the liquefied digested materials. The surface-dwelling species appear to be exclusively predators or scavengers, but a few of the submerged species may pipette tiny invertebrates, or even possibly algae, as mentioned in Chapter 1.

Larger and more prominent insects living on the surface tension layer of freshwater bodies include the members of six different families of Heteroptera: Gerridae, Veliidae, Mesoveliidae, Macroveliidae, Hebridae, and Hydrometridae. Two small families are also recognized and will be introduced in subsequent parts of the book. Both immature and mature instars live on the surface of the water, and nearly all are known to be primarily or exclusively predatory. Apparently, insects that have already drowned are also occasionally taken as food. However, because the mouthparts of these heteropterans are modified for piercing and sucking, the prey must be injected with digestive enzymes and the liquid contents sucked out, as mentioned above. During the digestive processes, it seems that one heteropteran sometimes steals the carcass of a prey species from another member of its own species. It is therefore often difficult to observe whether dead insects are taken initially as food or whether only insects still living are captured and killed. The answer may differ from species to species.

The question of whether or not only living insects are acceptable as food arises because of the apparent dependence of the insects on vibrations moving across the surface tension layer of the water for detecting food and for communication between members of the same species. Especially species in the family Gerridae are suspected of using the vibrations to orient toward terrestrial insects struggling in the water and for communicating with each other to find mates and possibly for other reasons.

Members of the family Gerridae are known as water striders. Both the larvae and adults live on the surface tension of the water. Most species prefer small standing water bodies, but a few live on the surface of streams and rivers, where the water flow is not very rapid. Exceptional species of the family are capable of surviving on the surface of the ocean, most near the seashore but a few on the high seas (Chapter 17). Like terrestrial insects, most exchange air directly with the atmosphere. However, their ventral surfaces are covered with a dense coating of setae capable of repelling water and holding a layer of air over the spiracles, preventing them from being blocked by the water, at least if there is no agitation.

The layers of setae coating the ventral surfaces of common species of *Gerris* inhabiting North German water bodies were found to consist of a dense coating of short, thin setae interspersed with a less dense coating of much longer and more robust setae. The long setae are curved near their apices so that the sharp ends do not puncture the surface tension of the water. The longer setae seem to have both a sensory and a mechanical function to keep the insects' bodies from coming into direct contact with the water. The shorter setae cause a layer of air to adhere to the body surfaces where spiracles are present whenever the insect is forced below the surface of the water (Heckman, 1983).

The underside of the surface tension of the water is typically the habitat of heteropterans belonging to the family Pleidae, which swim inverted along the underside of the surface tension layer. Like species in the beetle family Gerridae, these species simultaneously keep in contact with the epineuston community as well as the atmosphere above the surface tension layer of the water.

Coleoptera

Although the largest families of aquatic beetles encompass almost exclusively strong swimmers, several families of Coleoptera include species that typically walk on the surface tension of the water. These beetles are usually very small, typically no longer than about 6 mm, and often escape notice. Interestingly, several such species are known from the surface tension layers of fast-moving streams. The species belong to the families Dryopidae and Elmidae. Typically, the insects that live on the surface tension layer do not swim. They walk on the surface, and at times that they are forced into the water below, they can walk on submerged objects still carrying a bubble of air, with their ventral sides up. Some of these species move into and out of water, and they frequent places where there is an almost constant spray of water, such as beneath waterfalls or along the shores of streams, where there is a constant splashing of stream water on moss-covered rocks.

In addition, as already described, beetles of the family Gyrinidae typically swim almost as a part of the surface tension layer. The dorsal part of the adult body extends above the surface, while the ventral part remains below the level of the water. The compound eyes are divided into a dorsal and a ventral part with their separation being kept almost exactly at the level of the surface. At least some species of whirligig beetles are capable of taking off quickly from the surface of the water if danger approaches from below. When they break through the water surface and take to flight, all water instantly separates from the plastron, and the spreading of the elytra to release the metathoracic wings instantly liberates all water that might have accumulated along the edges of the wings. Apparently, most whirligig beetles have little trouble flying from one water body to another. However, they are seldom found beneath outdoor lamps where migratory beetles belonging to other families congregate during mating seasons. This indicates that they migrate during daylight hours, migrate seldom or not at all, or do not navigate using fixed lights.

Adults in other beetle families live on the underside of the surface tension layer. Swimming with the ventral side up allows them to keep in nearly continuous contact with the atmosphere while hunting for food on the underside of the surface tension layer. Adults in the families Elmidae and Dryopidae, as well as several other small families are represented in streams and standing water bodies, where they inhabit assemblages of detritus, mats of algae, or cling to other objects. Their behavior differs markedly from most of the insects that swim in the water column. Most of them creep on submerged objects or on the underside of the surface tension layer or on floating plants rather than swim. Some of them even appear clumsy when moving, occasionally causing them to make casual observers think that they are actually terrestrial species that fell into the water and are trying to escape to avoid drowning. In spite of the fact that they appear clumsy in the water, they are highly specialized members of their biotic communities, which are usually confined to rare and isolated habitats, such as waterfalls, the edges of rapidly flowing streams, or among rocks receiving continual spray from streams flowing nearby.

Some beetles belonging to overwhelmingly terrestrial families can also be found in wetlands, where they can walk easily on the surface of the many small assemblages of water and cross small water bodies without difficulty. Most species of the family Staphylinidae hunt small arthropods among the leaf litter in forests, on the soil in grasslands, and in other terrestrial habitats. In marshlands along the Elbe Estuary near Hamburg, several species appeared regularly in waterlogged marshes filled with the wetland grass, *Glyceria maxima* (Hartman) Holmberg. These species were *Stenus boops* Ljungh, 1810; *Stenus bimaculatus* Gyllenhal, 1810; *Stenus cicindeloides* (Schaller, 1783); *Stenus latifrons* Erichson, 1839; and *Xantholinus linearis* (Olivier, 1795).

Similarly, the same marshy drainage ditches supported a small population of *Bembidion semipunctatum* Donovan, 1806, a member of the family Carabidae, another family encompassing mainly predatory terrestrial species. This species is also known to be a marshland species. Whether or not these insects should be considered aquatic or semi-aquatic will not be discussed here. They are among the few species in these families associated with marshes.

Diptera

A great many species in the insect Order Diptera are associated with aquatic habitats of all kinds as larvae, but the adults of few species are seasonally abundant on water as epineuston. Two families including a few species that are well adapted to the surface tension layer of small water bodies are Empididae and Ephydridae. Most of the species in these families are terrestrial, but a few species appear in large numbers on the surfaces of small water bodies, often located in forests, during their mating seasons. They are sometimes joined by relatively uncommon aquatic species from other families of Diptera encompassing relatively few aquatic species, such as Dolichopodidae and Sphaeroceridae. In many parts of the world, there is not enough general information on the insect fauna to know for sure which of the dipteran species are aquatic during any stages of development.

During studies of the drainage ditches along the freshwater section of the Elbe Estuary (Heckman, 1990), large numbers of adult *Hydrellia griseola* (Fallén, 1813), members of the family Ephydridae, were observed on the shores near the water and on the surface of the water, as well. There were also large numbers of unidentified ephydrid larvae just below the surface of the water. It is fairly certain that after the identification of the larvae developing in the small water bodies, they will be found to be the same species that are frequently observed resting on the surface of the water. Chemical analyses indicated that there was a significant concentration of ammonium ions in the water, which had the odor of ammonia and appeared to be heavily contaminated with decomposing organic material.

There are a great many species of Diptera that develop as larvae in water or congregate near water bodies to mate and lay their eggs. Most of them do not join the epineuston community as adults and develop as larvae below the surface. They will be discussed further in the subsequent chapters.

It should be added that a great many species inhabiting small, standing water bodies as larvae shun the water as adults, except during spawning. These species include almost all species of mosquitoes in the family Culicidae, midges in the family Dixidae, species of Chaoboridae, and adult black flies in the family Simuliidae, to mention a few. Such species will either rest far above the surface of the water on emergent plants or will, as in the case of the species of Culicidae, travel far from water to seek out animals, including human beings, in order to find a meal of blood to provide the nutrients needed to produce their eggs.

Hymenoptera

Only in recent years has the abundance of species in the order Hymenoptera in the aquatic community become clear. The larvae of many species live as parasitoids inside of the bodies of other insects and are almost never seen during casual perusals of water samples. Other species develop inside of various aquatic plants, sometimes as parasitoids of other insects that develop as larvae boring through the stems and leaves of various plants. These will be discussed in more detail in subsequent chapters.

Adults of these parasitoid species are seldom encountered on the surface of the water in shallow aquatic habitats, often among aquatic plants. Most such species find terrestrial species as hosts. However, a large number of insect species that live on the surface of small, shallow water bodies are potential hosts for one of the hymenopterans that can develop as larvae inside of other insect species. Even in Europe, it is likely that more than half of the species that develop as parasitoids in other insects have not yet been described and named.

In addition, in tropical and subtropical regions with alternating wet and dry seasons, species of ant live during the dry season in burrows through the dried sediment, very similar to the nests of many terrestrial ant species. During the rainy season, when the lands on which the ants live become flooded and remain so for many weeks at a time, the ants move out of the burrows into the upper parts of emergent plants. At times, they remain together in masses on the surface of the water, rotating the mass slowly so that no ant at the bottom of the mass remains underwater for too long a period of time. Eventually, they find a place among the emergent parts of aquatic plants in which to remain until the dry season comes and the water level descends again to expose the sediment.

Aquatic ants are not found only on the surface of water bodies. There are species that also live underwater, both in freshwater and in the sea. Some of these will be discussed further in Chapter 28.

4

Herbivorous Insects on Aquatic Plants

Any aquatic habitat that is not extremely hot, cold, saline, or polluted and which does not flow very fast is colonized by a distinct aggregation of aquatic plants, typical for the geographical region. The plants will produce food on which terrestrial animals can subsist, and detritus, which will support other insect species through a complex food web involving bacteria, protozoans, and microscopic invertebrates. The herbivorous insects require only those modifications of morphology, physiology, and behavior of a typical terrestrial insect to utilize their food source and make themselves difficult for potential predators to capture. The parts of aquatic and marshland plants that grow above the surface of the water while their lower parts are rooted in water-soaked sediments or are submerged in deeper areas of open water can provide a habitat for insects with few recognizable modifications of basic morphology, physiology, or behavior of a typical terrestrial insect. Nevertheless, closer observations reveal that even hexapods from predominantly terrestrial insect families show subtle differences from their close terrestrial relatives. This gives the species with the suitable adaptations considerably better chances of success in the kinds of biotic community which encompasses the species of emergent aquatic plants on which it prefers to feed.

SECTION 1

Easy Transition from Terrestrial to Aquatic Habitats

Some herbivorous insects belonging to families consisting mainly of terrestrial species are specialized for living and feeding on the emergent parts of aquatic vegetation. Herbivores that do not specialize in feeding on any species or limited group of species may also feed occasionally on the leaves, fruit, or nectar of aquatic plants. None of these are considered further in this chapter, except for those species that have notable adaptations for surviving in or on water or which can survive only by feeding on specific aquatic plants. It is the species that rely upon one or more species of aquatic plant as the exclusive food sources for their adults or larvae, which can be considered aquatic species of special interest. Such species usually appear very similar to closely related species in the same genus, which feed mainly on terrestrial plants. Closer observation, however, almost always shows that the insects living in association with aquatic plant species show no distress if they fall into the water, and they may even enter the water voluntarily if danger threatens. Observation also reveals other behavior patterns very much different from those of closely related species that feed mainly on fully terrestrial plants.

An example of such an aquatic insect without direct contact with water is *Stenodiplosis phragmicola* Sinclair and Ahee, 2013. It is a gall midge in the family Cecidomyiidae, which develops as a larva in the reed, *Phragmites australis* (Cavanilles) Trinius ex Steudel. The reed often covers large areas along the shorelines of lakes and rivers throughout much of the world. Typically, such stands of reeds cover both submerged areas and areas usually just above the water line. The gall midges were collected and described from Ontario, Canada (Ahee et al., 2013).

In standing and slowly flowing water bodies, many species of aquatic plants with gas-filled structures permitting floatation are encountered. In tropical and subtropical regions, lakes often become covered by large beds of floating plants possessing root systems entirely submersed in water and without any contact with solid substrate. These plants form dense beds, which are moved around on the surface of the lakes according to the direction and velocity of winds and water currents. These "floating islands" of vegetation have been referred to as pseudoterrestrium because they are so dense that the surface of the water is barely visible anywhere through the parts of the plants above the surface of the water (Heckman, 1994). In eastern Africa, these floating islands include relatively tall plants, such as *Cyperus papyrus* Linnaeus, and they are locally called sudd. Such aggregations of floating vegetation take on characteristics of real islands, developing rudimentary soil out of humic matter that forms from dead leaves and stems of the plants, in which some of the component species may be rooted. The islands might even appear sturdy enough to support the weight of a person, although the water beneath them may be quite deep. For small wading birds, amphibians, and aquatic snakes, the pseudoterrestrium is dense and buoyant enough to support movement and activity similar to that in actual terrestrial habitats. For many species of terrestrial insect, there is no disadvantage at all of living among the emergent parts of the aquatic plants, and no morphological or physiological modifications are necessary to adapt to life on the floating islands. Most, however, differ in behavior from terrestrial herbivores by entering the water when they see a predator approaching and hiding among the roots or rhizoids of the plants until the danger passes.

Unlike the insects they provide with a habitat, the plants themselves are certainly well modified to form floating islands. Appearances can be deceiving, and almost all plants that thrive on the pseudoterrestrium have their roots fully submerged in water at almost all times. The roots of most do not reach the sediment. In many cases, the plants require aerenchyme tissue inside of the stems and roots to convey atmospheric oxygen to the submerged parts of the plants. Aerenchyme tissue is even more of a necessity for plants rooted in sediments beneath a shallow layer of water, especially where the sediments are completely anaerobic. Mangroves, for example, develop particularly complex systems to supply the roots with atmospheric oxygen, including vertical branches of the roots, which grow upward from beneath the sediment and through the water to reach the atmosphere. These are open at the top and conduct air downward into the roots deep within the underlying sediments. Mangrove forests will be discussed further in Chapter 18.

The relationship between the aquatic plants and their habitats sometimes require insects that live on the plants to develop subtle adaptations to their floating habitats. Careful observation reveals that insects living on plants in the pseudoterrestrium display morphological, biochemical, and behavioral characteristics not at all typical for herbivorous, terrestrial insects. Although herbivorous insects can and do live on floating and emergent aquatic plants without developing specific adaptations for their habitats when not under stress from predators or extreme weather conditions, many species can substantially improve their survival chances through special adaptations for survival in the presence of predators or during the heavy rainfall associated with tropical storms.

Because there is always water beneath the floating islands of vegetation, and the water can be deep and inhabited by fishes and other predatory animals that include insects in their diets, hexapods that live among these plants are at an advantage if they have the ability to avoid drowning after falling on the surface of the water. Insects living in the pseudoterrestrium must beware not only of fishes and amphibians living at or below the surface of the water, but also of many insectivorous birds that land on the plants to capture any insects that they notice from the air. Dense root masses can conceal an insect from enemies coming either from above or below, but only if the insects can remain concealed under water without drowning. If a herbivorous insect is forced to retreat into the water to escape a predator approaching from above or if it was washed into the water by heavy rain, it may display behavior markedly different from that of species believed to be closely related phylogenetically but which are required to live only on dry terrestrial vegetation under normal circumstances. For example, an aquatic grasshopper in Brazil, *Paulinia acuminata* (De Geer, 1773), lives in close association with the water fern, *Salvinia auriculata* Aublet, which forms a dense, feathery mass of rhizoids beneath the surface of the water. This species, in the family Pauliniidae, retreats into the water and hides itself within the dense rhizoids of the water fern whenever it detects a potential enemy approaching. Barthlott et al. (1994) noted that the herbivore mimics its host fern not just in coloration but also in its ultrastructure, as observed under a scanning electron microscope. *Salvinia auriculata* and *Paulinia acuminata* both repel water due to a similar pattern of epicuticular wax arranged as pentagons that are of similar size.

Just living temporarily on the surface tension layer of the water would not substantially improve the survival of the hexapods. Predators are usually plentiful both above and below the surface of the water, and they could observe any potential prey from any direction if it were resting on a small patch of open water between floating plants. An insect concealed within masses of leaves, stems, and roots of the floating plants, on the other hand, could remain completely concealed for long periods of time as long as it refrained from moving around.

Most species of aphid, grasshopper, cricket, and fly in the suborder Brachycera will struggle to escape from the water if they fall in. Failure to climb out of the water promptly will often result in the death of the insect. Species in the same families adapted to life on an aquatic plant species with parts that grow above the level of the water typically enter the water deliberately to escape predators, swim or climb on the submerged roots of the plants, and sometimes hide for appreciable periods of time until the danger has passed. Quickly finding concealment among the floating plants is clearly the only option for hexapods in times of danger, which offers good chances for the insects to survive. This shows not only modifications in the behavior of the insect not displayed by terrestrial species in the same family, it shows that the insect also possesses morphological differences to facilitate respiration during limited periods of time while the insect is submerged and probably physiological and biochemical adaptations, as well (Heckman, 1998).

In many places, the floating plant masses are considered a serious problem because they block shipping on lentic and slowly flowing lotic water courses and canals. In such cases, herbivorous insects can be used to control the floating vegetation, if suitable species are available. The problem with biological control of the floating plants is that the plant species causing the most problems had been removed from their natural ranges and introduced to other continents. The animals that fed on them and kept them under control on their native water bodies were not introduced to the new continents, and none of the insects native to the new habitats are adapted to feeding on the invasive plant species. While the insect introduced from the native continent of the invasive plant may well reduce the size of that plant's population in its new home, such insects have frequently become problems just as bad or worse than the nuisance plant by finding a useful and economically important plant in the land to which it has been introduced and making that plant its preferred food. Thus, the insect imported to biologically control an unattractive plant growing in the local waterways may turn out to decimate the harvest of the most important crop grown in the country to which it was introduced. To date, no way has been found to predict whether an insect will find a different plant species on a new continent preferable to the plant species it lived on at home.

Moreover, many invasive plant species encounter no species in their new habitats capable of competing with them for the open spaces of water or for the dissolved nutrients. Perhaps the most notorious floating plant species world-wide is *Eichhornia crassipes* (Martius) Solms, called the water hyacinth (Gopal, 1987). In the nutrient-poor waters of tropical South America, where the plant originated, it is a small, floating plant with attractive flowers. Each year, many of these plants and several other species which share the surface of the water with them, perish during the dry season, when they are stranded by the receding water of the many temporary lakes and rivers on their native flood plains.

In many Asian water bodies, to which they have been introduced, the concentrations of dissolved nutrients are extremely high. Where the pollution decimates the populations of numerous plants and animals, the efficiency of the aerenchyme tissues in the roots of *Eichhornia crassipes* permits this species to fully utilize the available nutrients for growth and the atmospheric oxygen for respiration by their roots. Thus, the water hyacinths in Brazil usually reach a height of 25 to 75 cm above the surface of the water, while conspecific plants in Bangladesh often exceed 2 m in height and form permanent stands, which make it difficult for boats to use the water courses along the canals and rivers for transportation (Fig. 4.1).

Figure 4.1 A small waterway between two floating restaurants completely blocked by a floating island mainly of *Eichhornia crassipes* washed seaward from upstream with outflowing water in the Bang Pakong Estuary, Chacheongsao Province, Thailand, at the end of the rainy season.

Naturally, methods to control the water hyacinth have been sought in many parts of the world. In Bangladesh, a search for products that could be produced from the water hyacinth so that it could be harvested profitably has seen little success. While paper has been successfully produced from this plant, it is of low quality, limiting the profit that can be made from producing and selling it. The hope of feeding the leaves to domestic animals has also led to disappointment. It simply contains too little nutrient matter to be good animal food. This has led to a search for a method of biological control for the water hyacinth wherever it has become an invasive plant (Cordo, 1996). The best choice for a species to select for biological control of an aquatic plant would most likely be an insect because of the vast number of different herbivorous species from which to select. The danger of using an insect to control a plant species outside of its natural range is that this insect would be able to adapt to feeding on species of useful plants in its new location and wipe out their populations, as mentioned above. In South America, where there are a great many native floating plant species, some of the insects live mainly on only one or two, while others are able to subsist on most of the plants, which they consume according to preference and availability. None of the species in the first group depend upon *Eichhornia crassipes* as their chief food or only food, and in the second of these groups, some of the species will eat holes in the leaves of *Eichhornia crassipes*, but they prefer to feed on other species of aquatic plant whenever they are available.

The caterpillars of one moth in the subfamily Arctiinae, family Erebidae, *Paracles palustris* (Joergensen, 1935) reportedly lives on *Eichhornia crassipes* growing on the surface of water bodies in Paraguay (Drechsel, 2014). These insects do not have any noticeable impact, even on the small specimens of *E. crassipes* that grow within their natural range in South America. It is not at all likely that they would make any inroads into the dense beds of the large specimens that cover the water bodies in tropical Asia.

Many members of the less selective species belong to the order Orthoptera. Among the most remarkable of these species is *Cornops aquaticum* (Bruner, 1906). One of the most surprising sights in the clear water of the northern Pantanal during the early rainy season is grasshoppers swimming underwater among the tiny, newly hatched fishes, which appear in great numbers after flood waters again cover the formerly desiccated plains. During that season, floating plants are limited to small stands that managed to survive the dry season. Within a surprisingly short period of time, these plants spread over most of the vast floodplains.

In the northern Pantanal, *Cornops aquaticum* is associated with the floating vegetation, but it is certainly not limited in its activity to the parts of the plants above the surface of the water. At rest, these grasshoppers in the superfamily Acridoidea usually remain on the leaves of either a floating *Eichhornia azurea* (Swartz) Kunth or a rooted *Pontederia lanceolata* Nuttal. When approached, they quickly move to the opposite side of the leaf they are resting on, out of sight from the observer, and then move up or down for a short distance. When the perceived danger moves closer, the grasshopper will jump from the leaf and fly a short distance to another group of emergent plants, on which it moves to the side opposite the approaching danger and downward. If the danger approaches closer, the grasshopper enters the water and hides among the roots and rhizoids of the floating plants. If the plants are disturbed, the grasshopper often swims at a depth of 10 to 50 cm below the surface toward another nearby group of rooted or floating plants and hides within the submerged roots. When the danger seems to have passed, the grasshoppers cautiously return to the surface and climb out of the water. They seem to prefer *Pontederia lanceolata* and *Eichhornia azurea* for food, and they seldom consume *Eichhornia crassipes*. However, there have been no detailed studies to determine exactly how these preferences determine the percentages of each of the species of plant in the grasshopper's diet.

Another species of grasshopper, *Marilia remipes* Uvarov, 1929, also in the family Pauliniidae, displays different habits (Carbonell, 1957). Unlike *Cornops aquaticum*, which feeds by eating small holes in the leaves of *Eichhornia crassipes,* but relying more on *Pontederia lanceolata,* and other emergent tracheophytes, *Marilia remipes* seems to feed primarily or exclusively on the submerged parts of *Hydrocleys nymphoides* (H. and B. ex Willd) in the family Limnocharitaceae. These small, red patterned grasshoppers generally remain on the dorsal surface of their preferred host plant, relying on its color pattern to camouflage it from its natural enemies, presumed to be mainly species of insectivorous bird.

When danger approaches, *Marellia remipes* most frequently enters the water and seeks a hiding place among the dense rhizoids of an aquatic fern. *Marellia remipes* was not frequently seen swimming underwater. Instead, it moves by creeping among roots and rhizoids from one hiding place to another underwater. Because it is not as strong a swimmer and flyer as *Cornops aquaticum*, it prefers hiding below the surface of the water to flight.

There are many species of insect that live on the emergent parts of the pseudoterrestrium formed by the aquatic plants. These may or may not display special adaptations for life on the plants that float on the surface of the water. However, a close examination of the mode of life of these insects indicates that most of them do. First, the insects are usually adept at climbing back onto the emergent parts of the plants should they be swept from the plants into the water. Unidentified species of aphid with wings were observed on the surface of the water at times. They could leave the water without difficulty and shed tiny droplets of water from their wings. Terrestrial aphids which fall into the water can almost never escape drowning if they are not eaten by a predator first.

Another subtle adaptation to life on emergent vegetation is the utilization of its foliage as a shield to predation from below. This provides relatively small hexapods, including members of the Collembola and insects in the orders Coleoptera and Diptera, with resting places that are completely invisible from below. In the Pantanal, it is much more important for insects to conceal themselves from the large number of predatory fishes in the water below them than it is in the small water bodies of northern Europe, where fishes are not nearly as abundant. Fishes and large predatory insects hunting for food beneath the surface can easily observe the tiny insects that live on the surface tension layer of the clear water that is present on the flood plains of the Pantanal during the early part of the rainy season. Once seen by one of the fishes, the insect is usually captured in a matter of seconds. Certain floating plants have leaves that lie flat on the surface tension layer, and it is a simple matter for an inhabitant of water bodies covered by such emergent plants to climb onto one of the leaves and thereby keep from being seen from below.

A rich fauna of small insects inhabit shallow water bodies in the Pantanal of Mato Grosso during the late rainy season in spite of a large variety of small fishes hunting for them just below the surface. After the plains have been flooded for three to four months, the water is no longer clear but rather filled with much detritus from the plants and many microorganisms breaking the detritus down. As more inorganic nutrients are released into the water, microalgae proliferate, and the water becomes increasingly turbid. The insects present conceal themselves between the leaves of emergent plants or in the detritus from such plants to keep out of the sight of the predators beneath them (Heckman, 1998). Among the aquatic species that protect themselves in this way are collembolans of the species *Sminthurides aquaticus* Bourlet, 1843. This species has already been mentioned because it is also a familiar member of the community of species living on the surface tension layer in northern Europe.

A similar group of species was encountered on the surface of ditches draining orchards in the floodplain of the Elbe Estuary near Hamburg, Germany. Many of these ditches remained flooded during the entire year and sometimes had ice on the surface during winter. These ditches were partially covered by various submerged and floating plants, which included *Lemna minor* Linnaeus, *Lemna gibba* Linnaeus, *Spirodela polyrrhiza* (Linnaeus) Schleiden, *Azolla pinnata* Robert Brown, and several larger emergent plants (Caspers and Heckman, 1982). The predominant floating plants were almost all much smaller than the species present in the tropics, but the emergent grasses rooted in the sediment were taller. The Collembola species that congregated in large numbers on the surfaces of these ditches included *Podura aquatica* Linnaeus, 1758; *Isotomurus palustris* (O. F. Müller, 1776); *Isotoma viridis* Bourlet, 1839; and *Sminthurides aquaticus*, as discussed in the last chapter.

In addition to the collembolans encountered on the surface of the water, small insects were encountered on and near the small floating plants. In Europe, a tiny weevil, *Tanysphyrus lemnae* (Paykull, 1792), is often found on the floating beds of *Lemna minor,* which cover many small standing water bodies each year during the summer, especially in places where the water is rich in inorganic plant nutrients.

Relatively few species of Lepidoptera spend part of their development on aquatic plants, especially on their submerged parts. Apparently, most such species inhabit South America and eastern Asia, but they are not well studied. Only a few species are known from Europe, and they are the species which have been most thoroughly studied (Wesenberg-Lund, 1943). The habits of aquatic moths are also known from North America and Japan, but most information on the habits of aquatic moths comes from European studies. Most caterpillars specialize in feeding on one or only a few species of plant, and the moths are aquatic because their host plants grow in or on water. Unlike most of the species discussed in this chapter, the moths tend to feed on the parts of emergent plants growing on or slightly beneath the surface of the water bodies in which they develop. Only one of the species discussed below feeds on plants growing at a depth well below the surface on the *Stratiotes aloides*, a plant that rises to the surface to permit the upper, flowering parts of the plant to emerge above the water. More will be said about these plants and the caterpillars they host in subsequent chapters.

A few insect species were found at the surface of the water in the middle of the floating plant layer. One of these inhabiting small standing water bodies in northern Germany was the larva of an aquatic moth, *Cataclysta lemnata* Linnaeus, 1758. This is a species that constructs tubes as shelter using whole *Lemna* plants held together with silk. It carries the cylindrical case around with it, giving it perfect camouflage in the middle of a solid covering of *Lemna* on the surface of the water. The pupa remains in the case until it emerges as an adult.

The caterpillar of a second aquatic moth also builds its shelter using *Lemna*, which is attached to the leaves of larger aquatic plants on the surface of the water with silk threads. According to Wesenberg-Lund (1943), the larva of *Elophila nymphaeata* (Linnaeus, 1758) is the largest of the Central European aquatic moths, and it often develops in association with plants such as the native water lilies, *Nymphaea alba* Linnaeus.

A European moth that can develop underwater some distance below the surface is *Paraponyx stratiotata* (Linnaeus, 1758). Its host is a remarkable-looking aquatic plant found in northern Europe: *Stratiotes aloides* Linnaeus. It has stiff, spiny leaves, which make it resemble an underwater pineapple. The typical habitat of these plants is small, shallow water bodies, where it begins growing from the sediment as a small plant in spring and continues growing throughout the summer until it releases itself from the benthos and rises to the surface above all of the floating plants in the water at the time it blooms. In recent years, the species has become increasingly rarer due to changes in agricultural methods, especially because of the decrease in the number of drainage ditches used to collect excess standing water in fruit orchards (Caspers and Heckman, 1982). Along with the host plants, the moth populations have been declining locally.

SECTION 2

Aquatic Crops, Their Pests, and Their Insect Protectors

In Southeast Asia, the climate is strongly influenced by the Monsoon, a large circulation pattern of the atmosphere centered in the Gobi Desert. When this large desert region warms up during spring, a counterclockwise air circulation develops around the stationary low pressure system that develops. In the autumn, this desert cools faster than the adjacent land areas and surrounding oceans to produce a high pressure area that remains in place for approximately six months. The winds circulate in a clockwise pattern around the high. The annual alternation in the warming and cooling produces reversals in the wind direction twice each year and variations in its velocity according to the air temperature on a rather precise schedule.

The geography of Southeast Asia is responsible for distributing the rainfall that results from the monsoonal winds. Between Vietnam in the East and Thailand and Laos in the West, there are mountain ranges that cause the air to rise and release much of its moisture. During the Southwest Monsoon from late April through mid-October, the winds move counterclockwise around the low pressure area over the Central Asian deserts, bringing the warm, moist air over the Indian Ocean across Thailand, Laos, and Cambodia. When it reaches the mountain ranges east of the Mekong River, it builds large thunderstorms, which persist until mid-October. On the other side of the mountains, there is little rainfall, and Vietnam has hot, dry weather during most of that period.

By mid-October, the air pressure over the Gobi Desert has become approximately equal to that over the Indian Ocean, so wind becomes nearly calm. It then cools further and begins to develop into a high pressure area over the desert. The wind then begins to circulate in a clockwise direction and to carry cool, moist air from the Pacific Ocean to the Vietnamese side of the mountains, where fine, cool rainwater falls almost daily on the eastern face of the mountains. The air crosses the mountain range, and as it descends on the west side of these mountains, it has lost its excess moisture and brings no rain. By the time it reaches the Mekong Valley in Laos and the Korat Plateau in western Thailand, it has become cool and dry.

The monsoon is responsible for the weather over a large part of Asia, and because it is so large, the areas affected have a climate that remains rather stable from year to year. The stable climate has long produced good conditions for a productive agriculture. The staple crop, rice, is aquatic. Rice, *Oryza sativa* Linnaeus, is apparently native to Southeast Asia, and its culture in both the Temperate and the Equatorial Zone had to await the development of strains that produce rice grains without the stimulus of monsoonal seasonal change. In addition to rice, there are other cultivated aquatic plants in the western part of monsoonal Southeast Asia, such as lotus, *Nelumbo nucifera* Gaertner; water chestnut, *Trapa natans* Linnaeus; taro, *Colocassia esculenta* (Linnaeus) Schott; yam, *Dioscorea* spp.; sweet potato, *Ipomoea batatas* (Linnaeus) Lamarck; and locally eaten green vegetables, such as *Ipomoea aquatica* Forsskal. All of these plants have emergent parts, which could be attacked by terrestrial insects without very much adjustment of the physiology or growth patterns of the insects (Fig. 4.2). Among these herbivorous pests are species of aphids and other members of the suborder Homoptera, leaf-eating members of the Orthoptera, various species of Coleoptera, and species of Diptera adapted to feeding on the aquatic crops. In such cases, many agricultural experts, who were trained in contemporary methods of pest control, have encouraged the introduction of newly developed mixtures of pesticides. In theory, the introduction of pesticides permits cultivation of more productive varieties of the crops, which are considerably more productive but less resistant to attacks by various herbivorous insects. The pesticides not only kill off the herbivorous pests but also the predatory species that have controlled the economically harmful insects while traditional cultivation methods had been employed. As long as the pesticides remain effective, the new methods of farming seem to be an improvement. However, such innovations seldom provide long-term increases in productivity.

Undesirable insects have shown a remarkable propensity for developing resistance to toxic substances used against them. As the pesticides used initially become less and less effective, new substances are substituted. Many times, the new insecticides prove to be less effective than those initially used, and some of these are more harmful to human beings. Even if the newer toxic substances substituted for the older one are just as effective as the older ones used to be, the process of developing resistance against the

Figure 4.2 A flooded permanent pond behind a factory at the end of a normal dry season in Chacheongsao Province, Thailand. The pond is in a rural area and inhabited by a large variety of aquatic insects. In the fore-ground, the long, creeping vine on the surface of the water is a popular green vegetable, *Ipomoea aquatica.*

newer ones begins immediately, and the effectiveness of the newer substances begins to decrease at more or less the same rate as that of the previously used pesticide mixtures. In addition, each new substance has an impact on the natural biotic communities by continuing to reduce the natural protection of the crops by predatory insects.

This introduces the problem of discovering the most effective method of protecting large monocultures of aquatic crops, such as rice in Southeast Asia, from herbivorous insects. Such crops are still economically important, although not as important as they were a few decades ago. During the past 30 years, there has been considerable diversification in the agriculture of Thailand and neighboring countries. Many new crops have been introduced, and the economic importance of rice has decreased. The introduction of new fruits and vegetables increases the potential profits of the farmers because they often bring much higher prices on the local markets, but they also require the education of the local farmers in the cultivation methods for the species never previously raised in the region.

With diversification has come a reduction in the percentage of the arable land devoted to rice and other aquatic crops and an increase in cultivation of terrestrial plants. Nevertheless, insect pests on aquatic plants are still present, and their control remains an important factor in agriculture.

The introduction of new crops to Southeast Asia will definitely result in the eventual introduction of insects that eat them. However, most of the pests will very likely be terrestrial species that feed on terrestrial crops. Where aquatic crops are concerned, the introduction of new pest species from other parts of the world should be avoided. However, experience shows that preventing the migration and even intercontinental transports of dangerous pests is not always effective. As it becomes apparent that trying to use chemical pesticides to control insect pests that feed on the emergent parts of crops is futile and often dangerous, greater emphasis is being placed on using predatory insects to control insect pests. Some of the most successful introductions of predatory insects to control the pests in tropical agriculture involve members of the order Neuroptera, especially species in the families Chrysopidae, Dilaridae, and Hemerobiidae. Neither the larvae nor the adults of these insects are aquatic, although many could probably adapt to the emergent parts of marshland plants.

SECTION 3

Aquatic Insects Capable of Natural Pest Control

Archaeological discoveries in Northeast Thailand indicate that rice has been cultivated there for at least several thousand years, and it has been cultivated in southern China for more than one thousand years. Rice cultivation was introduced to Indonesia much later, only after rice varieties were found that would produce rice grains in the absence of alterations between wet and dry seasons.

Rice has also been cultivated in West Africa, where different species of the genus *Oryza* have been raised along the downstream sections of West African rivers. The species of African rice include the cultivated species, *Oryza glaberrima* Steudel.

During almost the entire period that rice has been cultivated and a large and ever increasing percentage of human beings on the earth have adopted it as their staple crop, the plants have never been subjected to catastrophic losses due to attacks by insects. Rice has also proven resistant to blights and serious plant diseases.

Most rice is grown in flooded fields that remain under water for approximately six months each year. This period of time is sufficient for the larvae of various aquatic insects to complete their development. These species include some that feed on other insects in the water as larvae, others that feed on insects near water as adults, and still others that feed on potential insect pests throughout their entire life cycles. In addition, the water in rice fields serves as the nursery for small crustaceans, fishes, and amphibians, which feed on harmful insects while they grow and serve as protein-rich foods for the local human population after they have grown. A general list of insect species useful for the protection of aquatic crops is provided in Table 4.1.

Table 4.1 Aquatic insects belonging to the following orders, suborders, or families inhabit cultivated rice fields in Southeast Asia, and most serve to protect the crops being cultivated in the tropical wet-and-dry climatic zone from herbivorous insects, most of which feed on the emergent parts of the plants.

Odonata		
	Zygoptera	Larvae – All known species of Zygoptera feed on insects under water as larvae
		Adults – Adult damselflies capture and consume most flying insects
	Anisoptera	Larvae – Anisoptera larvae inhabit water and consume aquatic prey
		Adults – Adult dragonflies capture and consume fairly large flying insects
Hemiptera/Heteroptera		Almost all aquatic species are known to be predators as both larvae and adults
Coleoptera/Adephaga		Families Dytiscidae *sensu lato*, Haliplidae, and Gyrinidae: Almost all aquatic species are known to be predatory as larvae and adults
Hymenoptera/Apocrita		Families Ichneumonidae and Braconidae: Parasitoids of insects

With the introduction of the rice variety called "miracle rice" at the beginning of the 1970s came the perceived need to introduce various pesticides. The new variety allegedly lacked the natural resistance to pests and diseases that the traditional varieties possessed. Thailand had long been a major exporter of long grain rice to other countries in Asia. However, after the rice harvest in 1971, the great increase in the rice yield brought about a great decrease in the price of rice on the open market. Although the Thai government maintained the price of rice by permitting the farmers to sell their crops at government purchasing agencies for a guaranteed price, most farmers had to depend upon private rice buyers who could transport the rice in trucks to the purchasing agents. Farmers who did not own suitable vehicles were fully dependent on the private buyers and had to sell their rice for whatever price the buyers were willing to pay. This began a diversification of the crops grown within the country, and much of the land that had been occupied by rice fields was quickly converted to orchards used for growing many kinds of fruit and fields for growing a great variety of vegetables. Many plant species were introduced, and many new foods became popular.

The rich diversity of the aquatic insect fauna may be reduced as the total area covered by flooded rice fields decreases, but extensive changes are still occurring, and time will tell how well the insect fauna adjusts to the new conditions and the newly introduced plant species.

5

Hexapods on the Roots and Rhizoids Beneath the Pseudoterrestrium

a. The Home of Remarkable but Poorly Studied Insects

Almost everywhere affected by a seasonal climate, various plant species have adapted to life on or in water bodies. In the temperate zones, these plants are a familiar sight growing in shallow water bodies, in which they develop below the surface or are rooted in the sediment and produce emergent shoots. In the subtropics, where the temperature remains above freezing throughout the year, the seasons are often designated as rainy and dry, according to a strict schedule during most years or a system that is highly variable from year to year. The aquatic plants grow and produce spores, seeds, or vegetative propagules during an annual rainy seasons and as long into the dry season as the water bodies remain flooded.

Typical of this kind of habitat in all parts of the world where it occurs are large areas of water completely covered by aquatic plants, often dense enough to obscure the surface of the water. As discussed in Chapter 4, such vegetation has been named the pseudoterrestrium because it resembles large terrestrial areas, especially on aerial photos (Heckman, 1994). Under these circumstances, only the species of plant dominating this biotic community reveal that this habitat is aquatic. In some places, the plants contribute detritus to the water body until it rapidly becomes a truly terrestrial one. At other locations, the water is surprisingly deep beneath the vegetation, and there can be a strong water current flowing under the roots of the floating plants. Water movement might be seasonal or present for the entire year, but the current prevents the accumulation of sediment beneath the plants, which keeps the water deep. In agricultural regions and places receiving domestic sewage, the eutrophication of the water through human activities greatly promotes plant growth. The spread of floating beds of vegetation on water bodies has therefore been accelerated greatly by inputs of wastewater.

In Chapter 4, the insects and collembolans inhabiting the parts of the floating plants above the surface of the water were discussed. Many of these are herbivorous with habits reminiscent of those typical for terrestrial insects. Below the surface, a large number of aquatic insects take shelter in the masses of roots and rhizoids of the vascular plants. These insects are not too different in habits from those that live in submerged masses of aquatic plants growing entirely beneath open water, and many species of insect may be found in both habitats. The main difference between these two habitats is that little light can reach the submerged parts of the floating plants, while much light can penetrate through the open surface of a water body to reach the submerged aquatic plants. As a result, the floating plants can exchange gases directly with the atmosphere, while the submerged species absorb CO_2 from the water and release O_2 into it during photosynthesis.

Most of the insect species inhabiting masses of roots and stems beneath floating plants seem to be predatory. They feed on small animals or various microorganisms. Most behave as typical aquatic species, but a few of them living for part of the time on the emergent parts of the floating plants belong to predominantly terrestrial families and have distinctive patterns of behavior that are otherwise little different from those in the same family. They are more often herbivorous rather than predatory.

This chapter focuses on examples of those insects that remain underwater on the lower side of the pseudoterrestrium, usually to find and capture prey. Examples of species from predominantly terrestrial taxa that prey on other insects include a firefly species in the genus *Photuris* Dejean, 1833, of the family Lampyridae, the larvae of which live among the roots of floating plants, where they apparently exchange gases directly with the water. The aquatic larvae feed mainly on fishes and mainly other insects as adults. For most species, this is not known with certainty. Observations on the feeding behavior of *Photuris trivittata* Lloyd and Ballantyne, 2003, in Mexico were reported along with the description of the new species (Lloyd and Ballantyne, 2003). Also included in the diet of adults and probably also the larvae are other North American fireflies, mainly in the large and widespread genus *Photinus* Laporte, 1833.

In order to attract males, female fireflies flash their lights while on the ground or on plants whenever they see the light from males flying nearby. Males respond by flying to the females. In order to assure that they are flying to females of the same species, the males look for a species-specific pattern of flashing by females ready for mating. Females of several *Photuris* species investigated wait at the surface of the water or on nearby plants and emit a sequence of flashes mimicking those made by the female of the same *Photinus* species to attract the male. The signal deceives the male into flying to the light flashing in the pattern characteristic of conspecific females. When he lands expecting to find a female, he is captured and devoured by the large predatory firefly in the genus *Photuris* instead. The nature of the differences between the light flashes used by many of the Japanese firefly species to identify themselves to mating partners of the same species were shown in detail by Ohba (2004). He did not provide information on any species of *Photinus* and did not indicate any species with aquatic larvae, so it is likely that most or all of the species he investigated are terrestrial during all life stages.

Surprisingly, few studies reporting which species feed on which have been conducted, and some species in other genera may also employ this tactic to feed on other firefly species. Most species of firefly seem to have larvae that are terrestrial and develop in moist soils or on forest floors. However, many of the known species of *Photuris* are probably aquatic as larvae. More must be learned about the larval development of the individuals to prove or disprove this assumption.

Most species of *Photuris* apparently inhabit soils, wetlands, or water bodies in the Nearctic and Neotropical Zoographical Regions, where many species of *Photinus,* mostly terrestrial as larvae and adults, are also found. These genera are being reported from the Oriental Region, as well.

b. The Spread of an Unusual Habitat

In tropical and subtropical regions, the spread of species of floating and emergent plants and their associated fauna from one relatively isolated part of the world throughout other warm geographical regions has often been documented. Some of the plants were transported by human beings as crops, while others do not supply human beings with food or other useful items. A few of the plants have created problems because of their rapid growth in parts of the world to which they have been introduced. One of these, *Eichhornia crassipes*, the water hyacinth, has already been discussed in Chapter 4. It was transported to all parts of the world by people who thought its flowers were attractive, and it found its way out of botanical gardens to become a large aquatic weed in tropical water bodies in all parts of the world. Like its emergent leaves and flowers, its roots have also been colonized by aquatic insects, usually predatory species native to the location in which the water hyacinths are growing.

Reviewing the information provided in Chapter 4, the water hyacinth is undoutedly native to the wetlands influenced by the tropical wet-and-dry climatic zone of South America. There, it is a small floating plant, usually extending less than 50 cm above the surface of the water, with attractive flowers. Its appearance and capability of growing in nutrient-poor waters led to its being transported to all parts of the tropics. However, once it reached water bodies with large amounts of dissolved nutrients in the water bodies, it began to grow to heights greater than 2 m. In many places, floating beds of *E. crassipes* blocked waterways, interfered with local shipping, and appeared to threaten water quality. Attempts to control the plants have met with little success.

In some places, less creative ways were employed to control water plants imported from other parts of the world. For example, the water chestnut, *Trapa natans* Linnaeus, was the target of an expensive eradication program in the Mississippi River System at the same time that a small can of water chestnuts imported from Asia was selling in American supermarkets for more than $2.00, with the buying power that this money possessed during the early 1980s. Amazingly large amounts of money has been spent on herbicides to kill off the water chestnut plants in small water courses in the United States over the years.

Ignoring the potential economic value of some of the important water plants and falsely viewing them as menaces to boat traffic, research was initiated for the purpose of finding ways of eradicating many species. Naturally, the search for biological enemies of the fastest growing water plants motivated a search for animals that would consume the offensive plants. Of course, various insects were quickly found to be candidates for controlling the plants. The problem with importing insects to eat unwanted plants is that some of the insects might also consume important local crops and become worse problems than the plants they were supposed to destroy.

SECTION 1

Herbivorous Species

The search for insects that could exert biological control over floating and emergent rooted plants did yield some information about these insects in their native regions of the world. Of course, within the natural ranges of the plants, none of the insects could be expected to become their major natural enemies. These insects had been living in equilibrium with the plants for many millennia without eliminating them or even reducing their numbers from year to year. On the other hand, species native in the region into which the offensive plant had been introduced would never have fed on the plant before, and whether or not any insect would accept the new plant as a food would not be known until a period of adaptation had passed.

Not only do herbivorous insects inhabit the emergent parts of floating plants, especially within the original ranges of the aquatic species, predatory insects belonging to typically terrestrial families also seek prey among the herbivorous insects that feed on the emergent parts of the plants. Aphids and other members of the suborder Homoptera can quickly produce dense populations of small insects which pierce and suck the juices from floating tracheophytes. They provide excellent sources of food for the predators specialized in feeding on members of the Homoptera. Among the most obvious predators on aphids in tropical and temperate agricultural regions are species in the following orders and families: order Neuroptera, families Chrysopidae and Hemerobiidae; order Coleoptera, family Coccinellidae; order Diptera, family Syrphidae.

It can be concluded that floating aquatic plants show special adaptations for utilizing the surface of a water body, including an adaptation of some of the species to grow in water very poor in inorganic nutrients, at least during their growing seasons. It is unlikely that any of these invasive plant species can be controlled in eutrophic waters by insects that feed on the emergent parts of the plants, although control by other animals, such as grass carp, living on the submerged parts of the plants, may prove to be possible. At the present time, the grass carp themselves are regarded as a major problem in many parts of the United States. However, no insect to date has shown promise of controlling floating aquatic plant species, such as *Eichhornia crassipes*, that have become a noxious species outside of their natural ranges, by feeding on the roots or other submerged parts of the plants.

SECTION 2

Predatory Species Living Beneath the Islands of Floating Plants

The absence of obvious adaptations to life above the surface of water bodies does not mean that the species are not especially adapted to survive the small catastrophes that occur when terrestrial insects fall or are forced into a water body. In the Pantanal, for example, heavy rainstorms continue to fall for several months after the wetland has been flooded long enough to have developed areas covered by floating plants. Near the floating islands of pseudoterrestrium, it is not unusual to encounter drowning terrestrial insects, which have been obviously washed into the water by the rain. However, it is not at all usual to see any of the insects associated with the emergent parts of the floating plants, including the aphids, drowning in the water, as described in Chapter 4. These insects can be observed moving to sheltered parts of the leaves and firmly taking hold of their surfaces to avoid being washed off the plants by heavy rains. If deliberately removed from the plants and dropped into the water, they are also able to quickly either fly or move along the surface of the water to again reach the sheltering buoyant plants. This takes them out of the reach of the many fishes and predatory insects that move beneath the surface of the water. Terrestrial species drowning in the water, on the other hand, provide a rich source of food for the insects living on, in, or beneath the surface tension layer.

Obviously, some aquatic insect species have adapted to life among the submerged parts of the floating plants to take advantage of any potential prey on the surface of the water among the plants, whether adapted to the water or not. The predatory insects that patrol the surface of the water, such as members of the water bug families belonging to Gerromorpha, were discussed in Chapter 3. In this section, the species of insect living below the surface of the water and feeding on other insects which come into their reach will be discussed.

a. Neotropical pseudoterrestrium

Not only are the above-water parts of emergent plants the natural habitats for insects closely related to typically terrestrial species, their submerged parts are also habitats for the larvae, and even the adults of insects belonging to families not generally considered to encompass aquatic species.

Some of the insect larvae that feed on other arthropods found in the floating islands of vegetation belong to typically terrestrial families. In the Pantanal, for example, the larvae and adults of firefly species in the family Lampyridae develop among the roots and rhizoids of the floating plants. They remain beneath the surface of the water during the day but emerge on parts of the plants above the water during the night to glow or flash their lights. A species of *Photuris* is common in the Pantanal of Mato Grosso, Brazil. As described in the last chapter, species of this genus flash their lights in a sequence similar to that used by terrestrial firefly species to signal that they are seeking a mate. In this way, they lure the male fireflies of other species to the plant masses, in which they live in order to capture and feed on them.

Several other species belonging to more typically aquatic beetle families hide among the submerged parts of the floating plants. In the Pantanal of Mato Grosso, there is a distinct seasonal climatic cycle, which includes a period of frequent, heavy rainfall, during which the wetland becomes flooded, followed by a rainless period, during which the rainfall tapers off, and the northern part of the Pantanal drains through the system of the Rio Paraguai and gradually dries out. Meanwhile, the water level in the southern part continues to rise as the water from the northern part crosses it in its seasonal movement toward the sea (Heckman, 1998).

Because the downstream flow of water from north to south is only gradual, the highest water level in the Pantanal occurs simultaneously with or shortly before the last rainfall at its northern border, but not until late in the dry season in the southern part. Thus, the highest water level moves like a wave from north to south across the entire wetland. This movement creates significant differences in each part. In the northern section near Poconé, the water level starts to sink as soon as the dry season begins. By the end of the dry season, the wetland has become a large, desiccated plain covered by baked mud and small, short-lived terrestrial plants. The rainy season begins with large, heavy thunderstorms. As these storms approach, large numbers of aquatic insects of many different species take to the air and fly in large swarms around street lights. This is remarkable because aquatic insects are only rarely observed in the water during the annual period of flooding, yet these swarms are large and rich in species that live under water as adults.

When mating occurs farther south in the wetland requires more studies in both the central and southern parts. Because studies of insects in the Pantanal are still focused on basic taxonomy and systematics, there is still a great deal to learn about the ecoclimatological demands of the insect species. Comparing the mating activities of the adult insects in the Pantanal with those found in European water bodies and in Southeast Asian rice fields leaves a strong impression of a much greater species diversity in the South American water bodies but far fewer individuals belonging any one species within any given area of the wetland.

In general terms, such circumstances indicate that the Pantanal provides relatively unstressful habitats for the insects. This stands in strong contrast with the deaths of massive numbers of fishes which perish when the water dries up late in the dry season. The streets of the town Barão de Melgaço just before the arrival of the first thunderstorms in 1992 were covered by a dense carpet of aquatic insects, predominantly beetles, which had just completed their annual mating flights. Similarly, two years later, personal observations made under a street light just south of the town of Poconé, indicated that aquatic heteropterans, including many large belostomatids, take part in similar flights. At the times of these mating flights, a line of large storms produced a considerable show of lightning and thunder but no rain near the main road leading eastward. The first rains fell a few days later. Mating flights at such a time produce excellent conditions for survival of the offspring of the insects. If the mating is completed just before the arrival of a large thunderstorm, it is immediately followed by the production of a new seasonal habitat for the insects. The delay of the rain for just a few days may reduce survival somewhat. Some of the beetles had small patches of mud clinging to their elytra, indicating that they had been encased in dried mud shortly before they begin their flight around the street lights. The exact weather conditions that promote the emergence of the aestivating adults and the methods employed by the insects for sensing when the time is right to prepare for the arrival of the first rains provides a wide field for future studies.

The aquatic insects that leave the dried mud to mate and enter the water as it falls with the first rainstorm are predominantly those that live in or among the submerged parts of the aquatic plants. For example, aquatic beetles in the families Dytiscidae, such as *Anodocheilus silvestrii* Régimbart, 1903; *Anodocheilus sarae* Young, 1974; and *Megadytes laevigatus* (Olivier, 1795); as well as Hydrophilidae, such as *Tropisternus collaris* (Fabricius, 1775), and the grasshopper, *Paulinia acuminata* (De Geer, 1773) in the family Pauliniidae, were later found among the rhizoids of aquatic ferns and roots of floating tracheophytes, which begin to grow as soon as the fields become flooded by the first rains. An especially large number of species in the dytiscid sub-family Notarinae, treated by many authors as an independent family: Notaridae, seek out the feathery masses of rhizoids beneath *Salvinia auriculata* Aublet to hide from predatory insects.

The species observed and indentified during the research in the Pantanal included *Suphisellus grammicus* (Sharp, 1882), *Suphisellus rufipes* Sharp, 1882; *Suphisellus rufulus* Zimmermann, 1921; *Hydrocanthus debilis* Sharp, 1882; *Hydrocanthus laevigatus* Brullé, 1836; *Hydrocanthus socius* Sahlberg, 1844; *Suphus minutus* Régimbart, 1903, and *Canthydrus ovatus* Régimbart, 1903.

The best hiding places for insects beneath the floating plants in the northern Pantanal are among the rhizoids of the water fern, *Salvinia auriculata*, which are lined with feathery secondary rhizoids. A beetle the size of *Tropisternus collaris,* that is, approximately 6 mm long, can swim into the middle of the rhizoid mass beneath *Salvinia auriculata* and make itself completely invisible from below, that is, from anywhere in the open water near the plant masses. At the beginning of the rainy season, before the plants of the pseudoterrestrium have begun to spread, *Tropisternus collaris* is sometimes found in the interstices between the coarse stones near the shores of small rivers. They can be seen from the shore as they quickly swim to the surface of the water to renew their air supplies and then swim to the bottom again and disappear among the stones.

Paulinia acuminata lives in association with *Salvinia auriculata*, on which it feeds. It can be observed on the upper fronds of the fern when not disturbed. However, if approached by a possible predator, it will enter the water and hide among the rhizoids of the fern. When the danger passes, it will eventually leave the water and return to the tops of the ferns.

Interestingly, the rhizoids provide ample shelter in times of dangers from both above and from below for the insects living in association with them. Their fronds are the food for some of the insects, and the underwater rhizoids provide particularly large areas for settlement by microorganisms, which provide food for some the insects that live underwater.

Apparently, those adult insects that have succeeded in surviving the dry periods in regions with tropical wet-and-dry climates produce eggs early during the rainy season, and their aquatic larvae develop in the newly flooded regions of the wetland. The alternative would be for the imagoes to deposit eggs at the end of the rainy season as the shallow water is receding from the wetland. In that case, the eggs would have to spend the dry season at locations sheltered from the strong sunlight. Because most insects that spend the dry periods as adults, either aestivating in the mud or remaining active on the dried flood plains, are likely to deposit their eggs early in the rainy season, the predatory insects certainly can expect to find ample prey at that time.

It is notable that active adult aquatic insects are not at all easy to find in the Pantanal. A great many of the species of Coleoptera and Hemiptera found in large numbers under street lights near the Pantanal during the freshet were never encountered either as adults or larvae in water and floating plant samples taken in various parts of the wetland. Much more study will be required to elucidate the life cycles of these aquatic insects. It is possible that the adults die after spawning, and the larvae are particularly skilled at hiding from predators and entomologists. It is also possible that the insects hide in unexpected locations during the long periods of flooding and are simply very hard to find. Unlike Amazonia, where the land is covered mainly by tropical rainforests, the regions in and around the Pantanal are covered by dry forests. The trees growing there lack epiphytic plants, especially bromeliads, which could provide a habitat for aquatic insect species that avoid water bodies on the ground. Because of the large number of insectivorous fishes, birds, and mammals inhabiting the Pantanal, it could be expected that the insects would have to find good hiding places to survive at all.

b. *Pseudoterrestrium in the Palearctic Region compared to that in the tropics*

The equivalent of the pseudoterrestrium in the North Temperate Zone is hardly comparable to that in pristine water bodies of the tropics. Plants must contend with hot and cold seasons, and in most of Europe and western Asia, where sufficient precipitation occurs to keep ponds, lakes, and rivers filled during the whole year, plants must contend with annual periods of freezing, which can be expected to destroy both the leaves and roots of floating vegetation.

Because of the climate, aquatic vegetation develops in a different pattern from that familiar in tropical water bodies. Rather than solid masses of floating plants capable of utilizing the maximum amount of the direct sunlight reaching the surface of the water, most emergent aquatic plants in the Palaearctic Region are rooted in the sediment. Therefore, in the spring, among the first aquatic plants to appear at the surface and soon after begin to produce flowers above areas of deep water, belong to such species as the water lilies, *Nymphaea alba* Linnaeus and *Nymphaea incana* Savigny, which have been planted in lakes and ponds located in parks and gardens throughout Europe. In deeper water, natural populations of plants are mainly submerged species, at most with flowers appearing above the surface, which dominate the water during summer. In shallower water, emergent plants rooted in the sediment are predominant, as well. Many of these plants grow so closely together that other species are crowded out. Most of those in the Palaearctic Region are in the families Gramineae, Typhaceae, and Cyperaceae.

In addition to larger emergent plants, such as water lilies, there is a distinct community of floating plants much smaller than those in the Pantanal, which was mentioned in Chapter 3. In eutrophic waters, these plants can completely cover the surface of the water, blocking out so much light that most underwater plants cannot grow. This community includes species congeneric with those in the Pantanal, as shown in Table 5.1.

Table 5.1 Examples of related species found among the floating aquatic plants at locations on three continents: the Pantanal, a vast South American wetland; drainage ditches along the Elbe Estuary near Hamburg, Germany; cultivated rice fields in the Mekong Valley, Udorn Thani Province, northeastern Thailand, and plants cultivated in artificial water bodies in the city of Udorn Thani.

Pantanal	Elbe Estuary	Udorn Thani	Community function
Azolla mexicana	*Azolla filiculoides*	*Azolla pinnata*	Small floating fern
Salvinia auriculata	*Salvinia natans*	*Salvinia cucullata*	Floating fern
Lemna aequinoctialis	*Lemna minor*	*Lemna paucicostata*	Tiny floating monocot
Nymphaea amazonum	*Nymphaea alba*	*Nymphaea alba*	Emergent rooted water lily
Utricularia foliosa	*Utricularia vulgaris*	*Utricularia flexuosa*	Submerged carnivorous plant

The small selection of congeneric species shown in Table 5.1 shows that seasonal changes can be adapted to rather easily by many plant genera, regardless of what the nature of the changes are. Supposedly, congeneric species are related, having descended from a common ancestor in the distant past. Taking the examples of the locations in South America and Southeast Asia, it is easy to see how the alternation between a rainy period and a rainless one can be conducive to the development of species belonging to the same genus and filling similar niches within a body of freshwater. After long periods of geographic separation, as well as periods of adaptation to somewhat different conditions, differences between the two populations will develop that makes them distinct from ones another. At some point, the differences between the two populations become great enough to make them two separate species. No periods of temperatures below the freezing point of water have ever been known to occur either in the middle Mekong Valley or in the northern Pantanal of Brazil. The fact that the same genera are represented by different species that fill somewhat similar ecological niches in both tropical and temperate water bodies indicates that a similar rhythm of sprouting from seeds or spores, growth and production of flowers or sori, production of seeds or spores, and inactivity of those propagules until suitable conditions for growth return can be developed by congeneric species as adaptations to seasonal changes that induce periods of growth alternating with periods of inactivity based on different sets of climatological factors. It is clear that unsuitable conditions for aquatic plants may be the results of either freezing temperatures or desiccation of the habitat. Do the same seasonal rhythms control the hexapod populations that obviously live on and control the aquatic plant species?

Often, congeneric species share important characteristics which contribute to the biotic community as a whole. For example, all species of *Azolla* listed in Table 5.1 provide growth chambers for a symbiotic species of Cyanobacteria, *Anabaena azollae* Strasburger, which has the ability to fix atmospheric nitrogen. This makes all of these species of *Azolla* capable of enriching the water and soil of rice fields during the rice-growing season in Southeast Asia, enriching the nutrient-poor rainwater that floods the Pantanal in South America, and fertilizing the water and soil of the fruit orchards in northern Germany.

Observations of *Azolla* provide some interesting insights into the adaptability of this tiny water fern. In the Pantanal, not only *Azolla mexicana* Presl. but also *Azolla filiculoides* Lamarck is widespread during the rainy and early dry season. *A. filiculoides* is also the species common in northern Germany. It may be that *A. filiculoides* has been transported from one continent to the other, but the fact remains that its seasonal cycle makes it able to grow each year whether the season of dormancy is brought about by the drying out of its habitat or by freezing temperatures.

The species in Thailand, *Azolla pinnata* Robert Brown, is also found in Africa, where it was known under the names *Azolla africana* Desvaux and *Azolla imbricata* (Roxb.) Nakai, which were found to be synonyms of *A. pinnata*. The alternation between the African wet and dry seasons is considerably less reliable and regular than that in Asia because the size of the high and low pressure systems responsible for the speed and direction of the wind are small and more irregular from year to year, yet the same species inhabits the small water bodies under both weather systems.

The same kind of relationships are evident among the water fern species of *Salvinia*. The species in different parts of the world can be distinguished by the sizes of the fronds and their shapes, but they are similar enough to leave the impression that they are closely related. Of the two species that become abundant on the surface of tropical water bodies, *Salvinia auriculata* Aublet and *Salvinia cucullata* Roxburgh ex Bory, each have an aquatic grasshopper that accompanies them and apparently depends upon them for food and concealment. The species *Paulinia acuminata* (DeGeer, 1773) is the grasshopper that lives in association with *Salvinia auriculata* in South America. In Southeast Asia, the grasshopper living in association with *Salvinia cucullata* is an unidentified species in the family Tetrigidae.

Like the species of *Azolla*, those of *Utricularia* contribute to the amounts of inorganic plant nutrients, some of which they use for themselves. They do this by capturing tiny organisms in small traps and absorbing the chemicals they release as they decompose. In this way, *Utricularia foliosa* Linnaeus can grow well in spite of the intense competition for inorganic plant nutrients in the Pantanal during the rainy season. *Utricularia vulgaris* Linnaeus is able to do the same during their summer growth period in the small water bodies of the fruit growing district along the Elbe Estuary, just as *Utricularia flexuosa* Vahl is able to do in the rice fields of Southeast Asia. The prey of the species of *Utricularia* are frequently species of Cladocera as well as hatchling insects, often small species of such families of Diptera as Culicidae and Chironomidae.

The species of *Utricularia* can also be considered submerged aquatic plants because their growth occurs mainly under water. However, they produce flowers and seeds above the surface of the water, so for the purpose of this description, they can be classified as partially or seasonally floating plants. They are not the only aquatic plants that start out as submerged species and then move above the surface of the water. The host plant of the caterpillar of the moth, *Parapoynx stratiotata* (Linnaeus, 1758), is *Stratiotes aloides* Linnaeus, which begins growing as a submerged plant rooted in the sediment but rises to the surface during the summer, where it floats with the upper leaves and flowers above the surface of the water. The plant resembles an aquatic pineapple with stiff, spiny leaves. Both the plant and its accompanying moth have become rare in Germany during recent years.

The tiny plants in the family Lemnaceae have a very simplified morphology and are familiar in most parts of the world. They are not easy to distinguish because of their simple form. In Europe, one of the species, *Lemna trisulca* Linnaeus, grows below the surface of the water. The others float on the surface. The three species in

Table 5.1 can be distinguished by size and shape. *Lemna aequinoctialis* Welwitch and *Lemna paucicostata* Hegelmaier are both tropical species found in South America and Southeast Asia, respectively. They seem to survive in low-lying water bodies that contain water throughout the dry season and spread from there to the rest of the wetlands as they flood during the rainy season. *Lemna minor* Linnaeus is very common in Europe and North America, where its populations are greatly reduced when ponds and small artificial water bodies freeze over during winter. The populations begin to regenerate themselves in spring.

Another species of *Lemna* was recently introduced to Europe, either from North America or eastern Asia. *Lemna turionifera* Landolt was found in small water bodies along the Elbe Estuary near Hamburg (Heckman, 1984a). It is the only species of *Lemna* known to produce turions, which are vegetative propagules that can resist periods of drying. It may have originated in parts of North America with dry climates, but it has also been found in eastern Asia. After it was reported, it was found by researchers near Berlin and later throughout other parts of Germany. Since its introduction, probably with intercontinental ship traffic, it seems to have spread rapidly. Only one other species in the family Lemnaceae produces turions. That species is *Spirodela polyrrhiza* (Linnaeus) Schleiden, recognized immediately by its larger size. The insects associated with species in Lemnaceae will be discussed in subsequent chapters.

The familiar water lilies are presently found throughout much of the world, where the climate is not too cold or dry. *Nymphaea amazonum* Martius and Zuccarini is a tropical species found in South America, while *Nymphaea alba* Linnaeus is apparently native to Europe, from which it has been exported to many parts of the world. It was introduced to Thailand long ago, where it was planted in artificial ponds in gardens and temples. The water lily has since escaped the gardens and moved into natural water bodies, and today, it grows in most parts of the country.

There are also some large floating plants that fill nearly the same roles in all three parts of the world. However, they are excluded from one of them, the rice fields, because large groups of such plants would not only reduce the concentrations of the nitrogen compounds and phosphates available to the rice plants, they would also compete with the rice for the sunlight, thereby reducing the amount of rice produced. If no other processes excluded such large floating plants from the rice fields, the rice farmers would certainly remove it. In Thai water bodies not being used for growing aquatic grains and vegetables, many emergent floated plants, including *Eichhornia crassipes*, grow during the rainy season each year.

c. Seasonal succession of species in the wet-and-dry climatic zone

The life history of every species of Collembola and Insecta is distinctive, although there are many similarities that are used to classify various ecotypes. There have been many disputes about whether or not the hypothesis of the monarchial principle is valid. Proponents of this principle argue that every ecological niche can be filled by only one species. It two are found playing the same role in one habitat, it probably means that one of them was the only species filling the niche until a second one was recently introduced. The two species are in the process of competing for the niche, and before long, one of the species will take it over, and the second species will disappear.

There are basic problems encountered when it is attempted to prove or disprove this principle. There are simply too many factors in determining the parameters of a single niche. Such competitions for dominance could continue forever if, for example, one species required a relatively high concentration of calcium in the water, while the other required a relatively low concentration. The first species might prevail in mountain lakes, while the second might be common in bogs. Both species could migrate into nearby lakes with intermediate concentrations, and an endless competition for the same nutrient source could make it seem that the species can either coexist or that the competition will someday reach an end point. However, because water bodies containing much calcium will always be the exclusive habitat of the first species and bogs with acid water will be the exclusive habitat for the second, periodic conflicts will exist forever wherever intermediate or variable conditions exist.

Because water bodies in a wet-and-dry climatic zone typically show extreme physical and chemical changes over the course of a year, great changes in the parameters defining a niche occur during that same year. Not only that, in tropical water bodies located where the seasonal patterns of wind and rainfall are weak, the weather from year to year will also show considerable differences that impact the conditions under which the species present must exist. Most of the shallow water bodies within the tropical wet-and-dry climatic zone display typical characteristics of habitats for insect species living among the submerged parts of floating plants at least for part of the year. Deeper habitats, such as lakes, ponds, and oxbows, are far too large to be influenced much by relatively minor seasonal changes in wind and rainfall. Ice formation in winter might well eliminate some of the insect fauna in water bodies of the temperate zone, but shallow tropical water bodies do not have such problems to cope with. It is therefore evident that congeneric species can have seemingly minor differences in their adaptations to a whole series of factors which make it possible for one species to dominate one habitat, another to prevail in another, and a third species to prevail in a third. Before any species can be used as an example to demonstrate the validity of the monarchial principle, detailed studies must be conducted to find the usefulness for each adaptation by each species to the conditions in every kind of local habitat.

The most obvious factors that change during a seasonal succession involve nutrition, respiration, and reproductive opportunity. First, physical and chemical differences during the course of a year or from year to year have an effect on the growth of aquatic plants forming the communities discussed earlier in this chapter. Second, annual seasonal changes from wet to dry create conditions favorable for insects which obtain air directly from the atmosphere, while the return of the rains floods dry land and provides a habitat in which the water is relatively rich in oxygen, favorable for insects that employ respiration through gills or gill-like tissues. Third, the reproduction of insects and the development of their larvae is facilitated by a succession of other species in the habitat starting from few or ineffective predators at the time of mating and oviposition, progressing to a community of many small organisms suitable as food for the larvae, and ending with potential predators that become inactive at the same time as the other aquatic insects. Success in a habitat has a great deal to do with the timing of the appearance of the stages in the life cycles of both a particular species being studied and all other species in the environment with which it interacts.

The seasonal succession of aquatic insects in the tropical wet-and-dry climatic zone begins with all of the land being dry with the exception of a few deep spots and the beds of streams and small rivers. Even large rivers may become dry, except near the deepest part of the channel, where there may still be some water flow. At this time, local air temperatures are often the highest during the course of the year.

The activity of aquatic insects begins as the first thunderstorms develop in the vicinity of the water bodies. Before the first rains begin, there are evening flights of a great many adult insects, usually including beetles in the families Dytiscidae, including Notarinae, as well as Haliplidae, Hydrophilidae, and sometimes also members of smaller families. Together with these, adult hemipterans in the families Belostomatidae, Nepidae, Naucoridae, Notonectidae, Corixidae, and others appear flying around street lights. These insects often have patches of dried mud on them, showing that they have just emerged from the desiccated earth below.

Following the completion of the mating flights, it is not long before thunderstorms arrive and begin to bring about the annual flooding of the dry floodplains. Once water is there, the insect eggs can hatch, and the larvae can begin to develop. Where parts of the watershed are used for the culture of rice, the seed is planted in shallowly flooded fields during this period in the annual cycle.

Once the fields are flooded to a depth indicating that there is no great likelihood that they will dry up again before the end of the rainy season, the aquatic insect community will begin to appear in their full diversity. Apparently, the adults of some species and the larvae of others have spent the dry season encased in desiccated sediments. Those species that mated and spawned during the early part of the rainy season develop as larvae during this season. Still other insects develop rapidly and breed anytime the adults find assemblages of water large enough not to dry up before the larval and pupal stages have been completed. This strategy has been used successfully for most species in the dipteran family Culicidae, that is, the mosquitoes. During this period, rice farmers in Asia remove the rice plants that have sprouted, tie them into bundles, and transplant them into the rice fields, which have been ploughed and raked to prepare them for the rice.

During the next period, the rains continue, and the water reaches its highest level of the year. Each plant species follows its own schedule of maturing, producing flowers and seeds that will sprout during the next rainy season, and then dying off. When the rains cease, the plants continue growing until the water level becomes too low, and then all of the aquatic species are harvested or die off and do not return until the next rainy season. Cultivated rice produces its grains and is ready to harvest before the fields completely dry up. Just how long this whole process takes depends on the geographical location of the field, local weather conditions, and the variety of rice being cultivated.

This kind of tropical water body is the habitat for a great many aquatic insect species, and each one has its own life history. The structure of the biotic community in such a water body depends on the geographical location more than the biological needs of the insects. Tropical water bodies in a wet-and-dry climatic zone, like their counterparts in locations with four clearly defined seasons differing in temperature, provide habitats in which insects with simple adaptations for living in freshwater can survive and compete for a niche in the community.

Most species that cannot find a niche in such seasonal, lentic water bodies are those that require fast-flowing lotic habitats, usually with clear water and high quantities of dissolved oxygen. This group will be the subject of Chapter 8.

6

Active Submerged Hunters of Microorganisms and Small Prey in Lentic Water

The most familiar aquatic insects are those which live in the water of lakes, ponds, slow flowing streams, and small pools of water on wetlands. These kinds of habitat are also the places where almost all species that inhabit the water column as adults can be encountered. Because there are so many species found in such habitats around the world, it is not at all possible to examine the structure of these biotic communities, except on a case by case basis. Many comprehensive works were written by August Thienemann, on freshwater bodies, mainly between the 1930s and 1950s. Many of his works in his series, *Die Binnengewässer,* provide information relevant to the ecology of the insects discussed in this chapter. His book on the Diptera family Chironomidae is a classic on this large and complex family of insects (Thienemann, 1954). It was frequently said that whenever a young student of hydrobiology made what he thought was a new discovery in the field, he soon found out that it had already been reported and discussed in one of the comprehensive works of Thienemann published many years earlier. Although the nomenclature of some of the insects is outdated, the observations made and reported by this author continue to provide basic information on the water bodies forming the habitats of insects living in freshwater and some atypical bodies of water throughout the world.

There are three major modes of life encountered in the water columns of freshwater bodies around the world. One is life on the underside of the surface tension layer, which permits easy access to the atmosphere for respiration. The second is life in the water column, either in open water or among masses of plants. The third is life on or in the sediment. Some of these habitats are shared by both larvae and adults, while others develop underwater as larvae but spend their time as imagoes flying and resting in terrestrial communities as various distances from where they spent their larval stages.

Only a few groups of insects have free-swimming pupae, which also maintain some degree of activity in their preferred aquatic habitats.

Most insects that rest and feed primarily in one of the three sections of the water column will flee from potential dangers to hide in one of the other sections. Inhabitants of all of these three sections usually show distinct preferences for water of certain depths. Most of those inhabiting the middle section of the water column prefer specific depth ranges, with most seeking relatively shallow water, where access to the atmosphere or oxygen-rich water is most convenient.

Most insects have solved the problems related to respiration underwater in one of four ways: the use of gills or other sections of the integument for gas exchange directly with the water, direct contact with the atmosphere by insects that move along the underside of the surface tension layer, respiration using air transported underwater with the insect as a "plastron", that is, a thin layer of air adhering to part of the external surface of the insect or beneath the elytra or other structures, and respiration through long breathing tubes kept in contact with atmospheric air at the surface or with short tubes that pierce underwater aerenchyme tissue of plants. Species of parasitoids in the order Hymenoptera, apparently obtain their oxygen from the host, in ways that have not yet been adequately investigated. Descriptions of the modes of life of these parasitoid insects are found in Chapter 22.

SECTION 1

Gas Exchange with the Water

Several orders of insect and various families of others have a life cycle beginning with hatchling larvae which remain underwater and breathe using structures on various parts of the body which function as gills. This is presumably one of the oldest methods insects have used to utilize aquatic habitats and one of the most successful. Some of the orders that are considered most primitive and have left some of the oldest fossils typically employ gills for respiration as larvae.

The problem with this method is that it functions well only in oxygen-rich water, and it functions best in fast flowing mountain streams, which will be discussed in Chapter 8. In standing and slowly flowing water, eutrophication makes survival harder for the insect species breathing through gills. So does contamination of the water with certain substances that interfere with the gas exchange. Such substances range from fine silt in sufficient quantities to physically coat respiratory surfaces to chemicals that can interfere with gas exchange chemically in such ways as reacting with the dissolved oxygen rapidly, eliminating it from the water before it reaches the gills of the insect. While survival in water bodies that are anaerobic most of the time is not possible for insects depending only on gills for respiration, physiological adaptations enable certain insect larvae with gills to survive in water containing only small quantities of dissolved oxygen, at least for a few hours each day. In spite of the difficulties of using gills in standing water, many insect are able to do so, and the problems of surviving in strong currents seem to be even greater than those of living in water with less than optimal amounts of dissolved oxygen.

Certain species of Ephemeroptera cope with periodic shortages of oxygen by using gills that create water movement. Most mayflies inhabiting fast flowing streams do not encounter this problem because the water itself is continually moving rapidly over the gills. In the lower sections of streams, however, the current may not be fast enough to replace the oxygen consumed by saprobic microorganisms when many organic substances are being broken down. The larvae of several European mayflies in the family Baetidae are able to inhabit standing water containing significant amounts of organic material by rapidly moving the paddle shaped gill lamina to create a continual current of water. One of these species, *Cloeon dipterum* (Linnaeus, 1761), is the most abundant mayfly species which develops as larvae in small standing water bodies along the freshwater section of the Elbe Estuary. The larvae behave somewhat like the larvae of other species during most of the year, but during warm periods, when much of the dissolved oxygen has been exhausted presumably by the bacterial decomposition of detritus, the larvae begin rapidly vibrating their gill lamellae, creating a strong current of water, permitting the larvae to survive until metamorphosis in water where other species of mayfly would perish.

The presence of certain chemicals with an affinity for oxygen permit other insects to actively absorb and bind oxygen dissolved in the water, even when its concentration is very low. The larvae of many species in the family Chironomidae contain such chemicals. Many larvae in the subfamily Chironominae have a bright red color, accounting for their common name: bloodworm. The red pigment they contain is a substance like hemoglobin, which reacts with oxygen and binds it until it is needed by the insects for respiration. In many eutrophic water bodies containing large amounts of nutrients, algae supply the water with large amounts of oxygen while the sun shines each day, but at night, saprobic organisms in the water consume that oxygen until its concentration in the water is almost completely exhausted before sunrise. Chironomine larvae live in flexible tubes that they construct of sand particles or other objects at the bottom of the water body and store oxygen in their red pigments during the day for periods of oxygen shortages at night or during times of high water temperatures in summer. When the oxygen concentration in the water become low, the larvae can be seen moving actively within their tubes to create water current that permits their red pigment to bind and store the small amounts of oxygen still in the water.

This system seems to improve the survival chances of the chironomid larvae significantly, permitting them to develop in large numbers in small eutrophic water bodies. The lack of oxygen in the water makes it unsuitable for most species of fishes, which find these larvae particularly attractive as food items and can easily break their tubes and consume them. The lack of dissolved oxygen drives away fishes and other predators, protecting the larvae from the natural enemies usually present in better oxygenated, mesotrophic habitats.

There are several orders of insects with larvae that typically exchange gases with the ambient water through gills or gill-like structures, species of which are found in water bodies throughout much of the world. However, the individual species in these orders are generally distributed according to the climate, trophic level of the water, and geographical region of the world. The taxa that typically use this mode of respiration as larvae and live as adults either in flight or resting on terrestrial plants or other objects include the following:

Ephemeroptera: with species occurring as larvae in both lentic and lotic habitats;

Odonata: with most species developing in lentic or slow-flowing lotic habitats, but only a few of which are capable of survival in strongly flowing streams;

Plecoptera: with almost all species confined to lotic habitats, but one developing in a moist terrestrial habitat;

Neuroptera: with species in the sub-order Megaloptera, family Sialidae, developing in lentic water, and those in Corydalidae usually developing in fast-flowing lotic habitats. Members of the sub-order Planipennia, family Sisyridae, develop as larvae in lentic habitats inside of freshwater sponges;

Trichoptera: with species living in a wide range of lentic and lotic habitats. Most are apparently herbivorous and consume primarily tracheophytes, but some may consume microorganisms, small prey, or dead animals, according to availability;

Diptera: with species in the sub-order Nematocera, families Chaoboridae, Chironomidae, Simuliidae, and apparently individual species in families of the suborder Brachycera developing as larvae mainly in lentic water;

Hymenoptera: with a few members of the superfamily Ichneumenoidea living as perisitoids inside of the larvae of various insect species. Most details of their modes of respiration are unknown. The preferred biotic community in which they are found depends upon the habits of the host.

SECTION 2

Insects Utilizing Atmospheric Oxygen on the Underside of the Surface Tension

The species that feed on prey at the surface of the water while remaining below the surface usually benefit from morphological modifications which give them maximal advantages of simultaneous contact with the water and the atmosphere. They are obviously able to catch prey at or below the surface while being able to obtain atmospheric oxygen almost continuously while hunting. Their presence at the surface also provides them with immediate access to the terrestrial insects that fall into the water and are doomed to perish. For this kind of prey, those that hunt just below the surface must compete with insects that move above the water on the surface tension layer, such as water striders in the families Gerridae and Veliidae and flies in such families as Empididae and Ephydridae. Some of these species have been found to be able to detect the vibrations of the drowning insects through the surface tension layer or through the water. Competitors for the drowning terrestrial insects include fishes, which can hunt by sight and smell.

Adult insects that swim just beneath the air-water interface as adults are typically predatory. Examples of such insects include almost all hemipterans in the family Pleidae, coleopterans in the families Dryopidae and Helodidae, and a great many individual species from large, mainly terrestrial families of Diptera. Many of the Diptera prefer eutrophic water bodies or even mixtures of water and manure from animal husbandry. Many such species belong to families such as Tipulidae, Limoniidae, Culicidae, Psychodidae, and Dixidae in the suborder Nematocera, or to Ephydridae,

Empididae, and Sphaeroceridae in the suborder Brachycera. Some of these species obtain their air from the atmosphere through structures at the caudal end of the abdomen. These larvae differ from those which live deeper in the water only by the absence of a siphon for breathing. By moving on the underside of the surface tension layer, these insects are less exposed to predation by large insects and vertebrates that hunt prey on top of the surface tension layer by sight. Many protect themselves from aquatic predators by excellent camouflage or by the construction of shelters.

Species in the family Pleidae have very straight margins on what would be the ventral side of most insects. However, species of pleids spend most of their time swimming inverted below the surface tension layer. The straight border of the ventral side of the body passes along the lower side of the water's surface, permitting the whole ventral margin to pass in direct contact with the atmosphere. The mouth also passes just below the surface, bringing it into direct contact with potential food items, apparently including tiny arthropods that live above the surface tension layer, as well as terrestrial insects that fall into the water and drown. Pleids can also move along the surfaces of submerged plants with the ventral side directly in contact with the fairly straight surfaces of the leaves.

Other insects live on the underside of the surface tension layer only as larvae. Many species in the order Diptera develop as such larvae, and after metamorphosis from pupa to adult, they fly away to mate and then return to lay eggs. If the adults feed at all, they usually hunt prey in a terrestrial habitat or feed on the nectar of flowers. Food of the larvae often consists of microorganisms. Prominent among the dipterans that remain on the underside of the surface tension layer as larvae are members of the family Culicidae, which are strong swimmers as both larvae and pupae. Larvae of species in the family Dixidae, on the other hand, are weak swimmers. So are a few of the many species of Tipulidae and Limoniidae, which are able to develop among the debris floating on the surfaces of water bodies, which they themselves appear to be. Contact with the atmosphere through a short breathing tube permits the larva to rest on the surface while directly breathing atmospheric oxygen. Dixid larvae typically lie in the horizontal plane below the surface with the body bent in the shape of a long, narrow U. Some culicids lie in a diagonal place, while others lie horizontally when at rest at the surface. When threatened by larger predatory insects or fishes, the larvae of most species in the family Dixidae seek hiding places among plants and detritus floating on the surface, while those in the family Culicidae flee into deeper water and conceal themselves among debris. When the danger passes, they return to the surface. Species of Culicidae are among the only aquatic insects that have completely active pupal stages. Although they do not feed during the pupal stage, they swim actively to avoid predators. A few species of Chironomidae are occasionally found attached to various kinds of detritus floating on the surface.

In addition, individual Lepidoptera species develop underwater, usually among aquatic plants at the surface. One species, *Cataclysta lemnata* (Linnaeus, 1758), which inhabits tubes made of species of *Lemna*, usually *Lemna minor*, were already mentioned in Chapter 4 and will be discussed further in the next chapter. They are typically found on the surface among masses of *Lemna* species, where they are almost perfectly camouflaged by the tubes they construct to live in. The aquatic habits of the larvae of this species and a few other aquatic members of the Lepidoptera are very unusual among a great many related species of moth, which are fully terrestrial in their habits.

SECTION 3

Predators in the Open Water Column

Among the most abundant aquatic insects that live as predators in standing water bodies are those that swim rapidly and are able to catch their prey while swimming in the water column. Many members of the Hemiptera and Coleoptera hunt in one or more of these divisions of the water column, both as larvae and as adults. Species that subsist on microorganisms, insect larvae, other invertebrates, and, occasionally, on amphibian larvae, all can meet their needs in such environments, as long as the water is kept unpolluted and free from insecticide residues. For this reason, water bodies that remain in a natural state typically display a notably high diversity of aquatic insect species compared to other kinds of habitats within the same geographical region.

Because of their need for atmospheric oxygen, many of those insects which seek prey in the water column are frequently observed resting motionless just below the surface of water bodies. Unlike the species that both rest and feed just beneath the surface, most insects that hunt in the water column swim quickly toward their prey once they detect it, usually by sight. Almost all such species must swim beneath the surface to capture suitably-sized insects. Those that conceal themselves in masses of aquatic plants or detritus and ambush prey are discussed in the next section. Insects that pursue and capture prey in open water include a broad spectrum of species arranged according to the size of the prey which they are capable of overpowering.

Although a few such species exchange gases through gills, such as larvae in the dipteran family *Chaoborus*, most such species utilize atmospheric oxygen obtained at the surface and carried with them while swimming well below the surface. Those that rely on gills for respiration while swimming in the water column are usually confined to very clear, oxygen-rich water. Larvae of European and North American chaoborids generally remain motionless, suspended in the water of lakes. Their bodies are almost completely transparent, permitting them to approach their prey until close enough to snatch it with one quick spring. The prey of each chaoborid species is typically one or more species of copepod. The family encompasses relatively few species. Many of them are eliminated by pollution, after which the water column is dominated by insects that carry a supply of atmospheric air with them. If the pollution becomes worse, even those that carry air with them from the surface move elsewhere.

Apparently, the selection from among many classes of prey in mesotrophic European water bodies permits a large number of heteropterans in the family Corixidae to co-exist without much obvious competition for the food. These small insect predators consume microorganisms, including protozoans and microscopic metazoans. There are conflicting reports concerning their diet, and there are reports of their sucking out the contents of the cells of filamentous algae. It seems likely that there are major differences among the diets of the individual species.

Aquatic water bugs in other families feed on tiny insects and other invertebrates, but a few subsist on large insects, other invertebrates, small fishes, and amphibian larvae. Research has not yet revealed what the differences in the diets of the individual species are and how they minimize competition for the specific items in their diets.

The relatively large species in the family Notonectidae are much easier to observe in the field than the small corixids, which seem to subsist on diets of microscopic animals and various kinds of algae. That makes it easier to determine how the prey is selected by the individual species. Members of the Notonectidae are frequently observed resting at the surface, replenishing their supply of air. Like almost all species that swim and hunt in the middle of the water column, respiration is facilitated by means of a plastron of air, which adheres to the surface of the body. Many insects hold part of this air supply beneath the elytra. When the oxygen in the air supply is depleted, the insects must go to the surface of the water and let the air adhering to the body come into contact with the atmosphere. The removal of carbon dioxide that has built up in the air supply is usually not necessary because it is very soluble and absorbed by the ambient water from the air held by the plastron.

The possession of a plastron permits the insect to remain in the water column for long periods of time without going to the surface. This air is held by a dense coating of short, hair-like setae, all almost the same length. They differ in structure and purpose from those that coat the bodies of insects living on top of the surface tension layer.

Electron micrographs published by Heckman (1983) show that the water strider, *Gerris lacustris* (Linnaeus, 1758), has two kinds of hair-like setae covering much of the abdomen. One kind is long, sparsely distributed setae, which are the first to contact the surface tension of the water when the insect is forced downward. Presumably, they have a sensory function as well as the obvious one of keeping space between the surface of the abdomen and that of the water. The second kind of seta is shorter and finer, and they are much closer together. In case the insect is forced below the surface, these setae keep a plastron of water in contact with parts of the body surface, preventing water from blocking the spiracles.

Almost all insects specializing in preying on animals inhabiting open water depend on rapid swimming for success. In contrast to those on surface-dwelling insects, like *Gerris* spp., the setae covering the abdomens of insects that carry a plastron of air while swimming underwater are short and strong, arranged in a regular pattern, and evenly curved toward their apices. Some ambush animals small enough to overpower, so the plastron must keep the abdomen from becoming wet, even when the adjacent water is made turbulent. The insects which remain in the middle of the water column for long periods of time often remain concealed or camouflaged until the prey comes close enough for it to be grasped after a brief spurt of rapid swimming. Others consume small, slow-moving invertebrates, which can simply be sighted, pursued for a short distance, and grasped. A strongly bound plastron of air gives these insects a great advantage in hunting prey.

In addition to the heteropteran family Notonectidae, various families in the order Coleoptera are prominent in the middle section of the water column. Unlike the species in the families Corixidae and Notonectidae in the order Hemiptera, the larvae of which carry a plastron of air, like the adults, the larvae of the aquatic coleopterans breathe atmospheric air at the surface, and only the adults carry plastrons of air while swimming below the surface. Many of the species increase the volume of the air they carry in the plastron by storing some of their air beneath the elytra. These include members of the families Dytiscidae, including Noterinae, and Haliplidae. All of these families are represented by a great many species found throughout most of the world, just as the family Notonectidae is. Typical species in these families are strong and fast swimmers, which are able to overpower relatively large prey. They often rest just beneath the surface of the water, where they replace the air in their plastrons and beneath their elytra by extending the apices of the abdomen just above the surface of the water and ejecting their air supply into the atmosphere and replacing it with fresh air. While doing this, the anterior parts of their bodies are held below the surface, where they can search the area below them for suitable prey with their eyes.

SECTION 4

Epibenthic Species Breathing Through Snorkel-like Organs

Some of the largest predatory insects in the water column of relatively small, shallow water bodies remain on the sediment. Most of these large epibenthic species must use atmospheric oxygen for respiration. While they can store some atmospheric air in their respiratory systems, most of these insects cannot easily get to the surface to replenish it. Therefore, their survival depends on the possession of long tubes to replenish their air. This, of course, limits the depth of the water bodies in which they can survive to the approximate length of the tube.

Some of the epibenthic species increase their choice of water bodies in which they can live by moving up into the submerged vegetation, thereby staying at depths from which their snorkels can reach the surface. Among the largest epibenthic species that breathe primarily from snorkels in the tropical and temperate zones are species in the Order Hemiptera, Suborder Heteroptera, and family Belostomatidae. In most cases, they can remain on the bottom of shallow water bodies because their snorkels are several centimeters long, which is enough to reach the surface. This family encompasses the largest species of aquatic Heteroptera. In Southeast Asia, the depth of the water in most rice fields while the rice is growing makes them suitable habitats for a variety of belostomatid species. Although they are large and remain among the detritus at the bottom of shallow water bodies, the adults undertake what are apparently mating flights shortly before the freshet in both North and South America and in Southeast Asia. Adults are sometimes found on the ground under street lights shortly before the freshet. Presumably, the eggs are produced soon after the first thunderstorms of the annual rainy season. Once the adults have flown to breeding sites that appear promising and the first rains have begun to fill the flood plains, males begin to signal their presence by agitating the surface of the water and sometimes making low pitch vibrations. After mating, females begin to produce eggs. The males in the subfamily Belostomatinae have the unusual habit of pasting the eggs they take from the female on their dorsal surfaces, so that the prothorax and elytra are almost completely covered by closely packed eggs, protected from predators by the large and formidable father. The eggs are produced a few at a time while the male and the female of the mating pair remain together, carefully pasting each egg onto the dorsal surface of the male. Apparently, females in the subfamily Lethocerinae attach their eggs to aquatic plants.

The most widespread genus in this family is *Belostoma* Latreille, 1807. It is found in most parts of the world in which the water bodies do not freeze over for long periods of time during the winter. A covering of ice on the water surface would obviously interfere with the use of a long snorkel for breathing.

In Southeast Asia, these insects are eaten, providing the rural population with a good protein source from the rice fields (Heckman, 1979). In defense, they can inject a toxic digestive enzyme into a person who is not careful in handling them, inflicting tissue damage and much pain. When they are eaten, however, the repellant chemicals in the insect tastes spicy, and apparently they produce no harmful effects when the chemicals are eaten and not injected into the blood.

The species of this important family of large aquatic water bugs have been assigned to three subfamilies: Lethocerinae, Belostomatinae, and Horvathiniinae. The subfamily Lethocerinae includes the genera *Lethocerus* Mayr, 1853; *Benacus* Stål, 1861; and *Kirkaldyia* Montandon, 1909. *Lethocerus* is found throughout the tropics and subtopics of the world, while the other two genera in the subfamily include one species, each. *Benacus griseus* (Say, 1832) is a species known from North America, while *Kirkaldyia deyrolli* (Vuillefroy, 1864) is native to the Amur Region of Russia and parts of China, Korea, and Japan (Perez-Goodwyn, 2006). It is nocturnal and feeds on many different kinds of vertebrate, including fishes, snakes, and turtles, as well as large insects.

In addition to *Belostoma*, the subfamily Belostomatinae encompasses seven additional genera: *Abedus* Stål, 1862; *Dipdonychus* Laporte, 1833; *Hydrocyrtus* Spinola, 1850; *Limnogeton* Mayr, 1853; *Poissonia* Brown, 1948; *Sphaerodema* Laporte, 1833; and *Weberiella* de Carlo, 1966. After it was found that several species in the genus *Sphaerodema* feed on freshwater snails where they occur in Africa and southern Asia, they have become of interest as agents of the biological control of liver flukes and other parasitic worms that use the snails as intermediate hosts (Younes et al., 2016).

Only one genus is included in the subfamily Horvathiniinae: *Horvathinia* Montondon, 1909. It includes only one species: *Horvathinia pelocoroides* (Montondon, 1911), which has several junior synonyms. It is widely distributed in eastern South America.

A second heteropteran family with habits somewhat similar to those of the Belostomatidae is Nepidae. These insects are moderately large but significantly smaller than most belostomatids. Like the belostomatids, male nepids paste eggs on their dorsal surfaces to protect them from predators. These habits are not known from aquatic species in any heteropteran family other than the two mentioned here. Because of their smaller size, species in the family Nepidae are not found on the bottom in water bodies as deep as those inhabited by belostomatids. More frequently, they climb on submerged plants until they reach a depth from which their snorkels can reach the surface of the water to renew their air supplies.

The obvious advantage to a predatory insect of searching through benthic litter is that many animals suitable as prey for aquatic predators seek shelter beneath pieces of detritus during the day. The obvious disadvantage is that the water at the bottom of a fairly slow flowing river or standing water body is subject to oxygen depletion and the periodic occurrence of anaerobic conditions. Corixids and predatory beetles can minimize the effects of depleted oxygen in mesotrophic water bodies by hunting during the day, when photosynthesis by aquatic plants elevates the oxygen concentration, permitting an increase in the amount of oxygen in the plastron without the need for exchanging the air in the plastron at the surface. Although many insects carry a plastron of fresh air with them, anaerobic reactions frequently produce gases, such as H_2S and CH_4, which are toxic to most animals. Usually, active insects can hunt actively during the day using the plastron for respiration, but when toxic gases begin to appear in the water, the species diversity begins to decrease. Insects that carry a gas supply with them can usually survive brief periods in anaerobic water, but if forced to tolerate such conditions for extended periods of time, they generally perish and are replaced by insects with other adaptations to supply their need for oxygen.

Many of the aquatic insects that search through the epibenthic material in search of prey or carcasses of dead animals during their imago phase belong to families that have developed various physical and chemical adaptations for survival in hostile environments. As the trophic state of the habitat become richer, old aggregations of species disappear and new ones appear. In Europe, various degrees or levels of the trophic state have been defined, but no definitions have proven entirely satisfactory. In the less trophic stages, as defined by the amount of foods of all kinds in the habitat, the oxygen concentration is sufficient for the aquatic insects to exchange gases directly with the atmosphere. As the trophic state increases, insects that carry a plastron covered by atmospheric air become more successful. In the eutrophic stages, the benthos is frequently poor in dissolved oxygen, and benthic species with long snorkels become predominant.

In tropical waters, relatively large insects with long snorkels in the families Ranatridae and Belostomatidae become common, while in temperate water bodies of Europe, flies in the family Syrphidae become predominant in the most eutrophic shallow water bodies. The snorkels of aquatic syrphid larvae in the subfamily Eristalinae, tribe Eristalini, are long and flexible, while those in the tribe Brachyopini are short and resemble spikes. This tribe currently encompasses the genera *Chrysogaster* Meigen, 1803; *Melanogaster* Róndani, 1857; *Orthonevra* Loew, 1843; and others. When driven into the roots of aquatic tracheophytes, the short snorkels reach the aerenchyme tissue, which conducts air through the inside of the submerged root systems of the plants to keep them from suffocating. By utilizing the atmospheric air inside of the plant tissues, the syrphid fly larvae are able to develop without any direct physical contact with the ambient atmosphere. This permits the larvae to utilize an almost perfect system for avoiding predation by other species of insect.

For the large, predatory insects that live among the benthic detritus and must rely on a snorkel or breathing tube to survive, going any deeper into the sediment would deprive them of their food supply. The species that must hunt to survive generally conceal themselves among masses of detritus and await the approach of suitable prey. Most species in the family Belostomatidae are colored yellowish brown, dark brown, or another shade of brown that tends to blend in with most sediments of small water bodies. Other benthic predators that ambush rather than actively hunt their prey are usually brownish with mottling that makes them hard for their prey to see as they move toward them. When dead leaves or other kinds of detritus are present on the bottom of the water body, many predatory heteropterans will conceal themselves beneath an object with just the anterior part of the body extending slightly from beneath it. When prey suitable as food approaches, the insect grabs it with a quick snap of the fore-legs and injects it through the mouthparts with a fluid that poisons and begins to digest it.

SECTION 5

Hexapod Predators and Their Prey that Survive in Masses of Water Plants

Similar to the larvae of benthic heteropteran and dyticid larvae, which develop while concealed and camouflaged in aquatic plants, the larvae of many different kinds of aquatic insect conceal themselves in floating masses of water plants and remain motionless until approached by a species of prey within a size range that can be overpowered and consumed by the predator. The microhabitat resembles the benthos somewhat, but it differs by giving insects that do not swim well a kind of ladder, permitting them to ascend from the benthos to the surface, or a little above, by climbing on the plants. Generally, which species is considered the predator and which the prey depends almost exclusively on size and the strength of the individual insect. The great majority of the insects present inside of the submerged water plants prey on other animals, mainly insects, and the smaller species are also preyed upon by larger ones. It is not uncommon for adult insects to feed on small larvae of another species, the adults of which prey on their own larvae.

Just a few of the families that take advantage of the plant masses to travel throughout the water column without being able to swim well, if at all, include larvae of all species in all or almost all families in the orders Odonata and in a few families of Heteroptera, such as Naucoridae, Nepidae, and Ranatradae, in both the temperate zones and in the tropics. Many, but not all species in the orders Trichoptera and Coleoptera, as well as Neuroptera, in the family Sialidae, are also predatory species that live among submerged vegetation.

The Trichoptera encompasses families, most of which include species with larvae that feed on plants and others that consume other insects. A few are thought to consume both, but much still has to be learned about the diets of the many species in this order.

Neuroptera in the sub-order Megaloptera and family Sialidae, includes species that are predatory or scavengers in the benthos of standing or slowly flowing water bodies. A few species in the genus *Sialis* in Europe occasionally hide in masses of submerged vegetation. Most New World species are in the genus *Protosialis*. The feeding habits of the larvae of most species in this genus have not yet been reported.

One family of Heteroptera, Ranatridae, includes species that breathe through snorkels in a similar way as members of the Belostomatidae and Nepidae. Unlike these mainly epibenthic families, species of Ranatridae, called water scorpions, are very long and slender, making them very hard to see while they remain motionless in the middle of a mass of submerged plants. They are formidable predators, which seize relatively large insects with their long and narrow fore-legs and inject them with poisons and digestive fluids through their sharply pointed mouthparts. The internal organs of the prey are then liquefied by the digestive fluids, and the predigested insides are then sucked out by the water scorpion.

Another family of Heteroptera with many species that often conceal themselves in submerged aquatic plants is Naucoridae. They are relatively small and slow swimmers, so they cannot rely on rapid swimming to the surface or a snorkel for respiration. Instead, they remain close to the surface inside of dense masses of thin-stemmed pond weeds.

The most conspicuous species of insect found among submerged plants in the temperate zones belong to the order Coleoptera, and the family best represented is Dytiscidae. There is still some disagreement about the classification of the Dytiscidae. Sometimes, the families Noteridae and Laccophilidae are considered to be smaller families with habits similar to those of most of the dytiscids but with some morphological differences. Other authors treat all of these species as members of the Dytiscidae but classify the two smaller groups of species as members of the sub-families Noterinae and Laccophilinae, respectively. The larvae of these species have distinctive habits and different methods of respiration, depending upon the usual oxygen content of the water. They usually avoid water that is poor in dissolved oxygen. Usually, submerged water plants keep the oxygen concentration in the water relatively high, especially during the daytime.

Many beetles in the Polyphaga, especially members of the family Hydrophilidae, seem to be omnivorous. They are frequently found hiding among detritus on the benthos or in masses of water plants, usually in lentic water bodies.

Most adult beetles also depend on air forming a plastron on its submerged body, and gas exchange between the gas in the plastron and the water can also contribute to the duration of their stay under water.

The feeding habits of aquatic beetles are usually reported in general terms, although careful observations of the insects often reveal species-specific diets. For example, members of Dytiscidae have been given the common name, predacious diving beetles, while species in the family Hydrophilidae are called water-scavenger beetles. Most species in both families swim rather than crawl over surfaces, but some are more graceful than others. Close observation shows that most of the diving beetles do capture prey, while the water-scavenger beetles do feed on dead animals. However, the diets of many insects in both families overlap, and many of both families are considered to be omnivorous. The diets of most of the species in both of these large families worldwide have not been precisely defined, and it remains to be discovered which foods each species will accept, which it will reject, and the reasons for the choices.

Although the family Dytiscidae belongs to the large suborder Adephaga and the family Hydrophilidae belongs to the suborder Polyphaga, members of one family may be mistaken for the other by the casual observer. Familiarity with these two families makes identification very easy, however. The antennae of the species in the family Dytiscidae are long and filiform, while those of hydrophilids are short and clubbed. The nervous system of dytiscids also differs from those of hydrophilids, as determined from how they coordinate the movements of the legs while swimming. In the case of the Dytiscidae, both legs of each pair move in the same direction during each swimming stroke, permitting straight and rapid movement through the water.

Hydrophilids alternate the movements of the legs of each pair, making the movement through the water at a slower but steady speed.

Both of these families encompass many species in a variety of size groups, allowing a large group of species in both families to utilize different food resources in the same water body without competing with one another.

These two families of beetles are well represented in almost all parts of the world. Except for the largest of the insects, most species in both families typically conceal themselves in plant masses growing in the water column or in the roots and rhizoids of floating plant species. Species of small beetles can conceal themselves completely in the fine roots of plants like the water ferns in the genus *Salvinia*. The monocot species, *Pistia stratiotes* Linnaeus, is a floating aquatic plant in the family Araceae that produces long, richly branching root systems, in which somewhat larger beetles can hide. In the small, shallow water bodies of the temperate zone, there are few aquatic predators that can prey on water beetles more than 2 cm long. Therefore, the largest beetles often keep watch near the surface for large birds or mammals that might pose a threat to them and quickly descend to the bottom of the water body to hide under large objects.

In addition to the two largest families of aquatic beetles, there are small families encompassing small to moderate numbers of species, which occupy distinctive ecological niches in various kinds of water bodies. There are also a few members of some of the large families of terrestrial insects, which have adapted to conditions in aquatic habitats and share them with the members of purely aquatic families. Most of these are associated with aquatic vegetation.

Finally, mention should be made of the aquatic species in the order Hymenoptera. Species in the family Braconidae and other species in the superfamily Ichneumenoidea, which develop as parasitoids inside of aquatic insects, mainly those in the larval stage. Apparently, most of these species are terrestrial and most remain undescribed and unnamed. Personal observations of adult braconids in shallow, vegetation-rich water bodies along the floodplains of the Elbe Estuary were made by scooping out submerged plant masses into large pans. The adult parasitoid wasps would leave the water surrounding the plant masses and fly away. It could be assumed that they had been underwater among the plants seeking suitable hosts on or in which to deposit their eggs.

Apparently, while most host insects are terrestrial, a few are aquatic and must be searched for by adult hymenopteran parasitoids with larvae that develop in the bodies of one or a few aquatic insect species. Even in the places where these insects have been known of and studied for the longest periods of time, knowledge about them is fragmentary, although they are now starting to gain the attention of scientists that they merit.

7

Insects that Construct Underwater Shelters

The larvae of many aquatic insects greatly increase their chances of survival during development by constructing characteristic shelters typically made from various foreign particles or objects, which are cemented together with secreted substances, such as silk. These foreign objects always contribute to the camouflage of the insect larvae because they are made from materials from the habitat, which are similar in color and consistency to other natural objects in the water body. They sometimes also provide physical protection from predators. Many terrestrial insects belonging to several orders also construct shelters, including those used by larvae to conceal themselves and some constructed by and for large groups of insects, usually for mutual protection and sometimes to provide walkways that lead the insects to the plants on which they feed. Larvae that live in underwater shelters almost always occupy their shelters alone. Among the aquatic insects that construct shelters for their lives underwater, larvae in the predominantly aquatic order Trichoptera and those in several families in the order Diptera are the most abundant and widely distributed. In addition to these, the larvae of individual species in a variety of typically terrestrial orders may also be encountered in streams or ponds.

The shelters serve several purposes, although each kind of shelter does not necessarily provide all of the advantages that shelter construction could provide. In lotic water, the shelters provide anchorage to the substrate, preventing drift with the current. They must permit the insects access to the rich oxygen supply in the water, and a few insects construct nets in order to filter edible particles from the surrounding water. In addition, shelters constructed from stone particles or secreted as tough cases securely attached to solid substrate make it difficult for predators to extract the larvae. Some kinds of case have openings that can be partially or fully closed. Finally, some cases are especially constructed to camouflage them well.

Cases are constructed partially from secretions of the insect. Some of them, such as the cases made of pieces of stone glued to the substrate, are usually constructed out of silk strands that securely bind the stones together. The silk bonds are typically very strong, and the cases feel hard and sturdy to the touch.

Portable caddis shelters may also be made of small stones or sand particles held together by silk, but they cannot be as robust as some of the anchored shelters because the larvae must carry them around with them. They are typically smaller than the ones anchored to the substrate with silk strands and designed in a conical shape to reduce the forces from flowing water on the larvae.

In water with many floating plants, a much lighter kind of caddis can be constructed out of plant material cut from leaves and stems of the plants by the insect. Generally, such a caddis has the advantage of providing the larva with excellent camouflage while affording it a lighter weight shelter than any made of stones. As the insect grows, it is possible for it to increase the size of the caddis by adding longer pieces cut from the aquatic plants. Caddis flies live in water courses and standing water bodies in all parts of the world. Many families are almost cosmopolitan and are found on every continent except Antarctica. Their way of life is obviously well suited for success in all kinds of water bodies that are not too eutrophic to provide a satisfactory amount of oxygen for the developing larvae, which breathe through gills. Caddis flies frequently share their aquatic habitats with predaceous insects and fishes, attesting to the success of their defenses against predation.

Dipterans in the family Chironomidae make use of tube-like tunnels spun from silk for shelter and protection. They can usually construct their tubes more quickly than other tube-building insect larvae. However, they are less robust and provide the insect with relatively little protection from relatively large predators, such as small fishes. As a result, the predators usually are able to extract the larva from its shelter, but not without some difficulty. However, the sticky tubes quickly collect large amounts of small, sedimentary particles, such as sand grains, which provide excellent camouflage. Chironomid tubes may be destroyed more easily than the more robust shelters of caddisflies and black flies, which are dipterans in the suborder Nematocera. However, they can be reconstructed much more rapidly, and some of them are quite long. One of the best ways to investigate chironomid tube construction and the structure of the tubes is to collect substrate samples containing the larvae and allow them to remain in shallow plastic basins of suitably oxygenated water for one or more days. After the water and loose sediments are poured off and replaced by clear water from the same source, the tubes of the chironomids will remain attached to the bottom of the container. Obviously, chironomid larvae are able to flee from natural enemies some distance through their tubes.

Generally, it is apparent that the chironomid tubes are rather long, allowing the larva to move back and forth and avoid predators that are trying to extract them from their shelters. Many larvae in the subfamily Chironominae are bright red in color. This color indicates the presence of an oxygen-binding substance, which permits the insect to store a supply of oxygen in its body. This gives the larva the ability to inhabit water bodies that undergo periodic oxygen depletion. The shelters serve several purposes, although each kind of shelter does not necessarily provide all of the advantages that shelter construction could provide. In lotic water, the shelters provide anchorage to the substrate, preventing drift with the current. They must permit the insects access to the rich oxygen supply in the water and a means of filtering out food particles. In addition, shelters constructed from stone particles or secreted as tough cases securely attached to solid substrate make it difficult for predators to extract the larvae. Some kinds of case have openings that can be partially or fully closed. Long tubes resembling tunnels are sometimes constructed through benthic detritus, so that the larvae have physical protection from attack in some places and access to the overlying water to give them access to water that is relatively poor in oxygen during the night. Finally, some cases are well camouflaged in various ways.

Cases constructed by chironomids are usually constructed using secretions from the insect, which incorporate sand grains or other materials from the environment. The silk bonds are typically very strong, but the cases are more flexible that those of most other insects. They can be stretched but are hard to break.

Many are translucent and are hard to see. If the larva is forced to leave its tube, it will construct another one.

Some shelters constructed by dipteran larvae may be almost entirely secreted and much more sturdy than those constructed by chironomids. Very distinctive are the structures built by larvae in the family Simuliidae in lotic water. Gill filaments, through which the respiration of the larvae is facilitated, are allowed to trail in the flowing water through tiny apertures in the case.

A brief survey of the insects that construct shelters in water is provided here. Some will be discussed as insect orders because a considerable percentage of the larvae develop in tubes, cases, or other structures that they construct. Others are exceptional species that construct shelters even though most species in the higher taxon to which they belong do not.

SECTION 1

Trichoptera

The insect order Trichoptera encompasses species which typically construct one of two basic kinds of shelter. Some of the species that do not construct shelters for protection will collect food particles from the water by spinning nets which extend into the passing water and filter out material which adheres to the silk of the nets. They pull in the silk fibers from time to time and collect the particles that are adhering to them.

As stated in the introduction to this chapter, one kind of shelter is built on a firm substrate under the water, and it affords physical protection for the larva. The insect develops in one place protected by an immovable and strongly built shelter, typically constructed of stones or particles of sand. The other kind is a movable shelter, which the larva carries about with it. Such shelters are transported by the larvae as they move from place to place in search of food, which is frequently one or more species of aquatic plant, small prey, or tiny microorganisms, which are sometimes collected on the sticky silky threads described above. Many of the mobile shelters are shaped somewhat like quivers, in which archers carry their arrows. The name, caddis fly, is a medieval English name for the insects taken from the shape of the portable shelters. Members of each species construct shelters in the same manner using similar material, making it possible to identify many caddis flies inhabiting the same geographical areas from their shelters alone. Benthic species that construct portable shelters often use sand grains or small stones, while those that climb on submerged plants build lighter weight cases out of cut leaves of water plants or various kinds of detritus.

The portable shelters made out of plant material are usually held together with silk secreted by the caddisfly. The insects cut pieces of the aquatic plant they usually feed on. Many species cut each piece of leaf in a characteristic shape and fasten the pieces together into a typical shape for the species (Hickin, 1967).

Some caddis fly larvae that do not construct shelters, including certain species in the family Rhyacophilidae, typically inhabit fast-flowing streams, where they are able to anchor themselves to the substrate using silken threads. Like the portable or fixed cases, some of these nets are characteristic in shape, and the species can be recognized from their nets. Tiny particles, such as cells or colonies of microscopic algae, adhere to these webs as they pass by in the water. This method of fishing for their food seems to be an adaptation displayed almost exclusively by the larvae of caddisflies.

a. Adaptations to the habitat

Caddisfly species inhabit either lentic or lotic habitats, and some confine their development to lotic habitats within a range of water velocity. The preferred speed of the current differs from species to species. The typical caddisfly begins its development as a larva inside of a case it constructs shortly after hatching. The case is either firmly affixed to the substrate or portable, constructed in a manner characteristic of the species. The larvae may subsist on plant or animal material, depending on its species. Outside of Europe, little is known about the diets of most species. Relatively small food particles may be consumed, including organisms living as aufwuchs on objects upon which the caddisfly larvae creep. Larvae of other species consume the leaves of aquatic vascular plants by eating along their edges or biting holes in them in a manner not different in principle from that of lepidopteran larvae consuming terrestrial plants.

After completing the larval instar stages, the caddisflies metamorphose to the pupal stage. They emerge as adults, which leave the water and rest for much of the summer on plants or objects near the water. The time of the emergence of the adults and the duration of their mating and spawning activities vary from species to species.

Because typical caddisflies use gills for respiration, they depend upon water that usually contains high concentrations of oxygen. Some species live in relatively fast-flowing streams that almost always contain high concentrations of dissolved oxygen. Other species feed on submerged plants and prefer to live among the plants on which they subsist. During the warmer seasons, such plants typically produce large quantities of oxygen as a waste product of photosynthesis on every sunny day. The caddis larvae of species living in such habitats do not require the fast water flow needed by the larvae developing in mountain streams. Some of these larvae are able to develop in beds of submerged aquatic plants in standing water, and many of these are familiar among the aquatic fauna in lowland rivers of Europe.

b. General and ecological classification of the Trichoptera

Only one European species in the order Trichoptera could be classified as terrestrial during all of its life stages: *Eniocyla pusilla* (Burmeister, 1839), in the family

Limnephilidae. It is known to inhabit the Wyre Forest in England (McLachlan, 1868; Wallace et al., 2003). Other terrestrial species have been reported from other parts of the world, including New Zealand, and it is very likely that the larvae of several other species will also be found to be terrestrial. However, it is not likely that the total number of caddis fly species with larvae that are not fully aquatic will ever reach 0.1% of the total number of Trichoptera species in the world. At the present time, the larval development of most species in this insect order throughout the world are unknown or based on cursory reports.

Many descriptions of caddisflies from parts of the world in which entomological studies are few do not include mention of the larvae and fail to report whether they develop in lentic or lotic water. Where some studies have been conducted, the individual species typically develop in places where the water flow is limited to a speed optimal for them. The larvae of species that prefer lotic water often use pebbles or sand grains to construct their shelters. Species of some families build shelters that are firmly attached to large stones, allowing them to be anchored in one place. Such shelters are strong enough to prevent large predators, especially species of fish, from breaking them to get at the larva. The species inhabiting them must, of course, develop in streams or ponds containing stones of suitable size and shape for the shelters.

Species with portable shelters can also be found in lotic water, where many inhabit masses of submerged tracheophytes. Although larvae that construct portable shelters out of sand grains or small pebbles can sometimes be found among submerged plants in some ponds, they are most common in slow-flowing streams in which submerged aquatic plants are growing.

The great variety of caddisfly larvae found throughout the world makes the study of this insect order a field in itself. Zoogeographical studies of Trichoptera are still required to provide a world-wide inventory of the habitats utilized by the species in each of the families as well as the kinds of water body preferred by each of the larvae. The cases constructed by each caddisfly species is distinctive. At locations where at least a modest amount of research has focused on the Trichoptera, most species can be identified fairly reliably by examination of an empty case (Pescador et al., 1995).

A survey of the common families of Trichoptera and their general ecological characteristics is provided in Table 7.1. The known terrestrial species are omitted from the table and treated as exceptional and beyond the scope of this work.

The information in the table is tentative because the cases and habitats of many known species have never been described. In addition, there are still many species of Trichoptera that have yet to be discovered and described. There are certainly undescribed species in the same families as common ones that are already well known, which differ considerably in their case-building habits and preferred habitats from other members of their family. Similarly, species remain to be discovered in regions of the world far from the regions in which all known species in their families and even their genera are already known.

Table 7.1 A general survey of the families of Trichoptera inhabiting freshwater. The information provided is tentative because it applies only to larvae that have been studied, and in many parts of the world, there is little information available about the habits of the larvae. The reports concerning the diets of the individual species often seem contradictory and unreliable, and the families that have been studied was obtained from publications by Dumbleton (1963), Hickin (1967), Riek (1968), Wallace and Malas (1976), Cowley (1978), Neboiss, A. (1978), Stoltze (1989), Urbanič et al. (2003), Holzenthal and Oliveira Pes (2004), Kehl (2005), Holzenthal et al. (2007), Flint et al. (2008), Oláh and Johanson (2010), Hamilton and Holzenthal (2011), Ivanov (2011), Pes et al. (2013), Mundahl and Mundahl (2015), Vilarino and Calor (2015), Wiggens (2015), and Holzenthal and Calor (2017).

Family	Nature of shelter	Nature of habitat	Range
Stenopsychidae	No case, nets spun	Freshwater	Russia, East, South, & Southeast Asia
Rhyacophilidae	No case, net, or shelter	Benthic, lotic	Northern Hemisphere
Brachycentridae	Stones, plants, or only silk	Water bodies on coastal plains	Holarctic, Oriental, Indonesia
Glossosomatidae	Large stones	Mountain springs, lotic	Cosmopolitan
Helicopsychidae	Stones & sand in helical cases	Usually in lotic water	Cosmopolitan
Helicophidae	Tubes of sand grains or leaves	Slow lotic	South Temperate Zone
Hydropsychidae	Silk nets only	Lotic water	Cosmopolitan
Hydroptilidae	Very small size	Various	Cosmopolitan
Dipseudopsidae	Long, narrow tubes in substrate	Marine	Cosmopolitan
Leptoceridae	Various	Lotic or lentic	Cosmopolitan
Psychomyiidae	Silk trumpet nets & tubes	Lotic waters	Holarctic
Limnephilidae	Cases of various materials	Lotic or lentic	Northern Hemisphere
Philopotamidae	Long silk mesh finger nets	Lotic waters	Palearctic
Phryganeidae	Large cylindrical cases	Lentic water in cold lakes	Holarctic
Polycentropodidae	Silk trumpet and tube nets	Lotic waters	Cosmopolitan
Ecnomidae	Tubes and fixed silk shelters	Lentic & slow lotic	Southern Hemisphere, one Holarctic
Xiphocentronidae	Various tubes with sand grains	Littoral in fresh water	Tropical regions worldwide
Hydrobiosidae	No case	Lotic water, predatory on insects	Palearctic, Oriental, Australia
Oeconescidae	Cases of plants and small stones	Forest streams	Australia & New Zealand
Ptilocoloepidae	Cases constructed of liverworts	Hot springs with liverworts	Holarctic, Oriental
Phryganopsychidae	Elonate flexible cases	Pools beside springs & streams	East & Southeast Asia
Lepidostomatidae	Cases of bizarre shapes & materials	Springs & slow lotic	Holarctic, Neotropical, New Guinea
Kokiriidae	Sand cases	Sand banks of streams & rivers	South Temperate Zone
Plectrotarsidae	Cases of plant pieces	Shallow, lotic water among plants	Australasian
Goeridae	Traps from rock fragments	Most in cold lotic water	Cosmopolitan, except Australia & S. America

Table 7.1 contd....

...Table 7.1 contd.

Family	Nature of shelter	Nature of habitat	Range
Uenoidae	Sand & rock fragment cases	Lotic, most with rapid flow	Holarctic & Oriental
Apataniidae	Short cases, sand & rock chips	Rivers	Holarctic & Oriental
Tasimiidae	Flat stone cases, opening ventral	Clear, fast lotic water	Australia, Chile
Odontoceridae	Benthic case of rock chips	Benthic in mountain streams	Holarctic, Southeast Asia
Atriplectididae	Larvae bore into dead arthropods	Benthic in large lake	South Temperate Zone
Philorheithridae	Tapering stone chip cases	Cool lotic streams and lakes	South Temperate Zone
Molannidae	Flattened, flanged, stone or plant cases	Lentic water	Holarctic, Oriental
Calamoceratidae	Live in hollow plant remains	Pools along streams	Holarctic, Oriental, Southeast Asia, Fiji
Sericostomatidae	Silk only or with sand	Lentic and lotic	Cosmopolitan, except Australia & New Zealand
Beraeidae	Curved, tapering case with sand grains	Slow lotic	Holarctic
Anomalopsychidae	Curved cases of sand	Lotic or under waterfalls	Neotropical
Chathamiidae	Cases of coralline algae	Tidal pools on Australian seashores	South Temperate Zone
Calocidae	Cases of sand or plants	Forest streams	Eastern Australia, New Zealand
Conoesucidae	Cases, sand or spirally cut plants	Cool lakes, clear streams	South Temperate Zone
Antipodoeciidae	Curved, tapering cases of sand	Small rapid, lotic streams	Eastern Australia
Barbarochthonidae	Long cases of sand grains	Fast lotic streams & pools	South Africa
Hydrosalpingidae	Thin, slender, tapering cases	Lotic water	South Africa
Limnocentropodidae	Cases attached to rocks	Rapid lotic water	Oriental Region
Petrothrincidae	Flat cases attached to rocks	Small streams	South Africa and Madagascar
Pisuliidae	Cases made from twigs	Slow lotic streams	Tropical Africa, Madagascar
Rossianidae	Cases of rock fragments in gravel	Lotic water	Western North America
Thremmatidae	Bulging case of sand and stones	Cold mountain streams	Holarctic

SECTION 2

Diptera

Trichoptera species are by no means the most abundant insects with larvae protected by shelters they construct. By far, the largest number of individual species that develop as aquatic larvae in their own shelters belong to the order Diptera. The larvae of species in the families Chironomidae and Simuliidae, both in the suborder Nematocera, encompass the greatest number of such aquatic species in most parts of the world.

Chironomidae is a family very rich in species. The tubes the larvae construct are usually soft and flexible. Some species build no structures at all. The tubes are sometimes camouflaged with adherent particles from the sediment, such as small stones, sand grains, or silt, and common methods of collecting the larvae destroy their shelters. Therefore, the tubes constructed by many common species remain undescribed, and it is not known whether many of the species actually construct shelters or simply live concealed in detritus. Chironomids typically seek food only as larvae. Adults of each species usually emerge as adults at the same time of year, mate quickly, and the females deposit their eggs and die within a short period of time.

Species in the family Simuliidae, known as black flies, develop as larvae in water rich in oxygen. For most species, this means life in fast-flowing streams. Their shelters are anchored to firm substrates, and those from each species are all somewhat similar in construction. However, most species can be reliably identified from the structure of the shelter alone. Adult females of all or almost all species feed on the blood of vertebrates. Those living on the Arctic tundra, together with mosquitoes, can prove to be vexatious to human beings, just as they are to any birds or mammals that lack thick coats of fur.

a. Family Simuliidae

Members of the family Simuliidae, often called black flies in English, occur throughout the world. Most species develop in lotic water flowing at least with a moderate velocity. Many species show preferences for characteristic speeds of water flow. The family is very rich in species. The larvae feed mainly on microorganisms, which they filter out of the water as it flows by. The scope of any work designed to classify them according to their preferred habitats would be extremely great (McCreadie and Adler, 2006), although it has been begun in various parts of the world (Dumbleton, 1963; Minhas et al., 2005).

After hatching, the larvae find suitable sites to construct the structures which they will inhabit. The shelters of most species are similar in general construction, but that of each species is distinctive in ways revealing the identity of the insect larva it contains to the trained eye. Therefore, many of the larvae can be identified simply by examination of the shelter after the larvae have completed metamorphosis and the adults have departed. Most adults seem to live on the blood of vertebrates, and some are very vexatious to human beings, to whom they can transfer parasites.

The larval shelters of simuliids are similar in general appearance, but the features that give away the identities of the larvae which constructed them are finely structured short teeth and appendages, as well as other armaments and filaments. After the metamorphosis to the pupal stage, the pupae can remain in the cases until they undergo metamorphosis to the imago stage, after which they leave the water to feed and mate.

Simuliids are common throughout world, but they are particularly abundant on the tundra during the Arctic summer. In Alaska, they develop in massive numbers in small pools and rivulets of water above the permafrost, where, together with mosquitoes, they form large swarms near vertebrates that might be passing by in order to feed on their blood. The water in which they deposit their eggs is usually clear, cool, and oxygen-rich. It comes from the melting surface layers of the permafrost, and with the approach of the Arctic winter, it freezes again, encasing the resting stages of the biotic community in ice until the following spring thaw.

b. Chironomidae

As already stated in the introductory paragraphs, some larvae of chironomids construct long tubes in which to live. Many of these midges are benthic and live in tubes that appear flimsy. Most are simply thin silken tubes with scattered sand grains incorporated into the silk. However, they are strong enough to keep the larvae from breaking them by their movements, and they easily keep air inside from escaping, if the species is one that improves its oxygen intake by storing air. Those tubes constructed by some species in the subfamily Chironominae are transparent enough for an observer to see the larva moving inside, although it is usually obscured somewhat by sand grains or detritus. At times when the water contains only small quantities of oxygen, the chironomid larvae can be seen moving their bodies rhythmically to create currents of water through the inside of their tubes.

As mentioned in Chapter 6, larvae of many species in the subfamily Chironominae not only store atmospheric air as a source of oxygen inside of their tubes, many of them can also store oxygen bound to pigments in their bodies. Like the blood of vertebrates, the pigments are usually red, and the oxygen deficit of the insect can be judged by the color of its hemoglobin.

The large family Chironomidae encompasses species that are modified in a wide variety of ways to colonize habitats that are poorly suited for insects, in general. The larvae utilize silk and various objects to produce shelters that assist them in their various adaptations to harsh environments. The family will be mentioned again in subsequent sections. Because of the small size and inconspicuous appearance of their larvae, as well as the short life span of their imagoes, many species are poorly known, and it can be assumed that a great many of those living in parts of the world remote from institutes of scientific research remain to be named and described.

Species in the subfamily Orthocladiinae are notably smaller than those in Chironominae, making it easy to overlook them when examining water containing much detritus. However, some of the most enigmatic species of the family belong to that large subfamily. This family, like the Ephydridae, has been able to invade habitats that are hostile to most other insect species. However, all chironomids are aquatic, while only a minority of species in Ephydridae have fully aquatic stages during their development.

SECTION 3

Lepidoptera

Individual species in most insect orders may construct shelters in which they can conceal themselves or utilize the abandoned structures constructed by other small animals. These range from simple hiding places improved for use by individual larvae, tunnels dug by beetle larvae in the wood of dead trees, and traps into which the prey of the larvae might blunder into and be devoured. The great majority of such insects are terrestrial.

However, a few species from such terrestrial insect orders have adopted ways of life which involve adaptations permitting their larvae to develop under water or in waterlogged soils of temporary or permanent marshes. Such adaptations usually include modifications of the morphology, physiology, or behavior to permit respiration underwater, often including methods of carrying a supply of air with them beneath the water.

Prominent among those aquatic species of predominantly terrestrial insect orders are species belonging to the Lepidoptera, considered to be the closest relatives of the aquatic Trichoptera. The larvae of a few species of Trichoptera are terrestrial, just as a few species of Lepidoptera have aquatic larvae. Some of the families that encompass a few aquatic species include the Pyralidae, Crambidae, and a several others, mostly classified as moths. Almost all have specialized in feeding on the submerged parts of aquatic plants, which explains why their larvae must develop underwater.

a. Pyraloidea

Protective underwater shelters are not constructed only by members of families usually considered to be mainly aquatic in their habits. A few members of the Lepidoptera also have aquatic larvae, which construct protective tubes, often out of material taken from plants on which the caterpillars feed. A European moth in the family Crambidae, *Cataclysta lemnata* (Linnaeus, 1758), feeds on species of *Lemna*, predominantly *Lemna minor* Linnaeus, which form unbroken coverings on the surfaces of eutrophic water bodies and along the shores of ponds and lakes during the warm seasons. Sometimes, the thickest floating beds of *Lemna* form above water that is completely or almost completely lacking in dissolved oxygen. The larvae consume the *Lemna* but also incorporate some of the whole plants into a case spun with their silk. The case is tube-shaped and is completely camouflaged by the plants spun into its silken walls. Inside of the tube, the larva is protected from detection by predators, physically protected from attacks by small enemies, and supplied with atmospheric air stored in the tube. The insect can open the tube at the anterior end and extend its head and thorax out to crawl through the *Lemna* bed on the surface of the water to feed on the plants. If danger is detected, the larva retreats deeper into the tube and pulls the flap at the front closed after it. While the water beneath the *Lemna* is anaerobic, most potential aquatic predators are eliminated as dangers to the larval moths, which can access supplies of atmospheric air if the oxygen in the water is exhausted.

The larvae of many species of moth construct portable cases from material taken from the plants on which they live. Only the fact that the larvae of *Cataclysta lemnata* uses an aquatic plant as its source of material to build its case and lives in the case underwater makes it remarkable. There are a few other aquatic caterpillars found in various parts of the world, but *C. lemnata* is the best known species of Lepidoptera which develops in aquatic habitats because it occurs in parts of Europe in which the science of systematic entomology first developed during the 18th century. It was described as a species by Linnaeus.

While the cases of *Cataclysta lemnata* are constructed of whole plants in the genus *Lemna*, those of the caddis flies with portable cases are typically made of pieces of plants cut to the correct size and shape by the larvae. Therefore, the cases of the many species that are constructed in this way are often sufficient to recognize the species that made them on sight. They are not nearly as strong as those constructed by other species from stones, but they provide excellent camouflage, especially because the species that use them are often able to conceal themselves among dense beds of aquatic macrophytes.

Very few members of the Lepidoptera develop on aquatic plants. Most of those that do typically construct cases made out of parts of the plants on which they feed. These moths use the tubes for camouflage more than physical protection from predators. They may also serve as a storage place for air, which keeps them buoyant and helps the larva to remain just below the surface of the water. Apparently there are a considerable number of species in the tropics that have not been studied enough to know that they live entirely or partially on aquatic plants. These that feed on submerged plants apparently live underwater, and these are likely to build shelters out of parts of the plants, if for no other reason than to provide camouflage so that they are not easy prey for fishes and other predators.

In tropical South America, another species in the family Pyralidae *sensu lato*, *Samea multiplicalis* Guenee, 1874, consumes the floating aquatic plant, *Pistia stratiotes* Linnaeus, in the family Araceae. The caterpillars usually remain concealed within the plant masses and detritus at the water line during the day and come out to feed on the plant during the night (DeLoach et al., 1978).

The classification of this moth in the Pyralidae follows the classification system of Bachmann (1995), who considers Pyraustinae to be a subfamily of Pyralidae rather than an independent family of moths. Another South American member of this family and subfamily, *Sameodes albiguttalis* (Warren, 1989), was considered by DeLoach and Cordo (1978) to be a candidate for the control of *Eichhornia crassipes* (Martius) Solms, but it did not prove promising when it was tried experimentally. For details concerning other attempts to control *Eichhornia crassipes*, see Chapter 4.

b. Crambidae

The cosmopolitan family of Lepidoptera that seems to have the largest number of aquatic species throughout the world is Crambidae. Individual reports from various parts of the world provide examples of species in this family, the larvae of which consume water plants. Unfortunately, some of the reports fail to describe the cases constructed by the caterpillars underwater. It is remarkable, however, that the species in this family seem to be relatively unselective of the aquatic plants which they will accept as food.

An exception to this is the European species, *Parapoynx stratiotata* (Linnaeus, 1758), which feeds primarily on one species of plant, *Stratiotes aloides*. This second aquatic European moth is sometimes present in the same water bodies as *Cataclysta lemnata*, but it is much rarer because of the gradual disappearance of its host plant, *Stratiotes aloides* Linnaeus. Unlike *C. lemnata*, the moth, *Parapoynx stratiotata* (Linnaeus, 1758), lives on the host plant well below the surface of the water. This plant is usually found in water bodies that are only moderately rich in nutrients rather than in eutrophic habitats. In spring, the host plant, which is stiff, spiny, and reminiscent of a pineapple, sprouts at the bottom of the water body. As it reaches its full size, it breaks loose from the bottom and floats to the surface so that the upper part protrudes above the water while it is flowering. Its flowers and seeds are produced during the late summer and early autumn. The caterpillars of *P. stratiotata* remain on the host plant and grow as the plant matures. As the water bodies complete their processes of eutrophication and transformation to marshy terrestrial habitats, the plants disappear, and so do the moths. Where the water bodies are periodically dredged by local farmers to return them to oligotrophic or β-mesotrophic conditions, both the host plant and moth survive.

Another aquatic moth in the family Crambidae, called the pond moth, *Hygraula nitens* (Butler, 1880), from New Zealand and Australia, including Tasmania, is known to feed on a variety of aquatic plants and construct cases out of pieces cut from their leaves. Suitable food plants for them include various species introduced from other continents, as well as native aquatic plants. Redekop et al. (2016) tested seven plant species as foods and determined that the caterpillars were able to subsist on all of them. Of these, *Ceratophyllum demersum* Linnaeus, a species introduced to New Zealand from Europe, was the favorite food of the larvae, while *Hydrilla verticillata* (Linnaeus f.) Royle, *Lagarosiphon major* (Roxb.) (Ridley) Moss, *Egeria densa* (Willd.) Planch., *Elodea canadensis* Michx., *Myriophyllum triphyllum* Orchard, and *Potamogeton crispus* Linnaeus are also acceptable but less preferred. Only the last two are plants native to the region of Australia and New Zealand.

The caterpillars breathe through gills, and to the casual observer, they could easily be mistaken for members of the Trichoptera. However, in addition to morphological characteristics, the diet of the caterpillars indicates that they are moths rather than caddis flies. They consume aquatic tracheophytes rather than the microorganisms, small animals, detritus, or combinations of plant and animal matter on which most caddis flies seem to subsist.

c. Erebidae

The rich fauna of the Neotropical Region would be expected to have its share of moths with aquatic larvae that feed on aquatic plants. However, only the genus *Paracles* Walker, 1855, encompasses multiple species that develop on aquatic plants. This genus is in the subfamily Arctiinae of the family Erebidae, or, according to some taxonomists, what is regarded to be a subfamily should be elevated to the status of a family: Arctiidae.

Unlike the other insects in this chapter that build cases or shelters to live in during their larval development, those in the family Erebidae apparently live on the plants that cover the surface of the water in South America without building a shelter. This is hard to determine because there have been few reports itemizing the plant species that are acceptable as food for the caterpillars or describing any behavior of the larvae underwater. What seems apparent is that the species of *Paracles* reported to be aquatic spend at least part of their larval development on the surface of water bodies, either temporary or permanent. What can be assumed, however, is that the pupal stage of most or all of the species are spent in cocoons or other shelters prepared by the caterpillars. Most of these seem to be located in hollow stems of plants.

Larvae of *Paracles klagesi* (Rothschild, 1910) were observed feeding on its host plant, *Tonina fluviatilis* Aublet in the headwaters of a stream in Northeastern Brazil. It also fed on species in the family Nymphaeaceae. Specimens raised in the laboratory also accepted *Elodea canadensis* Michx. in the family Hydrocharitaceae and an unidentified species of *Cabomba* in the family Cabombaceae (Meneses et al., 2013). Although reports on the diet of some species include mention of a single plant species, most of the species raised in laboratories have been fed and consumed several species of aquatic plant. In one rearing experiment, Drechsel and Drechsel Garcia (2017) even successfully raised *Paracles aurantiaca* (Rothschild, 1910) on lettuce, *Lactuca sativa* Linnaeus.

In Paraguay, the larvae of the species *Paracles palustris* (Joergensen, 1935) have been found feeding on *Eichhornia crassipes* (Martius) Solms, apparently on the upper parts of the floating plants. Other foods were not mentioned. Observations of *Paracles laboulbeni* (Bar, 1873) in Amazonia near Manaus indicated that the early instar larvae feed on unidentified species of algae in the genera *Oedogonium* and *Melosira*, while the later instars consume Gramineae among the floating vegetation, including *Echinochloa spectabilis* (Nees ex Trin.) Link, *Hymenachne amplexicaulis* (Rudge) Nees, *Oryza grandiglumis* (Duell) Prod., *Paspalum repens* P. J. Bergius, as well as a submerged species in the family Lentibulariaceae: *Urtricularia olivacea* C. Wright ex Griseb. When he reported this information, Adis (1983) referred to the species as *Palustra laboulbeni* Bar, 1873, a junior synonym. The type specimens for this species were collected in French Guiana, where they reportedly were feeding on the plant, *Mayaca fluviatilis* Aubl., a submerged plant in the family Mayacaceae (Bar, 1873). Most of the known species of *Paracles* in South America were described from locations in Brazil, but a few were described from the Guianas, Venezuela, and Colombia, and others were reported from Argentina.

There is also a species of *Paracles* that feeds on aquatic ferns in the genus *Azolla*. It is widespread in the temperate southern part of Argentina and Chile. *Paracles azollae* (Berg, 1877) is an aquatic species (Drechsel and Drechsel Garcia, 2017), but there are no detailed reports on its larval development, and apparently the species of *Azolla* on which it feeds have not yet been identified. There are also no reports about the kind of cacoon its caterpillars build or whether or not it enters the water.

8

Ambushers in Streams

A specific feeding strategy used by predaceous insects is simply to remain motionless in a given place and seize any animal that passes within range. This strategy of ambushing is especially effective in moderately to fast-moving streams, where the predator can hide or even camouflage itself and wait until the water moving by delivers suitable prey. Some predators even construct shelters under rocks or in sandy sediments, in which the chance of being carried away by strong water currents is minimized.

The energy expended by the insect is greatly reduced because the flowing water delivers both food and oxygen, with the dissolved oxygen in the water usually at or near saturation, both day and night. The insect stage usually filling the ecological niche of an ambusher in a stream is the larva. Adults of such insects are usually found flying in the vicinity of the streams in which the larvae develop or resting on plants or inanimate objects along the shores. All species of many insect families are completely specialized for life in lotic water, but each of the species is specialized for life in a limited section or microhabitat of a stream. The females deposit their eggs in or near a section where the conditions are appropriate for the larvae. Often this is accomplished by laying eggs on plants overhanging the water so that the larvae can drop into the water immediately after hatching. Not surprisingly, the lower sections of streams with the slowest water currents are usually richest in species because the faster the current, the more difficult it is for the insect larva to resist the current to prevent downstream drift. Where the flow is slowest in the downstream section of the stream, the species often have habits similar to those inhabiting lentic water bodies.

When the eggs or hatching larvae must enter the water at or slightly downstream from the headwaters of a stream, the female must select a site at which suitable chemical conditions prevail. Different springs emerge from the ground with specific combinations of chemicals dissolved in the water. Water from melting glaciers or local rainfall usually displays less variability, but even water from these sources shows differences due to climate, geological history, and geographical peculiarities. In most cases, the first sections of streams flowing from the headwaters move rapidly downstream. The movement of the water can result in an increase in the oxygen concentration until it nearly reaches saturation, where it remains until the speed of the current decreases, usually far downstream, and the number of prey it contains increases. Usually the water temperature and the air pressure increase as the water moves rapidly downhill to lower elevations above sea level, so the water becomes supersaturated with oxygen as it flows.

The continual flow of oxygen-rich water in fast-flowing streams supplies the larvae with sufficient oxygen for respiration without elaborate modifications required to utilize atmospheric oxygen. Simple gills can exchange gases directly with the water flowing past. It is not even necessary for the gills to be moveable so that a flow across them can be created in times when the current in the stream decreases or the oxygen is depleted. Relatively large predatory insect larvae can survive in cold streams with rapid currents because of the ease with which the respiratory needs can be met.

Some of the insect families encompassing species which develop from predatory larvae typically inhabiting fast-flowing streams include Calopterygidae, belonging to the Odonata, suborder Zygoptera; Corydalidae, a family of Megaloptera, which had been regarded as an independent insect order for several decades but is now considered to be a suborder of Neuroptera; predatory beetle larvae belonging to the large family Dytiscidae, which is in the suborder Adephaga, and several small families of the suborder Polyphaga, also belonging to the Coleoptera, such as Elmidae, Dryopidae, and Hydraenidae; and members of several families of Diptera, such as Blephariceridae (Lutz, 1920) and Simuliidae (Hamada and Adler, 1998; Lane and d'Andretta, 1956). It is not certain that all of the small families of aquatic Coleoptera are predatory, nor is it certain that others are not. Several families include groups of minute insects most often found in the riparian zone of streams, which seem to be terrestrial as adults, and the feeding habits of their aquatic larvae remain unknown. If they are predatory, their prey would consist of microscopic invertebrates. Some of these species live on moss-covered rocks in the spatter zone near waterfalls and springs, which gives them the option of using atmospheric air for respiration. A few of them belong to typically terrestrial families, which give them some protection from dehydration in places where there are long periods without rain. The tiny species merit mention here, but enough information is not available at this time for further discussion in this chapter.

Almost the entire insect order Plecoptera, called stoneflies, develop as larvae in lotic habitats (Heckman, 2003). They require oxygen-rich water to survive and benefit from the constant water flow across the gills. Apparently, the first instars of all stonefly larvae subsist on microscopic plants and detritus, but the final instars of species in three families are omnivores, and they become hunters and stalkers of suitable prey that they can capture and overpower. These three families are Perlidae, Chloroperlidae, and Perlodidae.

A special niche is occupied by Heteropteran species in the genus *Rhagovelia*, which have conspicuously modified tarsi possessing a ring of curved setae specialized for very rapid movement over the surface of the water of fast-flowing streams. These species of the heteropteran family Veliidae prey on small insects, at least some of which are terrestrial species that fall into the water of the stream. They are remarkable for their ability to keep their positions while the water supporting them is moving very rapidly downstream. When observing their movements, it appears at times that the surface tension layer is anchored to branches and other objects extending into the water from the shore, while the water below the surface is moving rapidly downstream.

They function as ambushers of insect prey on the surface of fast moving water just as other lotic water predators capture suitable prey species as they are carried downstream below the surface.

Species in the genus *Rhagovelia* are common in the tropics, but they have no morphological advantages in streams in the temperate zones, where they do not occur. Their means of locomotion have not been studied in detail, but their skill in "skating" on the surface of fast-moving, lotic water is almost legendary. Like beetles in the family Gyrinidae, they continually move on the surface of the water, constantly changing direction. They rest in sheltered places along the shore of the stream. If they allow themselves to drift rapidly downstream with the current to avoid a predator, they return to their preferred site a few moments later, having moved upstream against the current.

The subsurface species discussed here are those which develop as larvae in streams with a strong current. They require an effective strategy to resist the strong flow of water, in which they must hunt their prey. The larvae of species of Odonata, suborder Zygoptera, which have the ability to live in lotic water, have a hydrodynamic body shape that is streamlined enough to permit development only in moderately rapid streams. Most of these belong to the families Calopterygidae, called Agrionidae in earlier literature, as well as a few members of the families Protoneuridae and Coenagrionidae that are able to live in streams by avoiding strong currents or clinging to rocks. It is probably because the gills of larval zygopterans are leaf-like appendages attached at the apex of the abdomen that these species are somewhat less successful than anisopterans in adapting to strong water currents. The internal gill chambers of the anisopterans are fully sheltered from mechanical forces from the flowing water and are not subject to injury from turbulent flow.

For this and perhaps other reasons, members of the Anisoptera adapted to mountain streams have larvae that develop in more rapidly flowing water. Rheophilic members of the family Gomphidae bury themselves in sediments as a strategy for resisting strong currents. The apex of the abdomen is narrow and elongate to form a breathing tube, which extends into the fast-flowing water above the sediment and conducts oxygen-rich water into the gill chamber. Larvae of other families commonly encountered in rapidly flowing water, including species in the family Libellulidae, have bodies that are widened and flattened enough to remain attached to the underside of large stones while the current flows over the dorsal surfaces of their bodies. The shape is such that the fluid dynamics acting on the body produce a force that presses the dorsal side of the insect toward the substrate, whether the larva is clinging to a rock above or below it. Their legs are strong enough to cling to the underside of rocks where the insect often remains in an inverted position, and it waits for prey to approach close enough to be captured by the mandibles on the mask, a structure formed by the specially modified mouthparts of Odonata species, which are used for the capture of prey, even when it is moving rapidly past the hiding place of the dragonfly.

Among the most conspicuous species found in the fast-flowing water of mountain streams are larvae of species in the family Corydalidae, members of the Neuroptera, suborder Megaloptera, called dobsonflies or fishflies. These are large enough for their activities to be easily observed, and they are able to overpower and consume small fishes and larvae of amphibians.

Such insects are best located by turning over large, flat stones in the streams. The larvae possess conspicuously, strong mandibles, which can seize any animal small enough to be overpowered by the large corydalid. These insects can remain in spaces below rocks and wait until species of prey seek shelter under the same rock. They are also capable of leaving their shelters at night to hunt, if necessary.

SECTION 1

Advantages of the Habitat

Mountain streams give aquatic insects the advantage of low water temperatures, permitting the water to maintain high oxygen concentrations. Such lotic habitats usually, but not always, have no more than low concentrations of pollutants, and they contain relatively clear water, permitting insects to hunt by sight.

The solubility of oxygen in water depends upon the pressure and temperature. Cold water can hold more oxygen, but at higher elevations above sea level, the percentage of oxygen saturation for a given quantity of dissolved oxygen is lowered by the lower air pressure. As the water flows to lower elevations but maintains its low temperature, more oxygen can dissolve because of the increasing pressure, and aquatic species that depend upon dissolved oxygen for respiration remain well supplied. In streams, there is usually considerable turbulence as the water flows, and this speeds up the dissolution of atmospheric oxygen in the water. Furthermore, the water at the bottom is constantly brought to the top, preventing a benthic layer of water with a reduced concentration of oxygen from forming. Therefore, for any insect larva capable of holding its place in a rapidly flowing mountain stream, problems obtaining enough oxygen during both day and night are seldom encountered.

While dissolved oxygen is plentiful, food is not so abundant. The numbers of suitable prey are usually lower than farther downstream, where the streams widen, the flow rate slows down, and the water begins to warm and lose oxygen. Those sections of streams closest to the headwaters usually contain fewer species, but the individual insects inhabiting them are often rather large. While they find fewer prey, they require less food because less energy is utilized in colder water. That leaves more of the nutrients in the food for growth of the larvae rather than for the energy needed for hunting and respiration. As a result, relatively large larvae are usually more abundant in streams with cold water and rapid currents. Some of the large insect larvae encountered include those that develop into dragonflies, dobsonflies, and large beetles in the family Dytiscidae.

Apparently, many of the predatory insect larvae wait for long periods of time for suitable prey to come along. Some seem to depend upon chemical and tactile senses to find insect prey nearby, while others seem to see the prey moving. As already noted, some species are large enough to capture small fishes and amphibians and do not have to depend upon other insects as prey. The problem is that there have only been a few studies on the behavior of the insects that develop in fast-moving streams, and it is not correct to generalize concerning the habits of all. In addition, while simple aquaria are suitable for raising and observing insects from lentic water, a rather complex system of pumps, channels, and thermostats are required to create a habitat in which stream fauna can survive and be observed under controlled conditions.

Many of the conditions that must be controlled in an artificial habitat simulating a mountain stream are difficult to reproduce in a laboratory. Insects occupying a niche in such a habitat usually die quickly in an aquarium containing standing water, and the conditions created for them in a laboratory are often not good enough to maintain the larvae all of the way through to metamorphosis. The water temperature must be controlled, and the water must be continually mixed to keep the amounts of dissolved gases and other physical and chemical conditions almost the same throughout the entire water column.

Observations in the field cannot provide the detailed information needed to determine precise information about the complete diet, growth rate, number of larval instars, and life span of the aquatic insects in lotic water. The hunting of prey cannot be observed as long as the larvae of the corydalids remain concealed under large rocks or gomphid larvae remain buried in sandy sediments, so differences in the bionomics of many individual species remain to be discovered.

Similarly, the habits of those dragonflies that develop in lotic water are less well-known than those of the species inhabiting lentic water bodies, which can be easily raised in aquaria. For the insects themselves, however, the life of predatory species in streams requires much less effort for seeking food, competing with other insects, and avoiding predators than that of those insects adapted to live in sanding water bodies. In most cases, there would also be fewer problems due to oxygen shortages and periodic episodes of water pollution in a well designed artificial habitat than there would be in a natural stream.

SECTION 2

Disadvantages of the Habitat

Obviously, fast-flowing streams have disadvantages that outweigh their advantages for most species. There are far fewer species in headwater streams than there are in lowland water bodies. In fact, moving downstream along most water courses reveals a steady increase in the number of species present as the speed of the current progressively decreases.

The advantage of high oxygen concentrations that fluctuate little over the course of a day or a season are more than is offset by the shortage of foods available. Lowland streams and lakes provide a surfeit of edible material for insects, while the clear water of a mountain stream provides little. While the low temperatures of the stream water reduces the energy requirements of an insect, there are more energy-rich foods available in the warmer lowland streams. However, lower temperatures can lower the energy needs for activity, but may also extend the time required for growth and development necessary to reach the adult stage. For a predatory species, like a dragonfly or a dobsonfly, a steady supply of food more than offsets the energy saved by lowering the metabolic rate needed at near-freezing temperatures. The energy saved by remaining inactive and waiting for prey to come along is much less than the energy that can be gained by actively hunting in warmer water.

These considerations are theoretical, however. A primary reason there are fewer species present in the fast-flowing waters of mountain streams is that few insects are able to contend with the current well enough to prevent rapid downstream drift. Dragonflies other than gomphids can only hope to colonize streams if their larvae have a broad and flat shape that causes the water flow to press them down against the substrate and legs and claws suitable to hold firmly to objects in the water. Fluid dynamics determine that the great majority of dragonflies could never hold their positions along a stream with anything other than a slow water flow. Dobsonfly larvae have suitable shapes and appendages to seek out and remain in their shelters under large rocks. They are also equipped with mouthparts that can seize, hold, and devour their prey with great efficiency. Insects without these features obviously required to hold their positions in fast-flowing water are not able to survive in mountain streams regardless of the advantages they present.

Another disadvantage of such streams is the shortage of mineral nutrients, which are necessary for the growth of aquatic plants, both large and microscopic. Without plants, there are no herbivores, which frequently are the food for the predators. Fast flowing water tends to move all soluble chemicals, as well as living organisms downstream, and there are few mechanisms in a stream system that move them back upstream. Streams in which populations of fishes swim upstream to spawn are exceptions to this. After tens of thousands of years of steady leaching of surrounding soils and the transport of their mineral nutrients downstream, the whole watershed of a stream system can become seriously depleted. Insects do have the means to move back upstream. If the larvae are washed far downstream during their development, as adults, they can fly back to the places of their own hatching prior to laying their eggs.

The exceptional way for nutrients to be transported upstream and replace depleted supplies is in the bodies of migratory fishes, especially salmon. While growing to a large size in the ocean, where all kinds of minerals are present in great abundance, such fishes can contribute a considerable amount of both organic and inorganic substances required for the growth and development of insects.

Salmon are the best known anadromous fishes throughout the world. They grow to a fairly large size in the ocean before returning to the streams high in the mountains where they hatched. There, they themselves spawn only once if they belong to one of the Pacific species in the genus *Oncorhynchus*, or sometimes twice or rarely more than twice if they are an Atlantic species in the genus *Salmo*, before they eventually die. The tissues of the dead fishes are broken down by a large variety of saprophytic microorganisms and larger scavengers, including many species of insect.

The great disadvantage of such a mutual dependence involving large fishes and aquatic insects is that if either suffers a decline in population, the other will, too. In this relationship, the role of the fish is to grow in the ocean and then return to its spawning stream with the nutrients stored up in its own body. It is the role of the insects to grow on or in the carcasses of the fishes and feed on the decomposing fish flesh or on the microorganisms that are decomposing it. With these nutrients, the insects must produce large populations, on which the hatchling fishes feed during their travels downstream to the sea. If anything interrupts the travels of the hatchlings or the adult fishes, such as a dam, the nutrients will not be replaced, and the insects will not develop large populations. Similarly, if the insects are eliminated by insecticides or temporary drying up of the stream, the hatchling fishes will lack the nourishment they need during the several weeks they require to reach the sea. In either case, both the fishes and the insects are in danger of perishing if the partner in this symbiosis is reduced in abundance.

In streams not used for spawning by anadromous fish species, nutrients must also be present for the insect larvae that grow in them. They may be provided by saprophytes breaking down leaf litter, aquatic plants growing in the water, aquatic mammals carrying their food into their underwater dwellings, or as a result of agricultural activities. In many places, the nutrient budget in the stream is determined by the kind of terrestrial communities present along the courses of the streams in the watershed. If a forest is cut down to create agricultural land, the streams flowing through it will almost certainly undergo major changes in the structure of their biotic communities, including their insect fauna.

The best known insects that play important parts in the food webs of streams around the world include the largest species, which are easily seen in water samples as larvae and flying along the streams as adults during the reproductive period. These include species as small as mayflies and stoneflies and as large as dobsonflies and caddis flies. These insects, while both larvae and adults, are well known as favorite foods of fishes, which is why the general appearance of the adults is imitated by artificial fly makers who sell their products to sport fishermen.

In addition to these large aquatic insects, a group of smaller invertebrates form the links in the food web between the saprobic microorganisms and the relatively large lotic-water insects, which the fishes depend upon for food. These links can include protozoans, rotifers, annelid worms, microscopic crustaceans, and various other animals, which can only be identified under a microscope using comprehensive literature on invertebrate zoology. The insects represented from among these tiny members of the food webs include species in dipteran families that develop in lotic water, such as certain species of Chironomidae, Ceratopogonidae, and Blepharoceridae. In most parts of the world, little can be said about the individual roles of the small species involved. The habits of some of the species inhabiting lentic waters are well known because they can be observed in aquaria, but those requiring strong water currents to survive are much harder to study because the larvae will die quickly if not maintained in a system with a rapid flow of oxygen-rich water.

Furthermore, the food webs possess complexities that are not at all well understood. For example, not only are substances released from the decomposing fish flesh made available to members of the food web, they also become available to algae, which are able to use the simple inorganic nutrients to grow and reproduce. Many of these have the ability to anchor themselves to substrates in the stream, where they produce foods by photosynthesis that can be eaten directly by certain aquatic insects or consumed by microorganisms, which are themselves food for different species of insect.

The main disadvantage to insects that must inhabit lotic water is that there are far fewer organisms that can potentially join in the food webs transferring nutrient substances from the saprobic breakdown of detritus, such as the remains of fishes, to rich and varied insect communities. Therefore, such food webs are not easy to bring into existence, and once formed, they can be easily destroyed through inadvertent harm done to one link in a complex web of diverse species.

SECTION 3

Zoogeographical Distribution of Lotic Habitats

The worldwide distribution of lotic streams suitable for a population of a rheophilic species is determined by the presence of mountain ranges in climatic zones suitable for providing enough rainfall to keep streams supplied with water and at prevailing temperatures warm enough and lasting long enough during the course of a year to permit each insect to complete its respective life cycle. Alternatively, seasonally melting glaciers or aquifers holding sufficient quantities of water to continuously feed mountain springs can replace rainfall as the source of the stream water.

All continents and many islands have locations with both mountains and water supplies sufficient to feed streams. Each species of dragonfly, dobsonfly, or other insect with the same ecological requirements for fast-flowing streams will find a local range of conditions on the continents and islands, some of which roughly meet their requirements. Prior to the beginning of the age of exploration and sea travel, only species already close enough to reach nearby habitats suitable for them could colonize them.

Large mountain ranges are drained by complex systems of streams, making them important locations for encountering insects in the taxonomic groups discussed in this chapter. Places at which many insects typical of mountain stream habitats are found include the Alps, Apennines, Pyranees, Balkans, Carpathian, and Ural Mountains of Europe, and these are the locations of the oldest sites of research on the insects of fast flowing streams. Other such rapid lotic water courses with rich insect faunas include those found in the Pacific coastal mountain ranges of North and South America, such as the Andes, Sierra Nevada, and Cascades, as well as the Rocky Mountains of North America. On the Atlantic coasts of North and South America, mountains, where they exist, are much older in geological terms, and most are lower and covered by forests. In North America, the Appalachians are drained by many mountain streams with typical aquatic insect fauna, while in South America, the eastern mountains are much lower, and much of the region in the tropics is covered by the vast lowlands in the Amazon Basin. The Guiana Shield is probably one of the least explored major mountainous region in the world today.

The highlands in Asia include the highest mountains in the world, but the scarcity of rainfall on many of the mountain ranges reduces the number of streams in many parts of these high ranges. Monsoonal winds blowing across such mountains continuously for six months at a time bring plentiful rainfall to the windward sides of the mountains, but on the leeward sides and the valleys beyond them, there is little or no rainfall at all during approximately half of each year.

All of the sets of local conditions have combined in the major zoogeographical regions to produce special combinations of species distinctly different from those in all other regions. However, during the past 500 years, increasing international shipping traffic has reversed the process of isolation of species with similar but not identical roles in lotic food webs, resulting of the spread of some species and the decline of others. Problems in studying the zoogeography of extant species and ongoing changes in their ranges is being made difficult, if not impossible, by the lack of fundamental information on their distribution and changes in their native ranges throughout the world in historical times.

While it is true that mountain streams are among the least likely to be colonized by species introduced from other continents because they not only have to cross large areas of the sea but also continental barriers, the process is ongoing and is likely to be accelerated as people continue to search for biological remedies to trivial problems.

9

Underwater Tunnel-Diggers, Aquatic Crickets, and Swimming Grasshoppers

Most members of the order Orthoptera are terrestrial. Many of them are restricted to arid or semi-arid habitats. In the temperate zones, they are familiar in fields, in the branches of trees, in bushes, under logs, and in kitchens. Usually, the emergent parts of large marsh plants are the closest these insects come to water.

In the tropics and occasionally in frost-free parts of the temperate zones, the order Orthoptera is represented by species that are obviously aquatic or semi-aquatic. However, most retain the ability to leave the water and function in ways similar to their terrestrial relatives. Among all members of the Orthoptera, there are herbivorous, carnivorous, and omnivorous species, and most have well developed jaws for chewing. Most are moderately large insects, and a few can reach lengths greatly exceeding 10 cm, but these usually live on forest floors or in trees and bushes, and they are not known to be aquatic. One, however, is known to hide underwater for extended periods of time. There are persistent stories, however, about giant members of the Orthoptera that do live in the water, but these are still in the class of legends, like Bigfoot or the Loch Ness monster. Only one of these stories has been confirmed.

One species in the family Stenopelmatidae, the species of which are called wetas, inhabits forests in New Zealand, where it hides from its enemies among the debris at the bottom of streams. Wetas are apparently the largest and heaviest of all insects. Perhaps further research in remote locations might confirm some of the other stories of giant aquatic orthopterans. More will be said about wetas in Section 2 of this chapter.

Although Orthoptera is a relatively small order of insects, and most of the species are large enough to be observed easily in the field, surprisingly little is known about the species that do not cause economic damage in agriculture or are thought to offer biological control of insect pests. Many species in this order have been domesticated to some extent, but much that has been learned about them has not been recorded in scientific literature. Crickets were domesticated in China centuries ago, and cricket fighting has been at the core of various gambling enterprises. The large praying mantis, *Tendora sinensis* (Saussure, 1871), which was introduced to the United States from China, can easily be tamed and will voluntarily stay with a person in return for regular feeding.

One of the most familiar insects throughout the tropics is the large cockroach, *Periplaneta americana* (Linnaeus, 1758). Although it has never been proven that this species spreads any harmful pathogens of parasites to human beings, it is combated using a wide variety of chemicals wherever it appears. The smaller German cockroach of countries in the temperate zones, *Blatella germanica* (Linnaeus, 1767), has been combated so successfully that it was placed on the red list of endangered species in Germany (Anon., 2010).

Although many species in the Order Orthoptera are well known, hardly any of the aquatic species have been given much attention. Most of those that have been mentioned in journals appear to be subtle advertisements for products containing microorganisms or chemicals designed to kill them off. To sell such products, assumed damage that they are doing is stressed to underscore the importance of eliminating them. Mole crickets, which are orthopterans in the family Gryllotalpidae, have been blamed for all kinds of damage, although little is known with certainty about their ecology. These will be discussed further below.

SECTION 1

Tropical Mole Crickets

When walking along the sandy or gravelly shores of ponds and lakes in the tropics, long lines of elevated sediments running from the shore near the water's edge down into the shallow parts of the water body submerged beneath a thin layer of water mark the locations of the tunnels dug by certain species of aquatic or semi-aquatic mole crickets, members of the family Gryllotalpidae. Like many other members of the Orthoptera, these are relatively large insects. Adults are almost always longer than 2 cm but seldom longer than 5 cm. Their generally cylindrical body shape is ideal for passing through the tunnels they construct with their powerful claws armed for rapidly digging subterranean burrows.

What is remarkable about the semi-aquatic mole crickets is their construction of tunnels through submerged sediments, including those that remain submerged for long periods of time. Members of the family Gryllotalpidae are characterized by prothoracic legs that are flattened and armed with strong setae to facilitate their underground construction activities. Many species are confined to relatively dry habitats that remain above the local water table. However, flat sandy areas kept moist by permanent or semi-permanent water bodies remain soft enough to facilitate tunneling by the insects but are firm enough to keep the tunnels from caving in. The aquatic and semi-aquatic mole crickets move rapidly through the tunnels and are not easy to locate by observing movements of the elevated sediments covering the tops of the tunnels. There is also at least one species that tunnels through marine littoral sand on beaches: *Scapteriscus didactylus* (Latreille, 1804), which has been reported to feed on the eggs of leatherback turtles, *Dermochelys coriacea* (Vandelli, 1761), in French Guiana by Maros et al. (2003).

In South America, several species of mole cricket have been found to be widely distributed in appropriate habitats. They were found in tunnels beneath the sand running from above to below the water level at the edges of rapidly drying pools of water during the dry season in the northern Pantanal (Heckman, 1998). Two species in the genus *Neocurtilla* were present, one relatively small and found along the Rio Bento Gomes near the town of Poconé at the northern limits of the wetland. The larger species was found only along the Rio Paraguai, deep inside the floodplains. Their burrows beneath the sand could be recognized easily at the edge of the receding water during the rainless season. Above the burrows, the sediment was obviously raised, and along the burrows that were already above the water level, their sediment was cracked laterally. They typically extended from the drying shore into the region still flooded. Mole crickets were often dug out of completely submerged burrows, and they frequently hid in parts of the burrows that ran beneath solid objects, such as stones or pieces of wood.

The adaptations of the semi-aquatic mole crickets to their habitat in the Pantanal raise a whole series of interesting questions that have never been satisfactorily explained. The first is simply how they are able to obtain oxygen for respiration. Is the air simply forced into the tunnels of the insects and perhaps stored under solid objects? If it is, this would seem to be a solution to the problem while the water is receding during the dry season. However, during the rainy season, many of the areas on which the mole crickets are found remain completely submerged beneath a layer of water at least several hundred centimeters deep. The larval stages either have a completely different survival strategy than the adults, or they have special adaptations to facilitate respiration. Apparently, the adults are active and reproduce during the rainless season, and the larvae develop while the Pantanal is flooded.

The species that was observed only in the Pantanal was found exclusively along the shores of the Rio Paraguai, in the central part of the wetland. The wetland drains slowly through the channel of this river due to a geological restriction to the depth of the river along the southern border of the wetland. This prevents its rapid drainage during the dry months each year. As a result of the restricted outflow, the Pantanal remains an ideal habitat for a large number of typical wetland plants and animals, even if there is no rainfall on the wetland for months. The presence of a mole cricket population along this river would suggest that it is adapted to the large fluctuations in the water level during each year. Studies of its behavior during its life cycle would therefore be especially interesting. Its mode of respiration during periods of submersion of its tunnels are yet to be determined. That some species may store atmospheric air in their tunnels or beneath submerged rocks or other objects is suspected but not yet demonstrated.

The question of how mole crickets survive along sandy marine seashores has only recently been addressed. A species of mole cricket has been reported from the State of Ceará in the northeastern section of Brazil (Bastos, 1977). Apparently, the habitat along the sandy seashores of Brazil is not too different from similar habitats in North America, where mole crickets have been reported near Florida beaches and other regions too far south to expect freezing temperatures during the winters. However, mole crickets could certainly survive light frosts lasting for short periods of time, such as those that occur in Florida from time to time.

Another part of Brazil noted for having mole cricket populations is Amazonia (Fowler and Fasconcelos, 1989). This is probably the part of the country with the richest fauna of gryllotalpids. Like the Pantanal, Amazonia undergoes considerable changes in water level during the course of a year. Large sections of the wetlands along the many rivers that are covered by a deep layer of water during part of the year are also dry during several weeks annually. The species diversity is very great there, and much of the region has never been visited by entomologists. Estimates of the number of undescribed insect species inhabiting the region vary considerably. It can be concluded that the Amazonian and Guianan regions of South America are among the last frontiers of basic biological research.

There is a considerable amount of literature in the field of pest control, which characterizes terrestrial mole crickets as pests in various kinds of habitat, such as Bahia grass pastures and marshy golf courses in Florida. Commercial products, including spores of fungi reportedly effective in killing the mole crickets, are marketed. This would account for the interest in finding new habitats in which it would be desirable to develop methods to control the insects using biological rather than chemical agents. It would also explain why studies to find various kinds of damage caused by mole crickets along beaches have been undertaken. Much of this literature appears more like marketing than reports of scientific research. The evidence that mole crickets are pests responsible for significant economic losses is not convincing. This is especially true for the aquatic species.

The results of one study indicated that the West Indian mole cricket, *Scapteriscus didactilus* (Latreille, 1804), also called the changa, may be tunneling into masses of sea turtle eggs in French Guiana and feeding on them (Maros et al., 2003). Another report indicates that the tunneling of these mole crickets is causing major damage to golf courses (Rentz, 1995). These reports have been used as a reference in order to justify the suspicion that mole crickets living beneath the sand along beaches in North and South America also feed on turtle eggs, although this conclusion was based on little more than observations of both turtle eggs and mole crickets on the same beaches in the United States. This mole cricket species was reportedly introduced to the coastal areas of New South Wales, Australia, in or before 1982 (Rentz, 1995). Certainly, in the interest of protected endangered sea turtle species, more detailed studies are required to determine whether the published suppositions are valid.

SECTION 2

Wetas that Hide Underwater

Members of the family Stenopelmatidae, called wetas, number about 70 species that appear to be large terrestrial crickets. In fact, they include the heaviest of all insects. They are all natives of New Zealand. One of the species employs a very simple method for utilizing water bodies as hiding places from natural enemies. It still requires a positive identification to species. *Paraneonotus* sp. normally lives on the floor of New Zealand forests, often near streams. When threatened by an approaching predator that consumes large insects, this weta simply jumps into the water of a nearby stream and sinks quickly to the bottom. It finds an underwater hiding place, where it remains motionless until the danger has departed from the vicinity of the stream. This semi-aquatic weta has been known to remain at the bottom of streams for as long as 12 minutes. It has not been classified as an aquatic insect (Wise, 1965), but it obviously has semi-aquatic behavior patterns, which substantially improve the survival chances of the weta in the presence of predators. To be successful, it must possess certain physiological or morphological modifications, which allow the weta to remain motionless at the bottom of small streams for more than 10 minutes without drowning. Whether any other wetas have these modifications, as well, remains to be determined.

Aside from its ability to hide for long periods of time at the bottom of streams, it has no recognizable morphological features or behavioral patterns that would distinguish it from terrestrial species of its family. To survive so long under water, it obviously requires one or more differences from purely terrestrial insects. Possible modifications of these insects which would explain their survival under water might be an internal respiratory system consisting of trachea that the insect can completely close off to the outside and hold a large enough amount of air to supply the insect with oxygen for a relatively long period of time, an internal supply of a pigment with an affinity for oxygen to store enough to meet the needs of the insect while under water, and a reduction of its physiological need for oxygen as long as the insect remains underwater.

SECTION 3

Pseudoterrestrial Crickets

In South America, there are also crickets in the family Gryllidae that live on the emergent parts of floating aquatic plants, and under normal conditions, appear to be typical terrestrial species with a taste for the emergent leaves of plants that float on the surface of the water. However, when they feel threatened by a predator or other large animal, they simply move downward on the plants they are resting on, enter the water, and hide among the submerged parts of the aquatic vegetation. These are mentioned in other chapters, where appropriate. In this section, they will be identified as well as possible.

Such crickets apparently feed on the emergent parts of the aquatic vegetation, but like most crickets, they also consume other kinds of foods when available. Unfortunately, many of the species that live among the floating plants have not been positively identified, so searches of the entomological literature will require much time and effort to determine whether these species have even been described, and the accuracy of the species identification would be only tentative in many cases. Two such species were encountered among floating plant masses in the northern part of the Pantanal. One was identified as *Argizala* sp., and the genus of the second has not been confirmed (Heckman, 1998).

There might be some debate about whether these crickets should be classified as aquatic or semi-aquatic. The floating habitat somewhat resemble that of terrestrial plants, which completely cover the substrate in which the vegetation is rooted. In some cases, when viewed from above while standing at the water's edge, the plants fully obscure the water on which the plants are floating with their lower leaves. This could easily lead the viewer to believe that the plants are terrestrial, while, in fact, the water below the plants could be one meter deep, or even deeper. For this reason, the name pseudoterrestrium has been proposed for such floating meadows of aquatic plants (Heckman, 1994).

The role of the crickets in the aquatic environment must still be determined. Many crickets are omnivorous, subsisting on fruits and vegetables as well as living or dead animals. Within the floating islands of plants, a variety of potential foods are available. Knowing which of these foods are preferred by each cricket species would go a long way toward understanding the roles of the crickets in their respective biotic communities.

SECTION 4

Swimming Grasshoppers

In the seasonally flooded Pantanal, a vast wetland fed and drained by the Rio Paraguai and its tributaries near the geographical center of South America, the inundations during the early rainy season produce newly flooded plains covered by as much as 1.5 meters of extremely clear water. After the first thunderstorms of the freshet, many species of aquatic plant begin to grow, and many tiny fishes appear in the water. This season is locally called the *enchente*, characterized by the development of communities dominated by floating or rooted aquatic plants (Heckman, 1998).

During this season, an observer can see many colorful tropical fishes swimming at various depths in the water, as well as typical species of aquatic insects. However, every so often, the surprising appearance of a green and yellow-brown grasshopper swimming together with the fishes well below the surface of the water catches the observer's attention. The species is *Cornops aquaticum* (Bruner, 1906), which feeds on the leaves of certain aquatic plants and travels from one group of the plants to another either by flying or by swimming underwater.

This was already described in Chapter 4, and what remains to be reported is the ecology of its congeneric species. Only one species of the genus, *Cornops frenatum* (Marschall, 1836), is not aquatic and feeds of terrestrial plants. These species belong to the family Acrididae, which encompasses mainly terrestrial species without any preference for aquatic habitats. The other aquatic or semi-aquatic species of *Cornops* are *C. brevipenne* Roberts and Carbonell, 1979; and *C. paraguayense* (Bruner, 1906), the habits of which are not known as well as those of *C. aquaticum,* according to (Adis et al., 2007).

Cornops aquaticum consumes the leaves of aquatic plants in the family Pontederiaceae, including *Pondeteria lanceolata* Nuttal, *P. cordata* Linnaeus, *Eichhornia azurea* (Swartz) Kunth., and *E. crassipes* (Martius) Solms. The species of *Pontederia* are emergent rooted plants, which grow in parts of the wetlands that are flooded seasonally. *C. aquaticum* may consume other plants as well, but it congregates mainly on stands of *Pontederia*. Grasshoppers could be preyed upon by relatively large fishes and by birds. Its two methods of flight from enemies give it a great advantage for survival. When the threat is from large, insectivorous fishes, the grasshoppers can fly appreciable distances to another group of plants to avoid these predators. If sought by birds, it can dive into the water and swim to a place of concealment, where it can remain submerged or partially submerged until the danger passes. Its green coloration with faint yellowish brown longitudinal stripes gives it excellent camouflage while the insects are resting on the plants.

Other species of aquatic grasshopper in the Pantanal include species of *Marilia*, such as *Marilia remipes* Uvarov, 1929, which dives beneath the surface when threatened and conceals itself among the dense, feathery rhizoids of aquatic ferns, such as *Salvinia auriculata* Aublet. Unlike the species of *Cornops*, they do not take flight by swimming underwater to another group of plants, but rather depend upon concealment to escape their natural enemies.

In order to take advantage of the conditions conducive to the development of members of the insect order Orthoptera in an extensive, seasonal wetland, a species requires little more than a way of securely closing off its spiracles temporarily to prevent drowning whenever the insect enters the water. This facilitates access to a rich supply of rapidly growing leaves and provides the insect with several alternative methods of hiding from, fleeing, and actively escaping from their natural enemies. It also permits the insects to live in a location with regularly alternating rainy and dry seasons, which excludes a great many potential predators. Many insectivorous animals have found it impossible to adapt to both aquatic and dry terrestrial habitats alternating roughly semiannually, but the life cycles of several species of Orthoptera are perfectly timed to the changes of the seasons in the tropical wet-and-dry climatic zone.

10

Insects Inhabiting Rainwater

In natural water bodies, insects must survive in the presence of many natural enemies, pathogenic microorganisms, chemical pollutants, and sometimes even acute shortages of oxygen. In most cases, however, these waters are also inhabited by many different organisms on which aquatic insects can feed, and the water contains various mineral nutrients, which most insects require. Ground water is typically rich in certain elements contained in the local geological formations. For the purpose of this chapter, rainwater is defined as precipitation that has not been in contact with the soil, ground water, or bodies of standing or flowing water, which are, in turn, in contact with a substrate that is not impermeable to water. The water from liquid or frozen precipitation may have been trapped in hollowed out rock, which is not porous enough to allow water to pass through and composed of minerals that are not notably soluble, or in parts of plants that are shaped to catch and store water, either through normal growth processes or after breakage and decay. Although many mosses and a few lichens form beds on the ground or in trees that fit this description of containers into which no ground water can enter, the discussion of the insects living in rainwater stored in such plants will be postponed until the next chapter. The rainwater described here is that isolated from all water that has been in contact with the ground or been in held in a pond, lake, stream, or river. Of course, water falling through the atmosphere often becomes contaminated with matter suspended in the air as dust or gases mixed with the air, but such contamination is almost always insignificant compared to the amounts of substances that enter the water immediately, once it has contacted surface water or moisture in soils. Just the presence of numerous actively growing bacteria, algae, and protozoa makes the water on the ground quite a different milieu than the falling precipitation, which might only have contact a few inactive bacterial spores on its fall through the air.

The fundamental problem encountered by insects that attempt to colonize pure rainwater is the almost complete lack of nutritious food organisms, active predators, and competitors. Although the search for food would have to be a primary activity of insect larvae developing in rainwater, they are relieved of the need to continually detect and escape hungry predators that prey on organisms in their size range. They are also relieved of the problem of competing with microorganisms for the available oxygen in the water.

Some species of insect do grow and develop best without a rich biotic community of aquatic organisms around them. For many of these species, survival is easier in rainwater trapped in depressions in the ground, rock formations, holes in trees, and large epiphytic plants. Some of these rainwater habitats dry up rather quickly; others are semi-permanent. For assemblages of rainwater to be useful as habitats for aquatic insects, they must remain in containers that do not subject them to complete evaporation faster than the aquatic insect can develop into a reproductive imago. The container must permit no more than insignificant leakage, almost no seepage, and have a shape that minimizes contact with the atmosphere. An additional factor to consider is the frequency of rainfall in the region. A site in a rainforest with daily rainfall would require a less well constructed container to provide a continuous supply of rainwater in which aquatic insects could survive than one in a region where only occasional rainstorms occur.

A few structures that catch and hold rainwater are found among natural rock formations. Naturally hollowed out rocks are found in many parts of the world, but they are not abundant. However, human beings have greatly increased the number of habitats similar in shape and size to such natural rainwater containers suitable for the larval development of aquatic insects preferring rainwater over the water in ponds and lakes. Today, old buckets, discarded earthenware, worn out tires, and plastic containers catch and hold rainwater near human habitations all over the world. The larvae of many kinds of mosquitoes that spread serious diseases are able to utilize such containers for their larval development and feed on the blood of human beings, who were kind enough to provide them.

Perhaps the most interesting but poorly known containers for catching and holding rainwater are certain plants that grow to form structures providing aquatic insects with a long-lasting supply of rainwater, which is subject to minimal evaporation. The best known of the plants that can do this are the bromeliads, which include many epiphytic species that grow attached to tropical forest trees. However, while it is known that they harbor aquatic insects, little is known about the symbiotic relationships between the individual bromeliad species and the species of insect which develop in the rainwater the host plants contain. Some of those relationships that have been discovered appear to be highly specialized and complex. The next chapters of this book will concentrate on such bromeliads and the insects they contain. These live in a biotic community known as the phytotelm (Corbet, 1983).

SECTION 1

Rainwater in Natural Rock Formations

Where the water collects temporarily on terrestrial soil rather than in hollowed out rocks or holes in trees, the amount of time the water will remain available for colonization by insects will also depend upon the porousness of the soil. While the amounts of minerals are usually much greater when the substrate sometimes acts periodically as a terrestrial soil, the water in hollowed out rock is typically clearer and richer in oxygen during most of the 24-hour day. Hollows in trees are usually intermediate in nutrient content between hollowed out rocks and flooded terrestrial soils. The oxygen content depends upon the time of day, the number of algae, bacteria, and other microscopic organisms found in the water, and the amount of free inorganic nutrients available to them.

Glaciers expand and push large rocks ahead of them. During periods of global warming, such as those which have prevailed since the end of the last ice age, glaciers melt away and leave a line of large rocks and coarse gravel marking the locations of the leading edges of the large masses of ice. These heterogeneous deposits of assorted rocks and minerals are known as glacial moraines. Over subsequent centuries, the material in the moraine is subjected to considerable abrasion and water flow, first from the melting ice and then from centuries of rainfall, which moves lighter material to places at lower elevations and erodes away larger rocks. Moraines in Europe and North America mark the borders of earlier glaciers that covered the region during successive ice ages.

In many places, large rocks left stranded by the retreat of glaciers are located in or along streams and lakes or beneath waterfalls. After the mechanical damage to the rocks from relocation when the moraine was formed followed by centuries of slow erosion, many large rocks found at locations formerly covered by glaciers have deep, wide holes and odd-shaped depressions, which can hold water at least for several weeks between rain storms. The water in such depressions is often populated by larvae of dipteran species that are apparently accustomed to such habitats. Because of their isolation, deeply eroded rocks are often overlooked by entomologists collecting aquatic insects.

The relocation of the larval habitats of species that have difficulties competing with large numbers of other aquatic insects inhabiting permanent water bodies to broken artifacts disposed of near human habitations requires little or no physiological or ecological adaptation. Taking the abundant disease vector, *Aedes aegypti* (Linnaeus, 1762), as an example, it is not difficult to envision a species that developed in rainwater trapped in holes in tree trunks and fed on the blood of arboreal mammals finding discarded pottery near human habitations containing trapped rainwater to be even safer from predators and more convenient for obtaining mammalian blood than the hollowed tree trunks. This species of mosquito has probably transferred more bacterial and viral diseases from other primates to human beings than any other insect. Several species of *Anopheles* have done the same for various species of *Plasmodium*, the protozoans that cause the various kinds of malaria. However, each of these *Anopheles* species has a more limited geographical range than *Aedes aegypti*, which is presently found near human habitations in most parts of the world not subject to seasonal freezing temperatures, where they breed in assemblages of rainwater, often contained in discarded artifacts.

SECTION 2

Man-Made Rainwater Habitats

Almost all towns and villages around the world have their own water sources. Few people still live so far from a well or a natural water body that they have to carry pails for miles to obtain water each day. One of the first goals of international aid projects is the construction of wells or pipelines to supply villagers in remote locations with access to running water. Saving the water that falls as rain is a popular method to minimize the need to transport water. Certain insect species have been able to claim a share of the rainwater saved by people throughout the tropical and subtropical regions of the world. Some are limited to certain regions or continents, but many have settled near human habitations wherever they occur in places warm enough for the insects to live over the past millenia.

Artificial containers to hold rainwater for long periods of time include various objects stored near human habitations. The development of several insect species in rainwater inside of objects used in human settlements has some important implications for public health and livestock rearing, just as it did when these diseases were first transmitted to man by the mosquitoes that are surmised to have obtained these pathogens from forest mammals, most probably other primate species.

The species *Aedes aegypti* (Linnaeus, 1762) has adopted itself to human habitations in tropical and subtropical regions around the world. This species is a major vector for two potentially fatal diseases: yellow fever and dengue fever. It also transmits several other less common or less serious diseases, such as the microcephaly caused in infants by the zikka virus. Although it is by no means the only mosquito that can transmit such diseases, it is probably the one that transmits them most often because of its close proximity to human habitations throughout the tropics.

In tropical countries, piles of metal and plastic containers, old tires, and temporary pools of water that dry up only during long rainless periods often contain rapidly developing nematoceran larvae, including the mosquitoes that spread contagious diseases, which will be discussed in Chapter 25. Old tires hold trapped rainwater for long periods of time without most people being conscious of the fact that they are filled with large numbers of mosquito larvae, often including those of *Aedes aegypti.* The people in such countries know that the elimination of trash that catches and holds rainwater makes an important contribution to preventing insect-borne diseases. In colder regions of the temperate zones, this problem is mitigated because the populations of most insect vectors that develop in stagnant water cannot survive in a climatic zone that has sub-freezing temperatures during winter.

Yellow fever was once a major cause of human fatalities in the tropics, but after it was discovered that it was transmitted by mosquitoes, removal of containers holding water near towns and villages greatly reduced its occurrence over the short term. This discovery permitted yellow fever to be controlled sufficiently to permit the Panama Canal to be completed. The disease nevertheless persisted until the development of a vaccine facilitated its eradication in many locations around the world.

Dengue fever, although not as dangerous to human beings as yellow fever is to unvaccinated people, can be fatal, especially after a person has already suffered once from the disease. The lack of an effective vaccine against dengue fever makes it a serious danger to public health, even now. These diseases and other insect-borne sicknesses will be discussed in more detail in Chapter 25.

Other mosquito species not so closely associated with human settlements may also be dangerous to human health and well-being. These include the species of *Anopheles* capable of transferring the protozoan parasites which cause malaria. The transfer of these protozoans, in the genus *Plasmodium*, is part of a very complex process involving multiple stages in the life cycles of the parasitic protozoans. Apparently, there are several processes that are still unknown or not understood. The fact that human beings frequently contract malaria in forests and wetlands far from human settlements demonstrates that there are other vertebrates that act as hosts of the protozoans in addition to human beings. These include primate species, which are known to harbor populations of *Plasmodium*. See Chapter 25 for a full discussion.

Mosquito eradication programs to kill off the larvae of the insects that transmit diseases has never been fully successful, although temporary improvements have frequently been reported. One reason is that there are too many accumulations of rainwater in places that escape the notice of human observers, and the species of insect vectors living in close proximity to human dwellings have long been resistant to all known pesticides and will most certainly quickly develop resistances to all new ones that are permitted on the market.

Other places that hold rainwater for long periods of time include various objects stored near human habitations. The development of several species in rainwater trapped in objects used in human settlements has some important implications for public health. The species *Aedes aegypti* has adopted itself to life near towns and villages in tropical and subtropical regions around the world. This species has fully adapted itself to life in close proximity to human habitations and has been spreading diseases from one person to another since prehistoric times. The location of its origin is still a matter of conjecture. There is no prospect of eliminating this mosquito by any known means, which is why vaccination has been found to be the only way in which yellow fever could be fully eliminated.

Its choice of habitats has proven to be excellent because the human population maintains an ideal habitat for it. People not only provide containers to hold water for their larvae to develop in, they also provide protein-rich food in the form of human blood. In addition, they also eliminate many of the natural enemies of the mosquitoes. Since pesticides were introduced, only a brief period of adaptation was required before *Aedes aegypti* had established populations resistant to all of the common insecticides that are used, one after another.

Aedes aegypti is certainly not the only aquatic insect capable of living in close proximity to human habitations. Species of *Anopheles* that transmit different species of *Plasmodium*, the genus of Protozoa which causes different kinds of malaria, has contributed to the loss of millions of human lives annually throughout much of recorded history. This will also be described in detail in Chapter 25. However, the species of *Anopheles* that transmit malaria do not always live in close proximity to human habitations all over the world, and in many places, malaria does not occur because suitable vectors are not present. Apparently, there are other important reservoirs of the *Plasmodium* species besides human beings, and these are generally assumed to be monkeys and perhaps one or more of the greater apes.

Aside from mosquitoes that are vectors for potentially fatal diseases, water trapped in artificial containers provide the habitats for other familiar insects. In the temperate zones, where the water outside freezes temporarily during most winters, the larvae of species such as *Culex pipiens* Linnaeus, 1758, replace *Aedes aegypti* as the most familiar summertime pest, although their bites do not constitute life-threatening incidents.

Insects other than mosquitoes are not nearly as common in water trapped in artificial containers as are mosquitoes. If they were, these man-made habitats would not be as attractive to mosquitoes, which any efficient predator could catch and eliminate in a relatively short time. However, certain species in the family Psychodidae might use such habitats when they are available. Their larvae certainly develop in polluted water near human habitations, such as that in outhouses, but some species might also be able to develop in water less rich in nutrients. The only species that occasionally utilize such habitats with certainty are species in the family Chironomidae.

11

Hexapods in Mosses and Lichens

Actively growing mosses and lichens are usually encountered at locations where the relative humidity in the microhabitat remains at or near 100%, except for brief periods during sunny days. Many species of lichens and some mosses have the ability to suspend their growth and development during periods of dryness and resume them promptly as soon as water is again available. In regions of heavy rainfall, both in hot and cold as well as temperate climatic zones, both mosses and lichens compete for suitable locations at which to grow, and small clumps of many species alternate along tree branches and on exposed surfaces of rocks. Other species are often encountered in dense patches of mosses on moist soils. This is one of the aquatic habitats designated by earlier authors as phytotelm (Thienemann, 1934). They are apparently the smallest aquatic habitats in which insects can be found, and the locations in which they can occur are limited to those with very frequent rainfall, such as both tropical and cold, temperate rainforests, certain hillsides exposed to onshore winds from oceans or large lakes, or seasonally on tundra in which suitable vegetation occurs.

SECTION 1

The Nature of the Habitat

a. Epiphytic mosses and lichens

Beds of mosses and patches of lichens become habitats for aquatic insects because they catch and hold rainwater and dew. Like sponges, beds of mosses absorb large amounts of water and offer suitably adapted insects a truly aquatic habitat, albeit a very small one. However, unlike most bodies of water, they permit the insects close contact with the atmosphere, regardless of their position inside of the saturated plant masses.

In addition to aquatic organisms, beds of water-soaked bryophytes and lichens offer many semi-aquatic hexapod species a suitable habitat, in which they can find shelter and concealment from small predators, as well as many kinds of food sources and ample water during periods when adjacent habitats have become completely desiccated.

Patches of mosses and lichens that provide reliable sources of water for longer periods of time than the rest of the adjacent vegetation are difficult to classify precisely as aquatic, semi-aquatic, or moist terrestrial. The habits of many of the hexapod species present in the waterlogged vegetation are not known well enough at the present time to adequately classify. Those species which exchange respiratory gases directly with the water can safely be called aquatic, while those which keep in contact with wet mosses so that they can conceal themselves if predators should approach are more likely to be classified as semi-aquatic. So many insect species have distinctive habits, however, that there continue to be problems in clearly distinguishing the ecological grouping to which an arthropod belongs.

Trees in rain forests are typically rich in epiphytic species of algae and bryophytes. The floors of such forests usually support other species of these plants. Many of these plant species form dense beds that soak up water and maintain a constant supply of fresh water for aquatic hexapods to inhabit with only infrequent interruptions during long periods of dryness. There are many rain forests in tropical regions, but rainforests also occur in cold temperate regions, such as those along the coast in the Pacific Northwest, including southern Alaska, as well as in New Zealand. The rainforests can be distinguished immediately from typical forests growing inland from the coast. The branches of the trees are thickly covered by mosses, lichens, algae, and other non-vascular plants, while those of the typical forests may support some epiphyte growth, but it does not cover most of the bark on the branches. These epiphytes often make it require some effort to discover patches of uncovered bark on the older branches.

Mosses form beds of closely packed plants, and many lichen species form folded or branching masses of thalli resembling large leaves. These plants retain water and maintain microclimates with a relative humidity of 100% and a thin coating of water on some of the plant surfaces. A search of these plant masses usually reveals the presence of various invertebrate species adapted to life in association with the plants. Most of these species are relatively small, but a few large insects are occasionally encountered, often with a coloration blending in with the background.

The plants in which the insects live do not necessarily supply them with food, but the foods for small, predatory species come to the moss beds. The moss plants are often located in forests, which are seasonally supplied with large amounts of wind-borne pollen. Most mosses only provide the substrate on which the insects live, moisture, and hiding places. The sources of pollen, microorganisms, and smaller arthropods, which are the foods of most hexapods, come from the environment. The exact relationships between a great many species of insect and the non-vascular plants in which they live still remain to be learned. Many of those found on the mosses and lichens on trees in tropical forests still need to be named and described. However, some of the most abundant species thrive in cold rainforests and even on tundra.

b. Mosses in bogs and raised bogs

A completely different kind of habitat is found in moors, bogs, raised bogs, and poor fens. The names of these habitats are basically derived from local names for basically the same kind of habitat. All are characterized by water bodies very poor in the mineral nutrients required by plants for photosynthesis. The water is typically yellowish brown and acidic from the humic acid it contains. The flora is characterized by plant species that are specialized at growing in poor soils, and it often includes species that trap and kill insects to augment their supplies of inorganic plant nutrients. Strictly speaking, it should not be classified as phytotelm because the water is in contact with the soil and surface water that has been part of a water body for a considerable amount of time. However, the water in bogs comes almost entirely from rainwater because there is no appreciable amount of water flowing in or flowing out of a bog.

The difference between bogs and raised bogs is mainly related to how they were formed. A bog formed mainly in lowlands, where a lake with no inflow of water carrying nutrients has been gradually filled in with the remains of plants that grew in nutrient-poor water. A raised bog develops at the top of a hill or on a plateau, and it has no input of water other than rain. The underlying rock is not porous and does not allow the water to seep out of the bog. Therefore the water can leave the bog only by evaporation or escaping over the rim, usually located at its lowest point along the circumference. Over many centuries, the water carries away the plant nutrients during periods of heavy rainfall. They are not replaced. This situation produces the same kind of nutrient-poor, humus-rich habitat as is found in a lowland bog. Moor is a British word for any kind of bog, but it is usually applied to a large lowland bog unless the raised bog covers an appreciable area.

A great many lowland bogs have been destroyed by attempts to convert them to farmland. These projects have failed because the necessary treatments with lime to bring the pH value up into the range tolerated by the crops and the large amount of fertilizer needed to permit ordinary crops to produce satisfactory harvests are too costly to make farming profitable. After this was discovered through bitter experience, farmers started growing those few plants adapted to bogs and bog soil. The most popular of these in North America both belong to the family Ericaceae. They are cranberries: *Vaccinium macrocarpon* Aiton, and the blueberry: *Vaccinium corymbosum* Linnaeus.

Because the soil is so poor in bogs, those in the pristine state usually support populations of insect-eating plants. In the water bodies, one or more species of *Utricularia* Linnaeus are usually present, and growing on the moist soils are many species of pitcher plants, several species of sundew plants, and other species more limited in distribution in many parts of the world. Examples of such local insectivorous plants include a member of the Lentibulariaceae, the butterwort, *Pinguicula vulgaris* Linnaeus, in scattered bogs throughout the Holarctic Region, and the Venus flytrap, *Dionaea muscipula* Ellis, in bogs of the Carolinas in eastern North America. The flora in any bog generally conforms to that of the zoogeographical region in which the bog is located. Generally, typical bogs are not encountered in the tropics, where the soils of most pristine habitats are typically poor in plant nutrients but not always acidic.

There is one characteristic genus of moss represented in bogs all over the world, and this is the predominant plant in the formation and evolution of each pristine bog habitat. These mosses are in the family Sphagnaceae, and their genus is *Sphagnum* Linnaeus. There are a great many species in this genus, and not all of them are dominant species in bogs. However, a review of the genus is beyond the scope of this book. What is important is that they contribute to the typical water quality of their habitat, and they absorb large amounts of water. In effect, they shape the physical and chemical characteristics of their habitat.

Since the last ice age, they have grown on the surface of water bodies in Europe, North America, and parts of southern New Zealand, Australia, and presumably South America. Each spring, they start growing and continue their growth and domination of the surface of the water body through the summer. In autumn, they die off and form a layer, upon which the plants starting growth during the spring rest. Each year a new layer of dead plants forms upon which the next generation of the moss grows. When cutting through the moss layers vertically, they reveal layers formed by each successive generation, which can be identified and counted. In this way it is possible to determine the year in which each layer of peat, as the dead mosses are called, grew and was buried under each more recent layer. In an undisturbed bog, the chronology of the peat growth can be learned, going as far back as the end of the most recent ice age. It can be determined by counting the layers of peat above it, the exact years in which artifacts, animals, and human remains were deposited in the peat. Such objects do not decay but rather are preserved in the peat because of its propensity for keeping the water acidic and free of oxygen.

The peat in bogs has frequently been used for fuel in Europe, and many bogs, especially those located in lowlands, have been eliminated or greatly reduced in size. Where the peat is left alone, geological processes slowly convert it into lignite. Over additional periods of millions of years, the lignites are converted to bituminous coal and even later to anthracite.

It should be clear that bogs are highly distinctive habitats, which stand out from the local climax community of forest or grassland. In forested regions, they stand out because of the paucity of trees. Everywhere, they can be immediately recognized by their low species diversity of plants and the presence of large masses of *Sphagnum* spp. and characteristic species of vascular plants, especially insectivorous ones. Chemical analyses of the water show it to be yellowish brown with a low pH and low concentrations of such ions as nitrates, nitrites, and phosphates.

Like the masses of epiphytic mosses and lichens, they support communities of aquatic and semi-aquatic insects, which have not been well studied in most parts of the world. It is not always easy to distinguish between aquatic and semi-aquatic species from their roles in these wetland communities. Most such species are not found in other kinds of habitat adjacent to the bog. Mention of some of these species from places where surveys of the fauna have been made will be found in the next sections.

SECTION 2

Collembola

Hexapod species in the order Collembola are often associated with mosses and lichens, as are tiny larvae of species in the order Diptera, mentioned in subsequent sections. Most of the species of Collembola in the bogs are very sensitive to dryness and require moist substrates that maintain the microclimate in their immediate vicinity with a relative humidity close to 100%. Although the number of collembolan species is not nearly as great as that of some of the largest insect orders, there are thought to be more individual collembolans on the earth than individuals belonging to all insect orders, combined. Many of the collembolan species produce large populations in aquatic habitats or in continually moist terrestrial environments. Examining primitive plants living in water-soaked masses can help convince skeptics that this hypothesis might well be true.

Not all collembolans die quickly when removed from the masses of wet mosses and lichens at room temperature and humidity. However, many aquatic species do. During periods when the natural microclimate at the surface of beds of lichens and mosses approaches the thresholds for survival of the springtails, these hexapods seek out the cooler and more humid spaces below or within the masses of plants, where they can survive until the more favorable conditions return, usually shortly before sundown or when rainfall begins.

In the laboratory, a Petri plate containing a small amount of water or moist material is sufficient to keep springtails adapted to continually moist environments alive for unlimited periods of time as long as the cover of the plate is not removed. If the cover is removed, the animals will live only as long as the microclimate remains moist, which can be as short as a few minutes. Apparently, the survival time is influenced by the relative humidity of the air in the room.

A few other hexapods that were formerly classified as insects belonging to Apterygota may also be encountered below beds of mosses that maintain conditions suitable for insects sensitive to drying. Most of these, however, can only be encountered among the detritus on the forest floor, most commonly in tropical rain forests, where mosses and other primitive plants are only sporadically distributed and the soils are wet but not submerged in water. This habitat is shared with subsurface soil species of Collembola, which have only vestiges of a furca and cannot spring.

As expected, species of Collembola can be among the hexapod fauna in water-soaked mosses and lichens, wherever they occur. Studies in southern Norway revealed two such species living in lichens growing on rocks in an alpine habitat between 1200 m and 1250 m (Leinaas and Samme, 1984). Such habitats are neither typically aquatic nor terrestrial. Like the springtails that are commonly observed living on the surface tension layer of water bodies, which remain in contact with the atmosphere as well as with the water beneath them, they have sought out a microclimate in which they are supplied with water and air with a humidity at or very close to 100%. As already mentioned, these hexapods die quickly if exposed to a relative humidity typical of most terrestrial habitats (Christiansen, 1964).

The two species observed among the lichens in southern Norway were *Xenylla maritima* (Tullberg, 1869) and *Anurophorus laricis* Nicolet, 1842. Their habitat provides them with a cool, moist environment and contact with water seasonally in liquid form or as ice sheltered by the lichens, which are mainly active during winter. During their aestivation, lichens become dry and sometimes stiff and hard. Upon exposure to rainfall, many species quickly become soft again and resume photosynthesis and growth. Obviously, their associated springtail populations must seek shelter and remain dormant until the humidity increases again. In the mountains of southern Norway, they are also subjected to being entrapped in the ice, during which their immediate environment can become anaerobic (Leinaas and Samme, 1984).

Experiments were performed to detect differences in the toleration of the springtail species to dryness. Other local species were included in the experiments, although they lived in somewhat different habitats, including soils and heath overgrown with other lichens. It was thought that some of the species from the lichens growing on soils were not able to survive among the lichens growing on rocks because of the lack of oxygen during periods the lichens and hexapods are entrapped inside ice crusts that cover the rocks during winter (Leinaas and Samme, 1984).

Species of Collembola also inhabit lichens and mosses that grow on tree branches. In the State of Washington on the western side of the Cascade Mountains, *Entomobrya nivalis* (Linnaeus, 1758) was found and identified inside dense masses of lichens and mosses on the branches of conifers. This species was frequently found on top of the clumps of ephiphytic mosses and occasionally on the lichens growing between them. Immature individuals were also observed between the branches of the mosses, but they were not identified.

The climate between the Pacific Coast and the Cascades is mild due to warm ocean currents running close to the coast. The warm, moist sea air rising as it meets the mountains produces a great deal of rainfall, producing a cool rainforest along parts of the Coast, especially close to the Olympic Mountains. The tree branches are covered with large coatings of epiphytes, which are produced each year, mainly during the colder seasons. Most species of Collembola present among these water-soaked epiphytes belong to species that are cosmopolitan in distribution, or nearly so.

SECTION 3

Insects from Mosses and Lichens Covering Water Soaked Soils on Forest Floors

Cold rainforests are fairly rare in the world. Some of the largest are found along the western Pacific seacoasts running from Alaska through the Pacific Northwest. A few other, smaller ones are located in New Zealand. However, cold rainforests cover a very small percentage of the geographical land area of the world.

Typically, cold rainforests have soils that are water soaked and covered by thick layers of forest litter with dense growths of mosses and many lichens covering the ground. The water supplied during the frequent periods of rainfall are held by the dense patches of these primitive plants, which act like sponges to hold water during rainless periods. They superficially resemble in their way of growth many of the epiphytic species covering the tree branches in the rainforests.

The habitat can be called semi-aquatic, although it shows fundamental differences from both water bodies and terrestrial soils. It is distinctive in having ambient rainwater trapped within dense masses of living plants, which form patches with only the lower parts in contact with the soil. Like vascular plants, which have roots in the soil but upper parts forming a different kind of habitat for insects, the upper parts of the clumps of moss offer a habitat of pure rainwater and dew, while the lower part keeps the rhizoids of the moss plants in contact with the ground. The upper parts appear to be similar to the phytotelm produced by bromeliads, but on a much smaller scale. The small hexapods capable of colonizing such habitats enjoy the availability of unpolluted rainwater trapped in dense masses of mosses and lichens, which conceal and protect the small invertebrates living in the water. The contact with the atmosphere and photosynthesis by the mosses keep the water well oxygenated. The underlying soils contain a slowly decomposing mass of twigs and dead needles from conifers. The fungi, bacteria, and other microorganisms provide food for a variety of tiny insects and other invertebrates that can live in the water-soaked habitat.

The animals that develop in this kind of habitat include a variety of invertebrates that are very poorly known due to a lack of research on both the plants and the animals they harbor. Poorly known species from the cold rain forest floor include protozoans, tardigrades, isopods, mites, spiders, centipedes, and salamanders, which share the habitat with many species of hexapod. In addition to the class Collembola, the beds of mosses and thalli of lichens harbor insects mainly in the orders Coleoptera and Diptera. Most springtails and beetles remain in the beds of water-soaked mosses and lichens throughout all of their life stages, while the flies and midges remain there only during their larval development.

The aggregations of species in the beds of non-vascular plants on the rainforest floor differ from those on the soils of typical terrestrial forests in subtle ways, which are nevertheless vital for determining the structure of the flora and fauna. The most obvious characteristic of the flora on the rainforest floor is its great abundance, the density of the beds of the individual species, and the large number of species present. More important for the semi-aquatic species present is the amount of water that the dense plant masses can hold. Several of the moss species form dense masses of closely packed plants, which act as sponges. When squeezed, they typically release a considerable amount of water, even during days without rainfall. Typically, the hexapods and other invertebrates which inhabit these beds of mosses are tiny, enabling them to develop in or directly in contact with ambient water. This gives them a certain amount of protection from terrestrial predators. Most such predators belong to the terrestrial predators in the order Coleoptera. The families Carabidae and Staphylinidae are best represented among the predatory terrestrial beetles inhabiting the forest floors in these rainforests.

Because these species are small and remain concealed among the waterlogged mosses, they have escaped the attention of systematic biologists. This explains why many of them have only recently been described and named by taxonomists. Some of them have been initially assigned to their own families before the characteristic features of an old, established families were noticed. Because it has proven difficult to assign several of these relatively new species to families, it is still uncertain to which they actually belong. It is fairly certain that careful studies of their waterlogged habitats will yield more species that are still unknown to science.

SECTION 4

Insect Fauna of the Micro-Phytotelm

Under a low power microscope, the shape of the leaves on many species of moss does not appear too different from the leaves on a species of vascular plant in the family Bromeleaceae. The leaves are concave on the dorsal side, and its lower part is adpressed to the stem. The hollowed spaces at the base of the leaves catches and retains rainwater. The surface of the trapped water has only a narrow surface in contact with the atmosphere, greatly reducing evaporation. This shape accounts for the large amount of water a clump of moss can hold. In the middle of the clump, water is held for a considerable amount of time, during which insects utilizing such habitats are able to remain sheltered from predators and find items of food that fall on the mosses, such as pollen grains.

While the structure and function of the leaves of many moss species is similar to that of the leaves of bromeliads, the size of the leaves are several orders of magnitude smaller. The moss communities could be called a micro-phytotelm, and they are far too small for most hexapods to fit into. The species best suited for this community are those reaching one or two millimeters in length, and the maximum size would not be much larger. The most notable hexapods observed living on or in the water trapped in moss plants include species of Collembola, tiny beetles belonging to families not among those most frequently encounted in aquatic habitats, and larvae of tiny members of the Diptera belonging to various families.

a. Coleoptera

When clumps of water-soaked mosses growing on the trees in a cold rainforest along the northern parts of the Pacific Coast of North America are examined under a low power microscope, it becomes apparent that many tiny insect species are adapted to living in this kind of habitat. There are few reports on the species present because they are tiny and conceal themselves well inside of the moss plants and because the cold rainforest is a habitat that does not occupy much of the total terrestrial area of the world. The species are not easy to identify, and their systematic position is not always agreed upon.

An example of an enigmatic beetle species is *Erpelus brunipennis* (Mannerheim, 1852), which inhabits the forest floors of cold rainforests from Alaska to California, apparently settling mainly in the clumps of actively growing mosses that cover the floors of the rainforests. In the past, it has been treated from time to time as a species belonging to several different families. It is now included in the large family, Staphylinidae. Very little is known about its habits.

Epiphytic lichens and mosses that produce semi-aquatic habitats in rainforest trees include numerous species. They differ from epiphytes that provide habitats for terrestrial species in their ability to absorb water and hold it like a sponge during a few rainless days. These species will be discussed further in Chapter 12.

b. Diptera

Many families in the suborder Nematocera encompass some of the smallest insect species known. They can be observed by microscopic examination of mosses and lichens growing on the branches of trees in cold rainforests. The larvae of most such species have never been mentioned because the descriptions provided by the authors who named the species were those only of adults. Some of the families with the smallest larvae include Chironomidae, Ceratopogonidae, Dixidae, Sphaeroceridae, Mycetophilidae, and Agromyzidae. Most members of the last three of these families are terrestrial. Other aquatic members of the Diptera are probably too large to successfully complete their larval development in an aquatic habitat as small as the one the mosses and lichens provide, although it cannot be ruled out that some small mosquitoes might successfully develop into imagoes in a habitat so well protected from large predatory insects.

The fact that so few aquatic hexapods living in cold rainforests are likely to have immediate importance for the economy, and the distance of these rainforests from large urban centers has made it unlikely that any large scale research projects will focus on them in the near future. There are a great many species of small midges and flies that have already been described only as adults, that there is a good likelihood of finding the larvae of some of them among the mosses and lichens that are so obvious when viewing a forest bordering on the Pacific Coast of North America between Oregon and Alaska. In addition, undescribed species are certainly still living in these habitats.

There are other locations in which epiphytic mosses, lichens, and vascular plants can be found growing on the branches of trees in great abundance within the temperate zones, and these also merit some attention. Cold rainforests also exist in coastal areas of southern New Zealand and nearby islands, as well as a few parts of Australia and South America. Somewhat similar conditions can be observed along the Gulf Coast of North America. Although it would not be correct to call the habitats "cold rainforests", the obvious existence of Spanish moss, *Tillandsia usneoides* (Linnaeus) Linnaeus, in the family Bromeliaceae, on the trees along the Gulf Coast of Louisiana, indicates that the moisture in the air is sufficient to keep the epiphytic vegetation soaked with enough water to facilitate the growth of non-vascular plants throughout much of the year. This kind of habitat is one that many species of tiny midges and flies have shown that they are able to thrive in.

In the next chapter, the kind of phytotelm providing the richest and most productive habitat for aquatic insect species of all sizes will be discussed. It is the habitat provided by the large epiphytic plants that grow in great abundance on trees in tropical rainforests. Associated with this habitat is that provided by holes in the trunks and branches of trees, which are able to catch and hold considerable amounts of water in locations where rain falls during most days, and when it does not, there is enough condensation on the surfaces of the trees themselves to provide a rich supply of water continuously to the true aquatic insects with larvae that must develop under water.

12

Insects of Phytotelmata

In addition to rainwater trapped in various inanimate objects, insects can find numerous small aquatic habitats that are associated with plants in the midst of various terrestrial habitats. Such plants form structures capable of catching rainwater and storing it for long periods of time. It was observed by pioneer researchers in tropical forests that certain aquatic insects utilized the water in these plants for their larval development (Alexander, 1912; Lüderwaldt, 1915), but studies by entomologists have been sporadic in the past, and only recently has sufficient attention been given to the phytotelm, that is, the community of species inhabiting small assemblages of water (Wesenberg-Lund, 1943). This is understandable because in the past, entomologists seeking aquatic insects would hardly think of seeking them among the upper branches of tall trees in equatorial rainforests or among mosses or lichens on the floor of a cold, temperate rainforest, as described in the last chapter.

Although aquatic entomologists long knew about the habitats in which aquatic insects deposit their eggs and develop high in the trees of rainforests (Thienemann, 1934), it was not until the last half century that the interest in the phytotelm as an aquatic community for insects began to rapidly increase. Various categories of habitats produced by plants for accumulating long-lasting supplies of rainwater for aquatic insect larvae have been enumerated. The parts of plants that have received the most attention for providing such habitats include bromeliad leaf axils, bamboo internodes (Fig. 12.1), holes in the trunks or branches of trees, and insectivorous pitcher plants. Much of the interest is explained by the presence of the larvae of mosquitoes that transfer the parasites responsible for causing malaria (Downs and Pittendrigh, 1946; Rattanarithkul and Green, 1986; Sinka et al., 2010) or the viruses that cause yellow fever or dengue fever (Forattini et al., 1998; Dahelmi and Syamsuardi, 2015).

Apparently, epiphytic plant species each have certain insect species that seek out one or more of them as places in which their larvae can successfully develop. In tropical rain forests all over the world, many such species are mosquitoes in various genera (Richardson et al., 2000). The introduction of the bromeliad, *Billbergia pyramidalis* (Sims) Lindl., to Florida turned out to also be the introduction of a new breeding place for mosquitoes (Frank et al., 1988). Another bromeliad in Florida,

Figure 12.1 Bamboo stems are often cut and used for small construction projects. The stems between the internodes are hollow, and when the bamboo is cut, the internodes can collect rainwater. The long and narrow internodes hold much rainwater, which evaporates slowly because its area of contact with the atmosphere is small. Cut bamboo used to hold down a wooden stairway in Gunung Leuser National Park on Sumatra, Indonesia, is shown with an internode filled with rainwater, in which various small insect larvae are developing. The small size of the openings makes it difficult for large predatory insects, such as dragonflies, to deposit their eggs in the internodes, making the smaller insects relatively safe during their larval development.

Catopsis berteroniana (Schultes f.) Mez., was reported by Frank and O'Meara (1984) to be the breeding place of two mosquito species in a genus encompassing only or predominantly mosquitoes that develop as larvae in the water trapped in bromeliads: *Wyeomyia vanduzeei* Dyer and Knab, 1906, and *Wyeomyia mitchellii* (Theobald, 1905). In Venezuelan national parks, the relationship between the phytotelmata and the mosquitoes that develop in them was shown by parsimony analysis (Navarro et al., 2007).

However, on the positive side, some of the insects that develop in the phytotelm serve beneficial purposes. One midge in the family Ceratopogonidae that develops in bromeliads is an important pollinator of cacao flowers (Fish and de Soria, 1978), which certainly benefits local agriculture. In addition, larvae of Odonata species in the suborder Zygoptera that develop in bromeliads feed on the larvae of the mosquitoes that develop with them in the phytotelm (Frank, 1983).

Another category encompasses the fluid-filled structures of pitcher plants, which contain a liquid in which insects are trapped and drown. There insectivorous plants then absorb the inorganic nutrients that are released by the decaying insects. The liquid they hold is inhabited by the larvae of several insect species, which feed on and digest the drowned insects trapped in the pitchers, thereby increasing the speed of release of the inorganic nutrients. Because it has not been determined whether the fluid in the pitchers is mainly rainwater or is secreted into the pitchers by the plants themselves, it is not clear whether all pitcher plant species proceed in the same way to catch and digest insects or whether there are major differences in how each species accomplishes the augmentation of its mineral nutrient supply for photosynthesis. Future studies will be required to determine which store mainly rainwater in their pitchers and which secrete fluids, which may contain enzymes to break down the bodies of the trapped insects. The group of insects that pass their larval development while helping the plants to digest their captured insect prey will be discussed in Chapter 13.

SECTION 1

Rainwater in Holes in Trees

Many insects find suitable water bodies in which to lay their eggs in holes in the trunks of trees. The importance of such habitats may be considerably underestimated in the large rainforests of South America, Africa, and parts of Indonesia. While rainwater trapped in rocks and artificial containers near human habitations is a medium from which predators are generally excluded, it may be poor in organic and inorganic nutrients that promote the growth of microorganisms on which many insect larvae feed. Holes in tree trunks located high in rainforest trees are usually relatively rich in detritus, which can yield enough organic nutrient material to promote the growth of microbiota. They also provide some safety from predators, although not as effectively as those in artificial containers provided by man.

When observing insects developing in rainwater, the advantages and disadvantages of specific kinds of container holding the rainwater are not considered by the insects depositing their eggs. Rather, it seems that each species has its own instinctive requirements for selecting sites to lay eggs. The female will deposit eggs anywhere the requirements are met, regardless of the kind of container holding the water. Much additional research will be required to determine why an aquatic insect will deposit its eggs in one rainwater habitat and not in another. It is also considered likely that some insect species will specialize in only one kind of habitat for its larval development, while another will accept a large variety of habitats for its larvae. However, many studies will be required before the choice of spawning sites can be fully explained. Where thousands of aquatic species in many insect families select their sites for laying eggs to assure the best conditions in which their larvae can develop are still completely unknown. Even in Europe and North America, the preferred microhabitats of many species of larvae have not been reported.

Because holes in tree trunks are not very important as habitats for most aquatic insects in the Holarctic Region, little work has been done to survey their aquatic insect fauna. Deforestation has greatly reduced the area covered by old trees in both Europe and North America, and this has eliminated countless habitats for the species preferring to develop in tree holes. Today, these small assemblages of water seem to be much more important in rainforests of South America, sub-Sahara Africa, and Southeast Asia than in places where much research had already been completed while forests were more plentiful. Systematic studies of the aquatic insects that develop in rainwater trapped in tree holes should therefore be undertaken as soon as possible to survey the aquatic insects as long as significant amounts of the pristine rainforests are left. Water trapped in tree holes is one kind of habitat that can be completely eliminated by cutting down tropical forests. If a species will only develop in such a habitat, the elimination of a rainforest will result in its local extinction. If the adults can adapt and spawn elsewhere, the species may survive (Fig. 12.2).

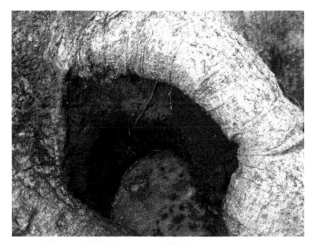

Figure 12.2 A typical tropical tree hole found in Chatuchak Park in Bangkok, Thailand, in which species belonging to various dipterans in the suborder Nematocera had probably developed as larvae. Photographing this small aquatic habitat in October, at the end of a normal rainy season, caused these insects to temporarily exit the holes. Tree holes are plentiful in the trees in the park.

Studies conducted to date have identified a few of the aquatic insects that develop in tree holes and similar habitats that are available in trees, hollow branches, and bamboo internodes. Table 12.1 provides a list of some of the species that may be encountered in such places (Kitching, 1992).

The fauna in both tree holes and in the axils of bromeliads are much richer than the tables in this chapter indicate. The problem is that most specimens collected cannot yet be identified to species as larvae, and even the species of adult are not easy to recognize without a considerable amount of effort. The tables are meant to provide the names of species that are known to develop in well defined phytotelm habitats, of which they could probably be considered indicators. It has been suggested that the larvae of at least some of these species develop nowhere other than the kinds of habitats where they have already been found. Of course, this remains to be demonstrated on a case by case basis.

SECTION 2

Rainwater in Epiphytic Plants with Structures for Holding It

Although there are insect species that can complete larval development relatively quickly, many of those living in rainwater prefer habitats in which the water remains for relatively long periods of time. In tropical rainforests, such habitats are frequently encountered in epiphytic plants. The shape of the stems and leaves assure that water will have places to collect, and these assemblages of water are sheltered enough by parts of the plants to delay evaporation. In tropical rainforests, daily rains are sufficient to replenish the water before it dries up. In Amazonia, through which the vast forested valley in which the Amazon River and its tributaries flow, epiphytic plants provide the normal breeding places for a large variety of insects, as well as other animals. The

Table 12.1 Species reported to develop in the water trapped in tree holes or in cut bamboo stems.

Species	Family	Kind of phytotelm	Author
Group 1: Species from the Americas			
Gynacantha membranalis Karsch, 1891	Aeschnidae	Tree holes	Fincke, 1992
Microvelia atrata Torre-Bueno, 1916	Veliidae	Cypress boles	Polhemus and Polhemus (1991)
Paravelia myersi Hungerford,	Veliidae	Tree holes	Polhemus and Polhemus (1991)
Paravelia loutoni Polhemus, 2014	Veliidae	Bamboo internodes	Polhemus (2014)
Aedes triseriatus (Say, 1823)	Culicidae	Tree holes	Williams (2005)
Culex mollis Dyar and Knab, 1906	Culicidae	Tree holes	Yankoviak (1999)
Toxorhynchites theobaldi (Dyar and Knab, 1906)	Culicidae	Tree holes	Fincke (1999)
Toxorhynchites (Lynchiella) pusillus Lima, 1931	Culicidae	Bamboo internodes	Frank and Curtis (1981)
Trichoprosopon (Shannoniana) schedocyclius (Dyar and Knab, 1906)	Culicidae	Bamboo internodes	Frank and Curtis (1981)
Trichoprosopon (Rhunchomyia) pallidoventer (Lutz, 1905)	Culicidae	Bamboo internodes	Frank and Curtis (1981)
Group 2: Species from the Palearctic Zone			
Aedes geniculatus (Olivier, 1791)	Culicidae	Tree holes	Williams (2006)
Anopheles plumbeus Stephens, 1828	Culicidae	Tree holes	Williams (2006)
Dasyhelea dufouri (Laboulbène, 1869)	Ceratopogonidae	Tree holes	Williams (2006)
Metriocnemus cavicola Kieffer, 1921	Chironomidae	Tree holes	Williams (2006)
Myathropa florea (Linnaeus, 1758)	Syrphidae	Tree holes	Williams (2006)
Prionocyphon serricornis Müller, 1821	Scirtidae	Tree holes	Williams (2006)
Group 3: Species from Queensland, Australia			
Aedes candidoscutellum Marks, 1947	Culicidae	Tree holes	Williams (2006)
Culicoides angularis (Lee and Reye, 1953)	Ceratopogonidae	Tree holes	Williams (2006)
Anatopynia pennipes Freeman, 1961	Chironomidae	Tree holes	Williams (2006)
Group 4: Species from Polynesia			
Aedes polynesiensis Marks, 1951	Culicidae	Tree holes	Bonnet and Chapman (1956)
Culex atriceps Edwards, 1926	Culicidae	Tree holes	Bonnet and Chapman (1956)

species are usually able to complete a life cycle of relatively long duration in clean and unpolluted water with relatively little threat from larger aquatic predators, except those few with larvae that also develop in bromeliads.

The fourth larval instar of the chaoborid, *Corethrella infuscata* Lane, 1939, takes one month to reach the pupal stage in the water trapped in bromeliads in the Brazilian states of Goiás and Mato Grosso, Brazil (Lane, 1953). The amount of time required for the first three instars to be completed has not been reported. Clearly, however, bromeliads must be reliable places of rainwater storage because the locations at which the chaoborid larvae were observed are located in a zone with alternating rainy and dry seasons.

Other places where rainwater collects are also utilized by insect species as nurseries for their larvae (Greeney, 2001). At certain locations in parts of the world with relatively regular rainfall and few periods of drought, rainwater will collect on low-lying soil that is relatively impervious to water, inside hollowed out rocks, and in the hollows of trees. Usually, the distribution of rainwater pools will be random, and the amount of time the rainwater is available for supporting aquatic insects will depend on how often rainfall can be expected and the shape of the depressions in which the rainwater collects. Rocks with deep hollows eroded in them, for example, will maintain the rainwater that collects for relatively long periods of time. Of course, breeding success in all kinds of standing bodies of rainwater depends upon the seasonal patterns of rainfall, the temperature and relative humidity of the surrounding air, as well as the shape and physical features of the depression.

Many species of insect are able to develop in temporary rainwater pools trapped in various kinds of depression. Their success depends upon how quickly the larvae are able to develop, at least as far as the pupal stage. The species most successful in placing their eggs in rainwater are therefore associated with specific habitats which tend to catch and store rainwater longer than the larvae require to complete their development. Among such suitable places are certain bromeliads in tropical rainforests, which can hold rainwater on a semi-permanent basis because rainfall is frequent enough to refill them before they completely dry out. Most of the bromeliads are capable of storing water for long periods of time in their leaf axils, making them ideal places for mosquito larvae to develop in the tropics (Haddow, 1948).

Many of the small insect larvae developing in bromeliads feed on algae, bacteria, fungi, or detritus. Nematocera, a suborder of Diptera, is probably best represented among such rainwater communities. These species are frequently overlooked, especially those of the tiny species in the family Chironomidae. However, even when their presence is reported, the species are almost never identified.

Among the predatory aquatic insects reported to develop in the bromeliads are damselfly larvae in the genera *Bromeliagrion* de Marmels and Garrison, 2005; *Leptagrion* Selys, 1876; and *Mecistogaster* Rambur, 1842, according to de Marmels (1985), Santos (1966, 1978), and Srivastava et al. (2005), respectively.

Presumably, the predatory insect larvae find a plentiful supply of prey, which also develop as larvae in the bromeliads. These insects can find much food in the form of moribund parts of the plants and other detritus, bacteria, protozoans, fungi, and small invertebrates, such as rotifers. In the rainforests of South America, observations of the larvae of dipterans found developing in species of epiphytic plants are commonplace. A few of these are included in Table 12.2. However, in their review of research and literature on the aquatic fauna of bromeliads, Frank and Lounibos (2005) showed that only a small percentage of the aquatic species actually inhabiting the plants have been identified, and much of the information provided is limited only to the insect families or occasionally the genera that are represented in phytotelm communities, including those found in bromeliads.

Not all aquatic insects found in a phytotelm community are present only as larvae. Several are also present as adults, although many have been observed but not identified beyond family or genus. One that has is *Tropisternus setiger* Germar, 1824, a beetle in the family Hydrophilidae, which was found living in water trapped by the species *Eryngium cobrerae* Pontiroli, in the family Apiaciae, in Argentina by Campos and Fernández (2011). The genus *Tropisternus* encompasses many species. Most of the others are encountered in ordinary water bodies rather than in the water trapped in plants. In contrast, *T. setiger* is encountered in a variety of aquatic habitats (Pereyra

Table 12.2 The larva of insects found developing in rainwater trapped in various kinds of plant, most commonly epiphytes, in the Neotropical and adjacent regions (Santos, 1966, 1978; Frank and Curtis, 1982; de Marmels, 1985; Lounibos et al., 1987; de Marmels and Garrison, 2005; Zillikens et al., 2005; Haber et al., 2015).

Species of larva	Family	Host plant
Bromeliagrion fernandezianum (Rácenis, 1958)	Coenagrionidae	Bromeliad
Leptagrion vriesianum Santos, 1978	Coenagrionidae	Bromeliad
Leptagrion siqueirai Santos, 1968	Coenagrionidae	Bromeliad
Mecistogaster modesta Selys, 1860	Pseudostigmatidae	Bromeliad
Erythrodiplax laselva Haber, Wagner, & de la Rosa	Libelullidae	Bromelad & *Cochliostema* sp.
Erythrodiplax bromeliicola Westfall, 2000	Libelullidae	Bromeliad
Mongoma bromeliadicola Alexander, 1912	Limoniidae	Bromeliad
Limosina bromeliarum Knab and Malloch, 1912	Sphaeroceridae	Bromeliad
Aedes albopictus (Skuse, 1894)	Culicidae	Bromeliad
Toxorhynchites (Lynchiella) solstitialis (Lutz, 1904)	Culicidae	Bromeliad
Toxorhynchites (Lynchiella) mariae (Bourroul, 1904)	Culicidae	Bromeliad
Wyeomyia vanduzeei Dyar and Knab, 1906	Culididae	Bromeliad
Corethrella infuscata Lane, 1939	Chaoboridae	Bromeliad
Corethrella downsi Lane, 1943	Chaoboridae	*Aechmea dichlamydea* var. *trinitensis*
Dixella wygodzinskyi Lane, 1945	Dixidae	Bromeliad
Fidena rufopilosa (Ricardo, 1900)	Tabanidae	Bromeliad

and Archangelsky, 2007), such as among submerged masses of floating plants, and visits to phytotelm communities are just part of their search for food.

Zaragoza (1974) recorded species in the order Coleoptera in epiphytes and reported 12 species new to the list of fauna of Mexico. Although all of them were not known to be aquatic as adults, the habitats of their larvae remain to be determined.

Among the other insects inhabiting phytotelmata as adults are heteropterans in the family Veliidae. Eight species in two genera have been identified, *Paravelia* and *Microvelia* (Polhemus and Polhemus, 1991). Insects in these genera are predatory, mainly or entirely on other insects (Heckman, 2011). The habitat of phytotelm communities is small, and there is considerable competition among the species that try to reproduce in those that are available. Predatory larvae are able to remain in these tiny bodies of water together with the larvae of their prey and maintain a food supply in close proximity to them that will last for their entire level development. Although mosquito larvae usually subsist on microorganisms throughout their larval development, certain species that develop in tree holes, such as *Toxorhynchites theobaldi,* prey on other insect larvae (Lounibos et al., 2003).

The question that presents itself here is the degree of specialization of the larvae. Are they capable of developing in more than one plant? If so, can they also develop in water trapped in objects other than plants if no suitable plants are available? If they can, do they develop better in plants, or do the adults leave their eggs in plants only if other spawning places are not available? Obviously, a considerable number of individual experiments would be necessary to determine whether the symbiotic relationship with the plant species is necessary for one or more of the insects that live in its trapped water to survive.

The other side of the question is whether the plant requires the insects to supply it with the inorganic nutrients it needs to survive. This will be discussed further in Chapter 13. What is important to consider here is how complex the relationships among the rainforest biota actually are. It is necessary to consider whether eliminating one species from a complex community displaying multiple symbioses will destroy the entire community or whether the elimination will simply remove the one species and leave the others to reorganize their interrelationships and survive.

An interesting field of research focuses on the relationships between the plants holding the trapped rainwater and the insects inhabiting that water. Such relationships may be far more complex than they seem upon superficial examination. There are certainly chemical relationships between rainwater, autotrophic plants holding the rainwater, small animals inhabiting it, and a variety of microorganisms interacting with the chemicals and the water containing the chemicals.

One important factor to be considered is the availability of the inorganic nutrients that autotrophic plants need for photosynthesis and to grow. Relatively large amounts of carbon from CO_2, nitrogen from nitrates, and phosphorous from phosphates are needed to produce structural proteins and carbohydrates to support the plants and chemicals that fix and store energy that the plants need. In addition, relatively small amounts of trace elements are needed to meet special needs of the plants, such as magnesium for the production of chlorophyll, zinc for its role in chlorophyll synthesis, iron for the transfer of energy through cytochrome complexes, and manganese for its role in oxygen evolution. The predatory insects living in the bromeliads feed on and digest the abundant larvae of such dipterans as mosquitoes and chaoborids, and this produces a supply of the inorganic nutrient compounds necessary for plant growth (Ngai and Srivastava, 2006). It is important to determine how common these trophic relationships are in aquatic bromeliad communities and how important they are for the survival of both the plants and the invertebrates living in the water they contain.

In most places, the local rainfall contains few nutrient elements needed to support plant growth and the production of spores and seeds. Only during severe thunderstorms do frequent discharges of lightning fix significant amounts of nitrogen and produce appreciable amounts of nitrates, but even this provides only temporary surpluses of fertilizer. Where the soil is packed down, denitrification occurs due to the activity of bacteria. Where the soil is also poor in inorganic plant nutrients, most plant species will fail to develop. For example, in bogs and moors, a typical flora consisting only of species tolerant of low quantities of nutrients in the substrate can develop. Many of them are successful in obtaining inorganic nutrients by possessing the means to capture and kill insects to obtain the elements they contain. Many nutrient-poor habitats are characterized by the presence of species of insect-eating plants, which derive their inorganic nutrients with the help of symbiotic insects, which help them break down and digest the insects they have captured. These will be discussed in Chapter 13.

In addition, certain insects come to the assistance of the plants by living on or in them and supplying them with some of the nutrients that they need after they have eaten parts of terrestrial plants or captured and killed prey from terrestrial habitats. Species in the family Bromeliaceae have large surfaces of their leaves that are not in contact with the trapped water but rather exposed to the atmosphere. These surfaces are available for the free movement of terrestrial insects seeking plants to feed on and water. The visiting terrestrial insects would not be expected to interact with the aquatic insects, although they might harm the plants which are harboring them. They do transport inorganic nutrients in their bodies in the form of wastes, which they leave on the plants they visit. The fact that in most rainforests, inorganic substances are in short supply, inputs of such substances, even in small quantities, determine the daily amount of photosynthesis. Considerably larger amounts of these substances can be provided to the plants by birds and mammals.

It should be remembered that like bogs in the temperate zone, tropical rainforests are typically dependent on outside sources of fertilizers to supply the plant nutrients that are scarce in the soil and in great demand. There is almost always excess energy available from sunlight in the upper canopies of the forest, and the limiting factor for photosynthesis is usually a compound containing nitrate or phosphate. Therefore, plants are dependent upon the activities of animals to supply them with the substances they need to grow.

This situation is considerably different from that in the temperate zones, where sunlight is the limiting factor for photosynthesis for most of the year. The seasonal cycle provides temporary niches for detritivores, herbivores, and predators of all sizes, and only the seasonal diminution of light and temperature limits it.

In investigating the bromeliads and the aquatic insects they harbor in the Equatorial Zone, it is important to remember that the basic conditions under which they live are fundamentally different from those under which aquatic insects live in zones with seasonal cycles, be they wet-and-dry or hot and cold. The phytotelm inside of bromeliads in an equatorial forest is a biotic community that defies comprehension when viewing it with the same principles that apply to habitats in the subtropical and temperate zones. Some of the new indications that the relationships between insects and plants, as well as between certain insect species and other species, are far more complex than previously thought possible. For example, in Chapter 28, we will return to the bromeliads to view their dependence on certain insects, just as we have examined the dependence of aquatic insect species on them in this chapter.

We can also see how important nutrients are in the equatorial rainforest, where there are no reserves stored in the soils. For that reason, mosquito species are frequently among the aquatic insects that live in the phytotelm. Meeting the nutrient budget of many of these insect species requires a search to find mammals, birds, or reptiles with rich supplies of blood to provide proteins and other vital nutrients for the production of eggs, which are placed into the low nutrient environment of a bromeliad phytotelm.

At this point, the nutrient budget shows complications that were not comprehensible to scientists of an earlier age. Prior to the general acceptance of a direct relationship between many fatal diseases that had afflicted mankind for centuries and contamination of the body with microscopic species of protozoans, bacteria, and viruses at the end of the 19th century, mosquitoes were considered a minor annoyance. Thereafter, it was recognized that mosquitoes were the vectors of these pathogens and that human health could be protected by controlling these insects. Because female mosquitoes are capable of replenishing their supply of blood for production of nutrient-rich eggs several times in their lifetimes, they could infect themselves by biting an infected human being or domestic animal, and by the next time she was ready to produce eggs, she could bite an uninfected person. By this time, the pathogen in the mosquito would have produced new, infectious life stages or new virus particles, and the mosquito bite would transfer the pathogen into its next source of blood, infecting him.

In this case, we have an example of a transfer of blood from a mammal into an insect, an infection of the insect with a pathogen, a transfer of the blood to a food-starved community of aquatic insects, a transfer of a pathogen from a human being into the insect, a second transfer of blood from the insect into the food-starved community of aquatic insects, an infection of an uninfected human being with the pathogen, and the release of inorganic nutrients to the bromeliad. During these transactions, the nutritional requirements of a mosquito, a pathogenic organism, two or even three predators on the mosquito larvae, and a bromeliad were met, and either one or two new generations of each insect involved would be provided for.

The complexities of the public health problems are added to these considerations, which are the most pressing problem the activities of the mosquito cause for human beings. Since historians first began recording the events that have had notable impacts on the course of human events, there have been numerous occasions when the transmission of diseases by insects have brought about great changes in the course of history. For example, one group of people tried to build a canal across the isthmus of Panama and failed, only because one species of mosquito needed to feed on the blood of the laborers. A second group performed research and discovered what was happening, and they succeeded in completing the canal. These discoveries and subsequent ones will be discussed in detail in Chapter 28.

13

Insect Fauna Living in Insectivorous Plants

The habitat of aquatic insects that live in water on or in terrestrial plants has been called the phytotelm, as described in Chapters 11 and 12. Thienemann (1932) defined two major categories of this kind of habitat. The first of these are communities in rainwater trapped in small inanimate objects or in suitably shaped parts of plants, as already discussed. The second category includes communities in liquids held in structures on plants produced to hold fluids secreted entirely or in part by the plants themselves. They are designed to trap, drown, and then digest insects that fall into the structure that holds the liquid in which the trapped insects drown. The liquids in pitcher plants are presumably derived entirely or in part from the plants themselves (Ellison et al., 2003), making the inside of their pitchers a separate category of habitat and placing the fauna they contain in their own distinct group within the phytotelm.

The other categories and subcategories of phytotelm are found in habitats in which the source of the water is mostly or entirely precipitation. These other categories of phytotelm were discussed in Chapters 10, 11, and 12. In most cases, however, it is still not known how much of the fluids held by the pitchers or other structures of plants that contain fluids in which their prey drowns are secreted by the plants themselves and how much of it is accounted for by trapped rainwater or condensation. The fluid in one species of *Nepenthes* that was investigated, *Nepenthes rafflesiana* Jack ex Hook f., is commonly described as viscous and elastic (Gaume and Forterre, 2007). It is also believed that it contains toxic substances and enzymes that kill and help digest the captured insects, and this was confirmed for an unidentified species of *Nepenthes* by Adlassnig et al. (2011). Some of the species have hoods that hang over the opening to the pitcher, which would presumably keep most rainwater out. However, during a windy rainstorm in Southeast Asia, a limited amount of rainwater would certainly enter in spite of the cover. In addition, the chemical composition of the fluid filling the pitchers of one species would certainly not be the same as that in all of the large number of other species of *Nepenthes*.

In contrast to species of *Nepenthes*, studies of the chemistry of the water inside the pitchers of plants in the family Sarraceniaceae, genera *Sarracenia* and *Heliamphora*, have fluid in the pitchers very similar to pure water. Even without toxic substances or enzymes, the death of the insects trapped in the pitchers is certain, but more cooperation from the insects that develop as larvae in the pitchers is required to digest the dead insects faster (Adlassnig et al., 2011).

In habitats where plants have difficulties obtaining enough nitrates and phosphates, such as moors, raised bogs, and sandy locations subjected to frequent rainfall, a few plant species have developed methods of capturing and digesting insects to obtain these inorganic nutrients. Among the strategies used by plants to capture and kill insects is the formation of a "pitcher", a structure formed by the leaves to hold water, presumably containing enzymes or other substances to hasten decomposition of insects which fall into the fluid and drown or are poisoned by toxic chemicals secreted by the plant. The pitchers attract insects to the vicinity of the opening at the top of a pitcher. The peristome of one species, *Nepenthes bicalcarata* Hook f., was examined and tested using five different species of ants. The peristome is temporarily covered by a waxy substance and is kept wet. The wet surface prevents the tarsi of the ants from getting a firm grip on the surface, causing them to slip and continue downward into the pitcher. Electron micrographs were taken of the surfaces to find out why the ant claws and foot pads could not keep hold of the surface of the peristome. Ants still could not get a firm hold of the rims of the pitchers that had lost their waxy coatings, as well as the rims of the peristomes of other *Nepenthes* species that lacked a waxy layer, as long as the rims remained wet. It is therefore not surprising that the greatest numbers of insects that fall into the pitchers do so during heavy rainstorms (Bohn and Federle, 2004).

The inside wall of the pitcher of most species is armed with closely set spines, which are usually long and pointed downward toward the bottom of the pitcher. Once an insect starts down, these spines prevent it from climbing upwards again. Once an insect falls into the liquid in the lower part of the pitcher, it cannot escape before it drowns.

Pitchers formed by the apical parts of the leaves of plant species dependent upon supplemental inorganic nutrients to survive are known from two plant families: Nepenthaceae and Sarraceniaceae. The family Nepenthaceae is confined to the Old World, while Sarraceniaceae occurs only in the Americas. All experiments described above were parts of studies made on species in the family Nepenthaceae.

Available information on the Sarraceniaceae has been compiled in two volumes: *Sarraceniaceae of North America* (McPherson and Schnell, 2011) and *Sarraceniaceae of South America* (McPherson et al., 2011).

The remains of trapped insects must be digested or decomposed before their phosphates and nitrates can be released and absorbed. Observers have reported the presence of larval insects within pitchers of various species from time to time, which do not die from exposure to the fluids in the pitchers. These symbiotic insects seem to feed exclusively on the arthropods that drown in the fluid, although some of the small larvae in the family Chironomidae, personally observed along the seashores of western Cambodia, are apparently nourishing themselves on the bacteria and other microorganisms that are digesting the remains of the trapped insects. However, the role of the larvae seems to be to break up the bodies of the dead insects in order to speed the process of their decomposition and release their inorganic nutrients to the plants.

Superficially then, it would seem that the symbiosis between the plants and the insect larvae is purely mutualism, that is, a relationship that brings only trivial benefits to both the plant and the larval insect. The plant benefits only from a somewhat more rapid release of nitrates and phosphates, due to an accelerated digestion of the drowned insect carcasses by the feeding of the larvae on them. The inorganic nutrients useful to the plants would, in any case, eventually be released. The role of insect symbionts is to assist the plant only by speeding up the digestion process. The insects benefit by enjoying protection from potential predators within the pitchers, where they usually find a continual supply of recently drowned insects to feed upon. On closer inspection, however, it appears that the survival of both the plant and the symbiotic insects are considerably improved by the relationship in subtle but effective ways. Most pitchers attract the symbiotic insects to lay their eggs inside the pitchers, and the insect species involved do not seem to occur anywhere else but in the pitchers.

The two families of pitcher plants are confined to distinct zoogeographical regions, which would suggest that the insects inhabiting the plants would also have distinctly limited ranges. Many insects inhabiting the pitchers have not yet been identified or even reported, but those that have been are predominantly members of the order Diptera. The Old World family Nepenthaceae, which encompasses species belonging to only one genus, *Nepenthes* Linnaeus, is noted for trapping their insect prey in modified, fluid-filled leaves with fine hair-like structures pointed toward the bottom of the "pitcher" to prevent the insects from moving toward the opening and escaping.

At the present time, the New World family Serraceniaceae encompasses three genera: *Sarracenia* Linnaeus; *Darlingtonia* Torrey; *Heliamphora* Bentham. The description of the insect-eating activities of Nepenthaceae in the last paragraph applies equally well to the Serraceniaceae. Only the part of the world in which the species of each family occur is different. Species in both families also receive assistance in digesting their captured prey from insect larvae that live in the fluid held in the pitchers. These larvae are obviously immune to any toxic substances produced by the pitcher plant and are also not affected by any digestive enzymes present in the pitchers.

The fluid filling the pitcher is water, which presumably contains substances to hasten the digestion of the insects and perhaps various kinds of microorganism (Wesenberg-Lund, 1943). The plants presumably are in need only of the inorganic molecules containing nitrogen and phosphorus compounds and also perhaps trace metals so that they can incorporate these substances into their tissues through photosynthetic processes. Therefore, organisms that can benefit by ingesting the remains of the insects and breaking them down into simple, inorganic compounds do not detract very much, if at all, from the nutrient value of the insect prey to the plants.

Upon examining the fluid inside of the pitchers, it seems surprising to observe tiny insect larvae developing normally while submerged in the fluid that has been lethal to many kinds of insect from which the plants derive their nutrients, presumably nitrates, phosphates, and trace elements. Upon the visible remains of those dead insects, both a community of microorganisms, including bacteria, protozoans, and rotifers, and the developing symbiotic insect larvae seem to thrive. The insect denizens of the insect traps of the pitcher plants most frequently belong to the order Diptera, but exhaustive studies may still bring some surprises concerning the structure of this aquatic insect community to light. From the general condition of the insect carcasses in the pitchers, it seems likely that the plant itself produces some of the enzymes that digest the material. Large pieces of chitin from the bodies of beetles and flies were personally observed in such pitchers near Koh Kong, Cambodia, during the 1970s. Among this debris suspended in the water were living larvae belonging to several families in the suborder Nematocera. The chironomids in the pitchers remained in simple networks of web-like fibers, probably secreted by the insects themselves.

Further evidence was provided by the pitcher shape, which included an umbrella-like flap of the leaf which covered the opening to the pitcher and kept rain that was falling from entering. Obviously, if water could enter the pitcher during every heavy rainfall, the overflowing fluid would carry away both any enzyme that had been secreted by the plant and some of the inorganic nutrients that had been released before the plant could absorb it.

The difficulty in determining what is actually happening inside of the pitchers is the large number of species of pitcher plant that exist and the very limited number of studies of their biochemistry and metabolism. There is no guarantee that a method of digestion employed by one species of plant together with its community of symbionts is the same as that of another species in the same family.

Much is to be learned about the fluid in the pitchers. It appears to be water filled with dead insects and pieces of chitin remaining from those in the process of decomposition. Also present are small invertebrates, including living insect larvae, which show no signs of being harmed by the medium in which they live. The fluid in the pitchers obviously contributes to the digestion of the dead insects, but how this fluid comes into being is poorly understood and will require much research. Is the fluid in all of the many pitcher plant species fundamentally the same? Does it always include toxic substances or digestive enzymes produced by the plant itself? Do the insects inhabiting each of the pitcher plant species contribute substantially to the survival of the pitcher plant, and if so, can the pitcher plant survive in the absence of such insects?

A toxic substance to kill the insects trapped in the pitchers would seem to be unnecessary. Without any way of escaping from the water, death of insects suited to survive only in terrestrial environments would be certain. The shape of the pitchers and many strong spines pointed downward toward the fluid preclude the escape of almost all insects that find their way into the traps. Even without any toxic substances or digestive enzymes in the fluid, water alone would be sufficient to assure that the plant's prey will drown. However, enzymes to facilitate complete digestion of the proteins contained in the insects would be helpful, especially if the pitchers did not contain living microorganisms with the capability of completely digesting dead insects.

Microorganisms smaller than insects are present in the liquid filling the pitchers, and some of them, especially bacteria and fungi, have the capability of at least partially digesting the dead insects that become trapped and drown. This raises the question of how they find their way into the pitchers. They may be species of bacteria, protozoa, and rotifers that are associated with the particular pitcher plant species, or they may enter on the bodies of the insects that drown. Other ways may also be involved in inoculating the fluid in the pitchers, but they are yet to be described. One hypothesis is that tiny midges and flies that develop as larvae inside of the pitchers remain associated with bacteria capable of promoting decay of chitin and proteins which are structural parts of all insect bodies. If that turns out to be the case, then the living insects inside of the pitchers would be essential for the well-being and perhaps the survival of the plants.

First, questions must be answered concerning how the insect larvae that develop in the pitcher plants reach the plant in which they apparently live as symbionts. A symbiosis seems to exist that is beneficial both to the plant and the insects that develop in the pitchers. The plants trap the prey and benefit from the release of inorganic substances they need for their own photosynthesis and growth, while the animal symbionts digest the prey and take a small amount of the substances contained in the bodies of the trapped insects for their own metabolism. If the plants are capable of digesting their own prey through the actions of enzymes they secrete themselves, then the role of the insects in the pitchers would seem to be that of parasites. Of course, the relationship between the plant and the animals found through studies of one species of plant and one species of animal in its pitchers would not necessarily be the same as those between other species of pitcher plant and insects living in the fluid contained in their pitchers. Addressing this question then would require repeated experiments, changing the combination of plant and insect each time. If it proves true that each species of pitcher plant provides a habitat for one species of midge or fly larvae, then it would be expected that the development of the pitcher plant and that of its insect symbiont would have to be coordinated in some way. At the end of a seasonal period of growth by the plant, the eggs of its symbiotic insects could well be deposited somewhere on the dying plant so that their hatching could be coordinated with the new growth of the plant during the next growing season. An alternative would be the survival of the species as adults throughout the winter or a dry season, depending on the climatic conditions at the locality. When conditions became favorable again, new pitchers would be formed, and the adults could deposit their eggs in them. The most likely way for the larvae to enter the pitchers would seem to be the transformation of the larvae to pupae at the end of the growing season of the pitcher plants each year. The leaves of the pitchers would then become detritus with pupae of the symbiotic insect species sticking to it during the cold or dry season, and the final metamorphosis to reproductive adults would occur as the new pitchers would begin their growth the following year.

In addition to the chironomid species that develop as larvae in the pitchers, Wesenberg-Lund (1943) reported the larvae of Culicidae, Ceratopogonidae, and other families in the order Diptera in pitcher plants from other parts of the world. Other insects belonging to the symbiotic community in pitcher plants will also require extensive studies to determine the ecological roles they play.

Larvae in the large midge family, Chironomidae, are often most prominent among the pitcher plant fauna. To benefit from the relationship with the pitcher plant, the adult insects would obviously have to survive entry into the pitchers to spawn and exit from the pitchers as fully developed adults of the next generation. This raises another question concerning whether adults appear only once per year or whether two or more generations appear during a single year. Adult chironomids that can be observed emerging from the pitchers when the plants are disturbed are very likely the same species that develop as larvae in the fluid the pitchers contain, but this must also be determined with certainty by direct observation. The community inhabiting the living pitcher plants could well be the subject of a study certain to yield interesting results concerning what appears to be a long-standing symbiotic relationship between an insectivorous plant and insect species which help digest its food.

SECTION 1

Diptera

As stated in the introductory paragraphs of this chapter, species in the family Chironomidae are best represented among the Diptera which develop in the fluid held in the pitcher plants. In addition, species in the families Culicidae and Ceratopogonidae have been reported from pitcher plants in many parts of the world. In addition, there are many reports of the larvae from insect families reported only once or twice in pitcher plants at isolated locations around the world. Therefore, species in the families Spaeroceridae, Syrphidae, Ephydridae, Empididae, and Micropyzidae are among the insects that can live as larvae in the pitchers and feed on the remains of the trapped insects.

One species that is especially remarkable is *Badisis ambulans* McAlpine, 1990, a fly in the family Micropezidae known only from a pitcher plant, *Cephalotus follicularis*, in Western Australia. The species is apterous, and the flies mimic ants in appearance and behavior (McAlpine, 1998). Both the pitcher plant and the fly are considered rare.

Of course, most of the species of pitcher plant have not been surveyed for symbiotic insects, so little definitive information is available about similarities and differences in various parts of the world. Especially great differences can be expected since pitcher plants belong to three different plant families, two widespread, and the third uncommon. Other plants that solve the problem of nutrient-poor soils by consuming insects, such as sundew plants and the venus flytrap apparently digest their prey using digestive enzymes they produce themselves and rely in no way on symbionts for assistance.

The significance of the question concerning the production of digestive enzymes by some or all pitcher plant species is a basic one for understanding why some species of insect can live in the fluids held by the pitchers and feed on the captured insects while others perish. If the captured insects simply drown and then decay through the action of microorganisms and insect scavengers, no special adaptations by the insects living in the pitchers would be necessary. Like many other aquatic insects, they would need only modifications for respiration under water. If the plants produced digestive enzymes for decomposing insects, then the symbionts would also need various biochemical defenses against being digested themselves.

Table 13.1 shows some of the species of Diptera with larvae that develop in the fluid held in the pitchers of species in the families Sarraceniaceae and Nepenthaceae. Not included are various species that have been reported but not identified beyond the level of genus.

Table 13.1 The larvae of midges and flies that have been associated with species of pitcher plants in the families Nepenthaceae and Sarraceniaceae. The publication reporting the relationship between the insect and the pitcher plant species is shown to the right.

Pitcher plant	Insect	Family	Reference
Nepenthes rajah Hooker f.	*Culex rajah* Tsukamoto, 1989	Culicidae	Tsukamoto (1989)
Nepenthes rajah Hooker f.	*Toxorhynchites rajah* Tsukamoto, 1989	Culicidae	Tsukamoto (1989)
Sarracenia purpurea Linnaeus	*Wyeomyia smithii* (Coquillett, 1901)	Culicidae	Paterson (1971)
Sarracenia purpurea Linnaeus	*Metriocnemus knabi* (Coquillett, 1904)	Chironomidae	Paterson (1971)
Darlingtonia californica Torrey	*Metriocnemus edwardsii* Jones, 1916	Chironomidae	Frank (2008)
Nepenthes sp.	*Nepenthosyrphus malayanus* Hippa, 1978	Syrphidae	Rotheray (2003)
Nepenthes sp.	*Nepenthosyrphus oudemansi* de Meijere, 1932	Syrphidae	Rotheray (2003)

SECTION 2

Lepidoptera

In addition to various species of Diptera encountered in the pitchers of plants in the family Nepenthaceae, the case-bearing caterpillar of a species in the order Lepidoptera was discovered in *Nepenthes destillatoria* Linnaeus on Sri Lanka. It was identified as *Nepenthophilus tigrinus* Guenther, 1913. Little information about the species has been published since its discovery in the pitcher plant was first reported (Guenther, 1913). It was initially thought to be a species of Trichoptera, but it would later be recognized as a moth in the family Psychidae. More precise information about its feeding habits would be useful to determine its role in the biotic community inside of the pitcher.

Other interactions between pitcher plants in the family Sarraceniaceae and caterpillars are more conventional in nature. Like almost all other caterpillars, these species are terrestrial and feed on the leaves of plants. They are species in the family Noctuidae. *Exyra ridingsii* (Riley, 1874) feeds on *Sarracenia flava* (Linnaeus), while *Exyra fax* (Grote, 1873) consumes *Sarracenia purpurea* Linnaeus, and *Exyra semicrocea* (Guenée, 1852) eats other species of *Sarracenia*. In all cases, these caterpillars eat the leaves, including the pitchers. Another species of Noctuidae, *Papaipema appassionata* (Harvey, 1876), even bores into the rhizomes of many North American pitcher plant species, killing the plants.

In the case of a Lepidoptera species in the family Tortricidae, *Endothenia hebesana* (Walker, 1863) destroys the fruit of *Sarracenia purpurea*. However, all of the moth species that feed directly on the pitcher plants are terrestrial and do not enter the pitchers. Therefore, they are not aquatic, and their ecology is beyond the scope of the book.

14

Aquatic Hexapods Active on Ice and Snow

A general rule has been proposed and is often discussed concerning the relationship between species diversity and the number of individual organisms in any given habitat. This proposed rule of thumb is based on the hypothesis that the less stressful a habitat is, the more individual species will inhabit it. As stress increases and survival becomes more problematic, the number of species decreases, but the number of individuals in each of the surviving species steadily increases to the limits that the resources of the habitat can provide. Studies on the distribution of the biota of the world tend to support this hypothesis.

For aquatic organisms, annual periods of subfreezing temperatures are certainly stressful. It would therefore be expected that the greatest species diversity would be encountered in places where the temperature never decreases below 0°C. Observations show very unequivocally that this is true. One look at the amazing species diversity in a tropical rainforest is demonstration enough. They also show that the numbers of individuals belonging to any particular species present in a habitat not exposed to freezing temperatures are low compared to the numbers of individuals belonging to a species filling approximately the same ecological niche present where temperatures below 0°C occur frequently during the coldest season.

Freshwater habitats in the coldest parts of the world, that is, the frigid zones and coldest parts of the temperate zones, as well as at the highest elevations in mountain ranges, are noted for their low species diversities and the vast numbers of aquatic insects belonging to those few species that are able to survive the long periods of subfreezing temperatures each year.

More simply stated, as conditions become more stressful due to low temperature, species disappear from the biotic communities. These species were preying on or competing with other species in the communities which can continue to survive in spite of the colder temperatures. As a result, the numbers of the survivors increase as their natural enemies and competitors are progressively eliminated from the community.

SECTION 1

Summertime Swarms of Adult Insects on the Tundra

Summer visitors to any of the regions of Arctic tundra in the world are often impressed by the large number of small flying insects they encounter. Some of these insects feed on blood, and people not well protected by thick clothing are often driven to seek shelter to escape the swarms of biting insects. The temperature can become uncomfortably warm on the tundra during the long summer days, so the visitor is sometimes forced to choose between uncomfortably warm clothing or being bitten by swarms of blood-sucking flies and midges.

The tundra is typically marshy during summer. Small streams and shallow pools of water cover the surface of the ground during that season. When disturbed, the vegetation of mosses, lichens, and small vascular plants are slow to regenerate. The vegetation that thrives on the cold side of the tree line has taken centuries to develop, and once damaged, it does not quickly grow back. Vehicle tracks across the tundra have remained visible in Alaska for more than half a century after they were made during the Second World War. In some places, the ruts formed by the tires expanded enough to form small bodies of standing water, which remain the same size or even grow slightly larger when the water freezes solid during the cold season each year.

Regardless of what happens to the surface vegetation, below the pools of water during the summer is a permanent layer of ice. No matter how warm the temperature becomes during the warmest part of the year, only the surface layer of the ice has time to melt before the air temperature drops below freezing again. Hence, during winter, the entire tundra is frozen, while during the summer, the ice at the surface melts, but the ice beneath the surface persists throughout the year as permafrost. Insects must adjust their life cycles to these harsh conditions, and members of the Diptera account for the overwhelming proportion of the aquatic insects present.

Although very abundant, the insects of the tundra have not been well investigated. The regions in which they occur are far from most research institutes, and human populations are not notably afflicted by diseases known to be spread by the insects, although some domestic animals are. Although human health can be negatively affected by some of the viruses and parasites these insects can transmit, most people are able to recover from the diseases without suffering from death or serious handicaps. More will be said about this in Chapter 25.

As in most stressful habitats, there are a great many dipterans, which belong to relatively few species. However, more needs to be learned about the geographical distribution of the individual species to know whether there are distinct geographical preferences of those species that occur and are abundant.

Tundra is a distinct kind of habitat known only from Arctic regions and high elevations in mountain ranges above the tree line. The Antarctic is not noted for the development of large regions of tundra because most of the land masses suitable for the development of such habitats are covered by glaciers. However, limited regions of tundra are present on small islands surrounding Antarctica, but lacking large numbers of vertebrates to supply blood, the kinds of insects encountered in northern Canada, Siberia, and Alaska are not abundant on most of the islands surrounding Antarctica.

North of the tree line in the Arctic and above it on mountains, tundra is characterized by large beds of mosses and lichens, which remain inactive on frozen ground during much of the year. The ecological dynamics of the biotic communities of high mountain tundra are complex and variable, and they are seldom noted for thick swarms of insects. Strong prevailing winds make it difficult for insects to fly and reproduce successfully. A few insects in the Andes Mountains find it easier to survive with wingless females among the imagoes rather than permitting some of the females attempting flight to be carried away by strong winds and not being capable of returning to their mating grounds. Several species of Blephariceridae, a family of species that develop as larvae in fast-flowing, mountain streams, are noted for having wingless females. The lowland tundra in the Arctic, however, is relatively uniform, and the same individual aggregations of plants and animals are seen over and over again on the Arctic tundra throughout large areas in the lands where it is present.

The physical environment of most regions of Arctic tundra is characterized by the permafrost, which, as already stated, typically remains frozen throughout the entire year. The insolation of such environments continues for most of the day in summer, and may continue for 24 hours daily for days or weeks before and after the summer solstice. During this period, the ice at the surface of the soil melts, saturating the mosses and lichens with a layer of well-oxygenated water. It is during this early summer season that the larvae of the insects, which remained dormant throughout the cold half of the year, complete their development and metamorphosis and seek food prior to producing eggs. The alternatives to leaving dormant larvae to continue their activity the following summer include leaving eggs that hatch after the ice thaws or spending the winter as pupae. Studies of the methods the many individual insect species employ to maintain their populations on tundra have seldom been attempted, and many questions remain to be answered.

In Alaska, the most notable of the flying tundra insects in the summer belong to the Nematocera, a suborder of Diptera. Mosquitoes and black flies, members of the Culicidae and Simuliidae, respectively, receive the most attention because many can live as adults on human blood. Also among the biting insects that develop from aquatic larvae are members of the Ceratopogonidae, which range from northern Canada and Alaska to Greenland and on to the northern regions of Siberia. For example, in the Krasnoyarski Territory of Siberia in the Russian Federation, Mezenev (1990) found four out of the eight species of the ceratopogonid genus *Culicoides* Latreille, 1809, known to inhabit the territory, feeding on the blood of reindeer, dogs, and human beings. The species he reported were *Culicoides chiopterus* (Meigen, 1830); *Culicoides pulicaris* (Linnaeus, 1758); *Culicoides fascipennis* (Staeger, 1839); *Culicoides alatavicus* Smatov and Isimbekov, 1971; *Culicoides obsoletus* (Meigen, 1818); *Culicoides punctatus* (Meigen, 1804); *Culicoides grisescens* Edwards, 1939; and *Culicoides helveticus* Callot, Kremer, and Dekeit, 1962. Several of these species have been investigated as vectors of viruses that affect people, and several others are suspected of spreading virus diseases to various other mammals and birds. See Chapter 25.

In the absence of human beings and domestic animals, all of these biting insects subsist on the blood of native mammals and birds, which pass close to the pools and shallow water courses above the permafrost of the tundra and in the taiga, where the larvae of many of the insects develop. Much has been written about the annoying insects that appear on the Alaskan tundra during the summer, but scientific studies of the insect species are few and confined to only relatively small areas of this kind of habitat. However, species diversity cannot be expected to be high relative to that in more hospitable zoogeographical regions, but the numbers of biting species of Diptera is regarded by many people as amazing and threatening. The principle that highly stressed biotic communities are usually characterized by large numbers of individuals belonging to few species seems to hold true for the tundra and much of the taiga.

Some mosquito species that develop in tundra pools are noted for their large size. Their larvae have the advantage of long hours of sunlight during Arctic summers, which can be assumed to promote the growth of algae and other microorganisms on which the mosquito larvae can feed. The black flies also have the advantage of high oxygen concentrations in the water, which the larvae of these insects can utilize for respiration without the expenditure of much energy. In warmer climatic zones, simuliid larvae are noted for inhabiting lotic water, usually with enough speed of flow to assure rich supplies of oxygen.

The tundra becomes a vast shallow wetland during the summer, in which water flows steadily from higher to slightly lower elevations. Thus, lotic waters interspersed with small standing pools are available for both larvae that require lotic water and those that can develop relatively quickly in oxygen-rich lentic habitats, such as the larvae of mosquitoes.

Not surprisingly, the insect species that develop as larvae in the tundra belong overwhelmingly to families regarded as aquatic, such as Culicidae, Simuliidae, and Ceratopogonidae. The substrate of the tundra offers the advantage of a thin layer of water distributed over large areas. It is clear, cold, and rich in oxygen. Predatory insects that feed on the larvae of Diptera species are not very abundant because of the long period of darkness and extremely low temperatures during winter. However, to take advantage of these conditions, each species must have the capability of coping with roughly half a year of continual sub-freezing temperatures and long periods of darkness.

In a paper published in 1954, Frohne provided the names of species in the family Culicidae found in Alaska. He stated at the time that it was premature to try to complete a full list of the species. To date, such a list of the common species on the Arctic tundra or in any other part of Alaska has not been completed, because many new species are being described and the identity of others is still not well defined. Taking the names of mosquito species known at the time from Frohne (1954), a provisional list of the Alaskan mosquito fauna with notes of their habitats can be provided with updates on their taxonomy in Table 14.1. It is notable that some of the species in southern Alaska are thought to range southward to parts of North America not regularly exposed to an annual frost.

The second important family of insects that feed on blood in the tundra and the Taiga of Alaska is Simuliidae. These also received much attention in Alaska because of their attacks on people during the Arctic summer. This family is also represented by a modest number of species on the tundra, but the populations of some of the species are also amazingly large. A list of the species compiled by Stone (1952) is provided in Table 14.2 with notes on their ranges in Alaska. As discussed below, many species have been described in recent years using various controversial characters to

Table 14.1 A list of species in the family Culicidae that were known from locations in Alaska in 1954, compiled from the publication of Frohne (1954).

Species	Nature of the Alaskan habitat in which the larvae develop
Aedes cinereus Meigen, 1818	Tundra and northern taiga; spend winter as eggs
Aedes communis (de Geer, 1776)	Tundra and northern taiga; spend winter as eggs
Aedes pionips (Dyar, 1919)	Tundra and northern taiga; spend winter as eggs
Aedes punctor Kirby, 1837	Tundra and northern taiga; spend winter as eggs
Aedes nearcticus Dyar, 1919	Tundra and northern taiga; spend winter as eggs
Aedes nigricans (Coquillett, 1904)	Tundra and northern taiga; spend winter as eggs
Aedes punctodes Dyar, 1922	Tundra and northern taiga; spend winter as eggs
Aedes flavescens (O.F. Müller, 1764)	Tundra and northern taiga; spend winter as eggs
Aedes cataphylla Dyar, 1916	Tundra and northern taiga; spend winter as eggs
Aedes diantaeus Howard, Dyar and Knab, 1813	Tundra and northern taiga; spend winter as eggs
Aedes excrucians (Walker, 1856)	Tundra and northern taiga; spend winter as eggs
Aedes fitchi (Felt and Young, 1904)	Tundra and northern taiga; spend winter as eggs
Aedes impiger (Walker, 1848)	Tundra and northern taiga; spend winter as eggs
Aedes intrudens Dyar, 1919	Tundra and northern taiga; spend winter as eggs
Aedes aboriginis Dyar, 1917	Tundra and northern taiga; spend winter as eggs
Aedes pullatus (Coquillett, 1904)	Tundra and northern taiga; spend winter as eggs
Culex territans Walker, 1856	Temperate Zone, feeds mainly on frogs
Culiseta impatiens (Walker, 1848)	Adults bite in winter in coastal forests
Culiseta alaskaensis (Ludlow, 1906)	Tundra and northern taiga, isolated in coastal areas
Culiseta morsitans (Theobald, 1901)	Tundra and northern Taiga
Culiseta incidens (Thomson, 1869)	Coastal forests, Temperate Zone species
Anopheles oxidentalis Dyar and Knab, 1906	Coastal forests, Temperate Zone species

determine their identity, and how many will be established as distinct will have to be judged as time goes by.

The species in the family Simuliidae that live on tundra and in other boreal zones have not been subjected to thorough taxonomic investigation, and the geographic locations and kinds of habitats they develop in are far from well defined. The information reported in the tables is considered fragmentary because vast areas of their potential ranges have never been explored by entomologists. Furthermore, the definitions and descriptions of the species must still be regarded as tentative until the reliability of the criteria used to distinguish them is established. Recent studies have attempting to delimit the species using such criteria as DNA sequences, which have resulted in the creation of "sibling species complexes" encompassing large numbers of tentative species-like nominal taxa, grouped under the name of a species, which was formerly treated as a conventional taxon at the specific level. Because the criteria used to delimit these "sibling species complexes" have never been tested until recently, their values for delimiting species are still uncertain. For that reason, the taxonomic identities and systematic positions used by earlier authors is maintained with the recommendation not to treat their classification as final until agreed upon by entomologists specializing in the family.

Table 14.2 A list of species in the family Simuliidae that were known from locations in Alaska in 1952, compiled from the publication of Stone (1952). Many more recently described species are omitted because almost nothing is known about their distribution other than locations where the type specimens were found. Ranges were provided, in part, by Rubtsov (1990, from an English translation of his work first published in 1956); Mezenev (1990), Adler et al. (2004), and Halgoš (2005). Most species with ranges listed as Nearctic have been found to be widespread in the region, indicating discontinuous distribution.

Species	Notes on ranges
Gymnopais dichopticus Stone, 1949	Palearctic taiga, Nearctic tundra
Gymnopais holopticus Stone, 1949	Palearctic taiga, Nearctic tundra
Prosimulium onychodactylum Dyar and Shannon, 1927	Nearctic, Yukon to Utah & Colorado
Prosimulium alpestre Dorogostajskij, Rubzov and Vlasenko, 1935	Palearctic taiga, Nearctic tundra
Prosimulium fulvum (Coquillett, 1902)	Nearctic, West Coast, Alaska to Arizona
Prosimulium orsinum (Edwards, 1935)	Nearctic
Prosimulium dicum Dyar and Shannon, 1927	Nearctic, West Coast of North America
Prosimulium travisi Stone, 1952	Nearctic
Helodon decemarticulatus (Twinn, 1936)	Nearctic
Helodon pleuralis (Malloch, 1914)	Nearctic
Cnephia mutata (Malloch, 1914)	Nearctic
Cnephia emergens Stone, 1952	Western North America
Cnephia sommermani Stone, 1952	Alaska
Cnephia minus Dyar and Shannon, 1927	Nearctic
Cnephia eremites Shewell, 1952	Holarctic
Metacnephia saileri (Stone, 1952)	Nearctic
Simulium aureum Fries, 1824	Nearctic
Simulium baffinense Twinn, 1938	Holarctic, Arctoboreal
Simulium gouldingi Stone, 1952	Nearctic
Simulium latipes (Meigen, 1804)	Holarctic
Simulium arcticum Malloch, 1914	Northern Holarctic Arctoboreal
Simulium vittatum Zetterstedt, 1838	Holarctic, from Greenland to Mexico
Simulium corbis Twinn, 1936	Nearctic*
Simulium decorum Walker, 1848	Nearctic
Simulium hunteri Malloch, 1914	Nearctic
Simulium malyschevi Dorogostajskij, Rubzov and Vlasenko, 1935	All of Siberia, Alaska
Simulium meridionale Riley, 1886	Nearctic
Simulium nigricoxum Stone, 1952	Nearctic, Arctoboreal
Simulium rubtzovi Smart, 1946	Holarctic, Arctoboreal
Simulium tuberosum (Lundström, 1911)	Nearctic
Simulium venustum Say, 1823	Holarctic, Arctoboreal

*Reports of *Simulium corbis* Twinn, 1936, in the Palearctic Region seem to refer to *Simulium murmanum* Enderlein, 1935, but the matter will require additional research before it is settled.

Omitted from the table are nominal species which appear to have been reported from more than one continent, although the species reported from each continent are thought to be distinct.

A comparison can be made between the biting insects of the tundra and the winter craneflies in the taiga. There is a fundamental difference between the life cycles of the winter crane flies in the family Trichoceridae, and the summertime insects of the tundra and northern taiga. The imagoes of the winter craneflies are more familiar on the taiga and in parts of the temperate zones, which, unlike the tundra, provide suitable habitats for trees. In this respect, they would appear to be ice age relicts rather than post-ice age colonists of cold wetlands, which resemble familiar insects of the temperate zones in their ecology, except for their particularly great tolerance for extremely low winter temperatures. The continual glacial coverage of the ground during the ice ages would have deprived the imagoes, which are so numerous on the tundra at the present time, of suitable places to spawn during both summer and winter. Not only would the ice of the glaciers not melt enough to leave suitable flooded areas for spawning, the water would form pools on top of the ice rather than in contact with nutrient-rich soils, which are needed to support the growth of bacteria and algae on which many of the insect larvae feed. In contrast, the trees covering the taiga and the epiphytes that grow on them would provide shelter and nourishment for the winter midges regardless of the conditions on the surface of the soil in which the trees are rooted. Therefore, biting midges are presently abundant only on the tundra of the North Frigid Zone, while winter crane flies can be found in both the North and South Frigid Zones.

The most striking difference between the insect fauna of the tundra and the winter crane flies of the taiga is the emergence of the adults of the tundra species during the spring and their persistence until the arrival of the low temperatures during the autumn. In contrast, adult winter crane flies are usually absent during the summer and first appear at about the time of the arrival of freezing air temperatures during autumn. Unfortunately, there are large gaps in the knowledge of the life cycles of the winter crane flies, so reflections on the survival strategies of most species cannot yet be based on clear and precise information.

SECTION 2

Winter Crane Flies between Water and Ice

The few insect species of the frigid climatic zones that appear as adults at temperatures at or below freezing escaped attention for many years because they are rarer and less conspicuous than species which appear in large swarms during Arctic summers. The typical species that emerge as adults during the seasons in which the temperature is relatively high and melts the surface of the permafrost force their attention upon the human population by their aggressive behavior in sucking the blood of warm blooded mammals and birds. In contrast, the imagoes of the winter crane flies typically appear in small numbers in seasons during which human beings feel uncomfortable outside of shelter, and none of these insects bite or otherwise annoy people.

An unusual complication in the life of insects adapted to develop in mosses and lichens or in flooded or waterlogged soils is the situation in the frigid zones and the colder parts of the temperate zones in winter. Only one family, Trichoceridae, the winter crane flies, is especially adapted to the extreme conditions in such habitats. The adults are usually encountered flying at temperatures slightly below to slightly above 0°C. Their imagoes are known from both the North and South Frigid Zones, as well as from the colder regions of the Temperate Zones in winter, and in habitats high in the mountains. Adults are not encountered during warmer seasons, and they are not active during periods when the temperature is much colder than 0°C. Because their periods of flight, mating, and egg laying can occur during periods when the temperature is usually lower than freezing, they are confronted with the problem of depositing eggs in places usually covered with ice. When their larvae hatch, they must also cope with the problem of maintaining activity in contact with ice, at least for a short period of time.

In this respect, their habits differ fundamentally from those of the abundant species of Diptera encountered as imagoes during the long days of summer in the Arctic. The known winter crane flies are certainly more characteristic of the taiga than the tundra, and the adults fly during milder periods of the colder seasons rather than during the Arctic or Antarctic summers. They are best known from the Boreal Zone of the Northern Hemisphere, but the ranges of some species extend into subarctic regions and high in mountain ranges at elevations where the climate is most similar to that in the Subarctic and Boreal Zones. Other genera of winter crane flies inhabit the South Frigid Zone, but because of geographical differences in the regions surrounding the two poles of the earth, they are limited mainly to relatively small islands. Because they typically wait to fly until windless periods, they are not observed in those high mountains where the prevailing winds tend to be strong.

Winter crane flies are also present in the Southern Hemisphere, and apparently there has been some transfer of species from the Arctic to the Antarctic Region. The species *Trichocera maculipennis* (Meigen, 1818) was found in a research station on King George Island, which is in the waters off Antarctica (Fig. 14.1). It is the first species in the family Trichoceridae found in the region of continental Antarctica, and it is known to be a common native species from the Arctic introduced with shipping traffic to the Southern Hemisphere (Volontario et al., 2013). In addition, *Trichocera annulata* Meigen, 1818, has been introduced to the Southern Hemisphere, and it has apparently been successful in establishing itself.

Species apparently native to the Southern Hemisphere have been identified on islands off the coasts of Australia and New Zealand, as well as in southern Chile (Alexander, 1926 and 1929; Dahl, 1970b; Krzeminska and Young, 1992).

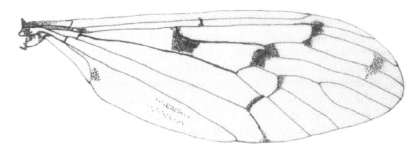

Figure 14.1 The wing of the common northern European winter cranefly, *Trichocera maculipennis*, which was reported to have been introduced to the Southern Hemisphere. Redrawn and modified from Volonterio et al. (2013).

During summer, winter crane flies typically remain inactive, and adults are not observed. Apparently larvae remain in an inactive or semi-active state whenever the substrate becomes too dry, but that has yet to be determined with certainty for most trichocerid species. There are many gaps in the knowledge of winter crane flies, including whether or not the species have similar life cycles and periods of development or whether there are significant differences in their periods of activity and diets. The larvae of most species still cannot be identified, but observations in the seasons during which adults are active have been recorded. Recent publications are adding substantially to the knowledge of both the larvae and the imagoes, but there is still a great deal to learn.

Observations in a fruit-growing district along the Elbe Estuary in northern Germany revealed the presence of four species, the adults of which could be found near small eutrophic water bodies during every mild period between late autumn and early spring (Heckman, 1982). All four species were most abundant above marshy ditches in the fruit orchards among the upper parts of the emergent aquatic grass, *Glyceria maxima* (Hartman) Holmberg. In order of abundance, the four species identified were *Trichocera regelationis* Linnaeus, 1758; *T. hiemalis* (De Geer, 1776); *T. saltator* (Harris, 1782); and *T. annulata* Mergen, 1818. Adults of a fifth species, *T. major* Edwards, 1921, were found near a retention basin in a wooded park in the northern part of Hamburg, several miles from the Elbe Estuary and uninfluenced by the tides or by agriculture. They were active only on cold days in winter, usually when there was very little wind.

To determine how insects could remain active at temperatures below freezing in winter, the heat generated by the sediments in highly eutrophic parts of shallow drainage ditches in the fruit orchards was recorded. The ditches were found to generate a considerable amount to heat during the winter due to the decomposition of large amounts of detritus by microorganisms. This was enough to prevent the shallowest parts of the ditches from freezing when the deeper parts had already frozen over. It was also enough to provide steady updrafts sufficient to support the flight of the winter crane flies for extended periods of time during windless winter days. Adults of none of the species were observed when the temperature was higher than a few degrees above 0°C. According to Coe et al. (1950), adult winter crane flies may be found throughout the year in mountain ranges high enough for the temperature to remain at or near freezing almost continuously.

In northern Germany, the adults were most often observed when weather conditions were very similar to those described here. Preferred periods of flight also seemed to be sunny days with air temperatures no more than a few degrees above or below 0°C. Often, the flights begin before light periods of snowfall. If the snowfall increases, they are interrupted but are resumed after the heavier snowfall abates or ceases. A small amount of snow is often on the ground, which makes the imagoes easier to see, since most of them are blackish or dark brownish. The humidity is usually high, and winds are no more than very light. Usually, few specimens are observed in any one place, although it was obvious that the imagoes of each species were emerging at about the same time for mating and deposition of eggs. Only occasionally were they somewhat numerous, but the mating swarms could never be described as dense. It seems likely that mating occurs quickly, probably after dark, and the females remain near the sites where they will deposit their eggs. The insects often fly near or beneath trees, and they may take shelter in or near buildings if precipitation increases. Because of the season, time of day, and weather during which their mating and egg laying occur, predation on the spawning imagoes must be negligible.

Like crane flies in other families, what is already known about the larval stages of trichocerids indicates that they develop in a variety of different habitats ranging from moist soils to muddy, marshy, or waterlogged substrates, possibly including mosses and lichens on tree branches. The wettest of these habitats are periodically beneath snow or ice. Reported observations suggest that one or more species inhabit rodent burrows, but these species have not been positively identified. The nature of these habitats during winter makes it difficult to classify the larvae as either terrestrial or aquatic. More research will be required to determine species-specific differences, but from observations and reports, the larvae of many species must be able to survive periods of immersion and freezing in ice (Alexander, 1967).

Dahl (1970a) reported the ranges of the known species and provided zoogeographical and habitat information on the winter crane fly species of the world. This can be supplemented with updated notes about more recently described species. Because of the lack of ecological information about most winter craneflies, the following generalizations are provided, which supposedly are common to all or most species without regard for whether they will be classified as terrestrial or aquatic after the habitats in which their larvae develop are known with certainty.

Trichoceridae is a relatively small family. There have been somewhat more than 100 species names assigned to specimens, of which several have been or are expected to be demoted to junior synonyms of other species. Those native to the South Temperate Zone are all presently assigned to the subfamily Trichocerinae, but all of those circumantarctic species originally placed in the genus *Trichocera* have been transferred to genera known only from the Southern Hemisphere. Two such genera are present in Australia, New Zealand, islands surrounding Antarctica, and in the Andean region of South America: *Paracladura* Brunetti, 1911, and *Nothotrichocera* Alexander, 1926. *Paracladura* is known only from New Zealand and nearby islands. All of its 12 species have been discovered in New Zealand, but one of them, *Paracladura antipodum* Mik, 1881, is also known from Auckland and Campbell Islands.

Four of the six species of *Nothotrichocera* were found in Tasmania and reported from a small area along the southern coast of Australia in the state of South Australia: *Nothotrichocera cingulata* Alexander, 1926; *N. tasmanica* Alexander, 1926; *N. terebrella* Alexander, 1926; and *N. tonnoiri* Alexander, 1926. The other two species were reported from islands near New Zealand: *N. aucklandica* Johns, 1926, from Auckland Island, and *N. antarctica* Edwards, 1923, from Auckland and Campbell Islands.

The seasons and activity patterns, as well as the natural history of these insects, suggest that they are relicts of the last Ice Age. Of course, the thick glacier covering the entire continent of Antarctica with ice easily blocks the activity of insects with habits similar to those of winter crane flies. Islands surrounding the coast, however, could have been more hospitable to these insects, but the species known today are limited to the southern seacoasts and islands to the south of Australia and South America.

In the Northern Hemisphere, glaciers were free to move back and forth across continental land masses several times during the millennia since the Miocene, allowing the family Trichoceridae to move back and forth along their leading edges, establishing populations of species in the northernmost parts of the continents. Near the leading edges of advancing glaciers, trees would still be present, and patches of vegetation, including trees and their epiphytes, would certainly have still existed on the sunny sides of ridges and hills. Such places would have been ideal habitats for winter crane flies. Much more study of the species in the family Trichoceridae will be necessary to confirm their origins and systematics, as well as their modes of life. In any case, the arrangements of the continents explains why there are far more species in the Northern Hemisphere than in the Southern.

The physiology of insects that can actively fly around during the hours of darkness at air temperatures below the freezing point of water requires some explanation. Similarly, it is still necessary to determine whether breeding and egg-laying are seasonal, that is, whether each species can be expected to undergo metamorphosis at a particular time of year, or whether the adults appear only when weather conditions are conducive. For insects that are active when all or almost all of their potential natural enemies are inactive, it might not be helpful to have one coordinated mating flight each year in order to minimize the impact of predation on the imagoes. In northern Germany, one of the common species of *Trichocera* can be observed flying during every month from November through March, and similar observations can also be made in Norway.

Further information about the bionomics of these insects would be very useful for a better understanding of the seasonal cycles of the species. For example, the diets of each species and the seasons during which the larvae of each actively develop would be very useful for understanding their life cycles. In addition, the habitats in which the larvae develop would provide additional information to assist in distinguishing the individual species. All of this information would be useful for developing plans for using these insects as indicators of environmental changes and to assist in protecting the species.

In addition, discovering the geographic ranges of each species would be very helpful for zoogeographers and ecoclimatologists, who could use this information for better delimiting the respective regions based on the conditions in winter. There are still a number of important and interesting questions to be answered with the help of the winter craneflies.

In spite of about 40 years of observation in Norway, Hågvar and Krzemińska (2007) were only able to speculate about the seasonal pattern of breeding and larval development of 15 trichocerid species they had identified in forested regions. Their specimens were collected on snow or flying at air temperatures as low as $-1°C$. In the State of Washington in the Pacific Northwest, adults of an unidentified species of *Trichocera* were personally observed flying during a light winter snowfall above a thin layer of snow on the ground at air temperatures below $-1°C$. As the temperature decreased, some of the insects landed to rest on the snow beneath trees or on branches of the conifers themselves. How they can avoid freezing and remain active at temperatures below the freezing point of water and how they shelter themselves from the effects of even colder temperatures and other kinds of bad weather is a question that still has not been answered satisfactorily for most species.

SECTION 3

Snow Fleas

Several species in the subclass Arthropleona, class Collembola, have been called "snow fleas" because they are small, jump, and live on the surface of snow. Four species with these habits were reported by Hågvar (2000) from Norway: *Hypogastrura socialis* (Uzel, 1891), *Vertigopus westerlundi* (Reuter, 1898), *Isotoma hiemalis* Schött, 1893, and *Isotoma violacea* Tullberg, 1876. In his discussion, it was pointed out that Hexapods trapped in ice would be subject to anoxic conditions for long periods of time in winter. For many species of insects, this could prove fatal. He found that the Collembola species active in winter could cope with this problem very well, and some would move up into the overlying snow layer to keep from being frozen in the ice.

In addition, two other species of Collembola were reported from among lichens on wet, exposed rocks during cold seasons in Norway (Leinaas and Sommer, 1984), as already mentioned in Chapter 11. The two species observed among the lichens in southern Norway were *Xenylla maritima* (Tullberg, 1869) and *Anurophorus laricis* Nicolet, 1842, noted for their preference for saturated substrates in cold, semi-aquatic habitats.

Apparently, five of these species can move about by orienting themselves to the sun and leaping using the furca. One of them, *Hypogastrura socialis,* in the family Hypogastruridae, has a vestigial furca and cannot spring very far. However, they can move in small groups over the snow and cover several meters per day.

15

Adults that do not Eat

Many aquatic insects develop as larvae in freshwater bodies, and the adult stages of some of these live only to reproduce as quickly as possible before dying. The imagoes of all species belonging to one insect order and one large family of another have lost the ability to consume any nourishment at all, and they must eat enough food as larvae to supply all the needs of pupal development, a brief period of dispersal, usually involving flight, and the maturation and release of eggs or sperm. An alternative strategy encountered in species belonging to a few insect families is for the male to consume no food and die shortly after mating, while females do consume food as adults and continue to deposit multiple batches of eggs as long as the local climate permits and they can find sufficient nourishment. This is possible because of the transfer and storage of enough sperm during one mating to fertilize all of the eggs the female will produce for the rest of her life.

There are a great many variations in this basic life cycle employed by insect species that do not eat after reaching the pupal stage. For example, *Culex pipiens* Linnaeus, 1758, is among the most successful aquatic insect species in the temperate zones. The most familiar reproductive strategy of such insects is to mate one time and store enough sperm for successive batches of eggs. The males do not eat or consume only fluids produced by plants and die soon after mating, but the females feed on blood and produce successive batches of eggs after each feeding as long as the seasonal weather conditions permit (Frohne, 1954). This kind of reproductive strategy, which is so successful in the temperate zones, is not at all suitable for the polar or tropical regions of the earth, where *Culex pipiens* is typically absent. In the Arctic regions, the great majority of mosquito species produce eggs only once during the course of a year, and the eggs remain dormant during the cold winter periods (Frohne, 1954).

The large number of aquatic insect species with either male imagoes or imagoes of both sexes that do not consume food indicates that this survival strategy is an effective one. The brief flight period is sufficient for a few of the females to reach neighboring water bodies, which the species can colonize, but the bulk of the eggs are presumably deposited where the most successful populations have lived as larvae from generation to generation.

Most insects with this survival strategy are relatively small. Presumably, this would make their females containing the masses of eggs they carry particularly nutritious foods for larger predators. The strategy of simultaneous emergences of adults ensures the survival of a maximum number of egg-bearing females. Flying predators have an easy time capturing prey during the emergences, but because so many of the prey appear almost simultaneously, the predators cannot capture too large a percentage of the breeding stock before the mating flights and oviposition of most have been successfully completed.

Species in certain families of Diptera are noted for having adults of both sexes which reproduce quickly and then die without having consumed any food. The largest such family which employs this strategy is Chironomidae, the adults of which often have only vestiges of functional mouthparts. Diptera families are not the only higher taxa displaying limited feeding capabilities as adults, and it is best to review this strategy of reproduction as employed by each individual insect order on a case by case basis. As far as known, Insecta is the only class of Hexapoda in which some of the reproductive adults abstain from all food after their metamorphosis.

SECTION 1

Ephemeroptera

Ephemeroptera species are noted for rapid reproduction, no consumption of food, and brief life spans as adults. This strategy for survival of the species is not one encountered in many insect orders, but it is assumed to be the strategy for all or almost all species of mayfly because these insects have only vestigial mouthparts (Heckman, 2002). Parts of the legs are often vestigial, as well.

Mayfly species are notable among the insects for having two stages bearing wings. After completing all larval instars, the mayfly molts, and a winged insect emerges. This stage, however, is not sexually mature, and shortly later, sometimes within a day, it molts a second time, and another winged state emerges, which is capable of taking part in mating flights and laying eggs. The first winged stage is called the pre-imago, and the second is the imago, the name given to the adult.

In the case of Ephemeroptera, breeding swarms of adults belonging to many species emerge at about the same time. When these swarms appear depends on the species involved. The number of imagoes involved depends upon the individual species and the size of the water body in which they developed. Massive numbers are encountered near many lakes for short periods of time, prior to the introductions of modern pesticides. Near the Great Lakes of North America, older residents remember that bulldozers were sometimes used to remove the carcasses of mayflies from roads after the completion of spawning. The advantages of such large mayfly populations in the lake for the survival and growth of the native fishes must have been enormous. The larvae of mayflies are a favorite food of some of the most popular sport fishes in North America, and their decline in numbers has been accompanied by considerable decreases in the populations of the fishes.

During the appearances of the breeding swarms of mayflies, trout are frequently seen leaping out of the water to catch them in the air before they can fly out of range after shedding their larval exoskeletons. For this reason, many of the artificial flies used to catch sport fishes using rod and reel are made to appear like mayfly imagoes.

As already mentioned, species in the order Ephemeroptera have only vestigial mouthparts during the adult stage, making it doubtful that members of either sex of any species are physically able to consume any nourishment as adults (Heckman, 2002). Of course, there are many species of mayfly around the world that still have not been studied well enough to rule out the possibility of their being able to feed as a pre-imago or as an adult at all, but it appears as if this condition is universal throughout the entire order.

Most species of mayfly require relatively clean water with a reasonably rich supply of dissolved oxygen. Many prefer lotic water, and some are limited to life only in fast-flowing streams. Exceptions to this include a few species in the family Baetidae, which can survive in α-mesotrophic standing water. One of these exceptional species that was found to be moderately abundant in northern Germany was *Cloeon dipterum* (Linnaeus, 1761). The larvae of this species are able to move the gill laminae rapidly to create a flow of water over the gills during periods when the concentration of oxygen in the water becomes low.

Cloeon dipterum is also exceptional in other ways. Lock and Goethals (2011) referred to it as the mayfly species most tolerant of unfavorable water quality in Belgium. This apparently reduces the number of predators they have to contend with in their habitat. Fishes are among the main predators of mayflies, but most fishes avoid the detritus-filled, oxygen-poor water in which the larvae of *Cloeon dipterum* develop. As a result, these mayflies do not fly only in swarms during a brief mating and spawning season. In Korea, *Cloeon dipterum* was found to breed during every season except winter, and single, mass emergences do not occur. Between April and October, four peak periods of breeding were observed (Lee et al., 2013).

Most other species of mayfly lack the movable gill laminae that give the larvae the capability of living in water bodies that become temporarily poor in oxygen. The typical habitat of most larvae is clear and cool water, most frequently in large lakes or streams. Observing mayflies emerging from a stream or lake demonstrates the strategy for their success. They leave the water rapidly, coming to the surface quickly and molting quickly at about the same time so that as many as possible can avoid being consumed by a fish. Many molt at about sundown, minimizing the chance of being caught by a bird. However, a few bird species have adopted crepuscular habits to take advantage of mayfly emergences, and bats are also able to exploit crepuscular and even nocturnal swarms of mayfly imagoes. The pre-imagoes fly to a tree or a fixed object high enough above the water to be out of the range of a leaping fish and molt again. As soon as that is complete, the imagoes mate quickly, and the eggs soon become evident at the apices of the female abdomens. This makes it easy for predators to capture the adults for only a brief period of time, after which all members of the entire generation have deposited their eggs and died, making imagoes unavailable as food for fishes and other natural enemies again during the rest of the year. Other mayfly species sharing the habitat may repeat this rapid reproduction performance during other times of the year, however. After hatching, the new larvae are very small, and most can conceal themselves among benthic detritus. With such a reproductive strategy, mayflies make it almost impossible for any one group of predators to depend entirely on them as a main source of nutrition.

SECTION 2

Diptera – Chironomidae

The largest insect family known to include a considerable number of species that do not feed as imagoes is Chironomidae, which belongs to the suborder Nematocera. Like mayflies, chironomids possess only vestigial mouthparts, indicating that they cannot consume food as adults (Heckman, 2002). Because there are no special problems caused by midges in this family, less is known about the species, and many are certainly still to be described and named worldwide. For that reason, it cannot be stated with certainly that the descriptions provided in this Section apply to all species in the family.

Mating only once after metamorphosis to the adult stage definitely provides advantages for survival and reproduction of chironomid species, as demonstrated by the fact that such a large number of species worldwide belong to this family. The female can store enough sperm during one mating to fertilize all of the eggs she produces until the end of her life. As in the case of mayflies, chironomids employ the strategy of emerging at the same time and forming mating swarms to minimize the exposure of their imagoes to predators. In northern Germany, some of the species formed these mating swarms in autumn just before sundown when the air temperature was only a few degrees above freezing, and all local predatory insects, including dragonflies, were grounded by the temperature.

At other times of year, the strategy for survival of those species that mated at such times is to emerge shortly before sunset, often during weather conditions that are not encouraging for predators to hunt them. Some species were observed swarming during cool evenings or in light rain. One swarm of males in the subfamily Orthocladinae in Germany was observed seeking people standing near a water body on a cold autumn day. The body heat from the observers created updrafts sufficient to keep the tiny insects in the air without expending very much energy, and they were able to keep their position near the water until a female rose from the vegetation along the water body to mate with one of them. In the absence of warm-blooded animals, the water itself frequently provides updrafts in autumn at times when the air cools quickly in the evening, while the water decreases slowly in temperature. It was also found that small water bodies containing large amounts of dead vegetation, including tree leaves, generate an appreciable amount of heat through the decomposition processes, causing updrafts of warmer air during cool evenings and even delaying ice formation, after it had already begun on adjacent water bodies. It was observed frequently how well certain chironomid species were able to utilize such local microclimates to their advantage for successfully completing of mating and oviposition.

Obviously, the reproduction must occur very quickly after metamorphosis, and the females have to live only a little longer than the males to complete their work of egg laying to provide for the next generation. The species of insect reportedly with the shortest life expectancy as an adult is a marine chironomid: *Pontomyia natans* Edwards, 1926. More information on this species is provided in Chapter 17.

The water bodies inhabited by the larvae of insects that must consume and store all nutrients for the pupa and adult stages must be sufficiently rich in organic nutrients to preclude nutrient shortages, which would limit the abundance of the next generation. Hence, there are usually large numbers of species that feed only during the larval stage in eutrophic water bodies and much fewer in nutrient-poor habitats. The adaptations of many chironomid species to water relatively poor in dissolved oxygen, including the presence of hemoglobin in larvae in the subfamily Chironominae, attest to the need for surviving periods of oxygen shortages due to eutrophication of their habitats. Apparently, this pigment, with its strong affinity for oxygen and ability to store it, serves to make it possible for the larvae that possess it to develop in water bodies rich in algae, which release large amounts of oxygen into the water during the day. The oxygen stored in the hemoglobin during the day meets the needs of the larvae during hours of darkness, when saprobic activity results in the nearly complete depletion of the dissolved oxygen before sunrise on the following day.

Richness of the food supply in the water provides an advantage for insects with very short reproductive periods. By emerging from the water for only short periods of time, usually in large numbers, the imagoes of such insects can avoid severe predation by terrestrial predators and quickly release their eggs into their aquatic habitats, in which the larvae of the next generation will presumably benefit from a large food supply. Naturally, the larvae will require such a generous food supply because they must consume enough energy to complete their development and emerge as adults with enough nutrient reserves in their bodies to produce gametes and undertake mating flights. In addition, the females must produce eggs and deposit them in places where the larvae will be able to conceal themselves from natural enemies and obtain sufficient food and oxygen to develop.

Obviously, a great deal is still to be learned about the very numerous species in the family Chironomidae. Apparently, neither males nor females in this family feed as adults at all, although enough of them have not been studied to determine whether there are any exceptions. From the development of the mouthparts, however, it can be assumed that exceptions, if any, would certainly be few.

The family Chironomidae encompasses a large number species, and the ecology of most has never been studied in enough detail to make final conclusions about their feeding through all of their life stages. Other taxa of aquatic dipterans that include species incapable of consuming food as adults belong to few families. In almost every case, adults that do not consume food are noted for being short-lived. Their stored energy is used for intensive mating activity, and their proteins are used for the production of gametes.

By minimizing the time the adults are available to terrestrial or littoral predators, they avoid becoming a food reserve for such natural enemies, promoting their population growth, and attracting them to the locations in which the swarms of mating adults appear. Hence, the faster the mating and deposition of eggs takes place, the fewer the losses due to predation will be. The more the period of reproduction is reduced in time for a given population, either seasonally or through local weather conditions, the less will be the impact of predators on the reproductive success. This accounts for the vast numbers of mayflies that sometimes appear during the reproductive period of local species and the dense swarms of chironomids that appear briefly during the specific mating period of each species present in a given water body.

Needless to say, the adults of species utilizing this kind of reproductive strategy must have ingested all of the necessary nutrients needed for successful reproduction prior to metamorphosis. Therefore, the larvae generally live in water bodies with rich food supplies, usually including microscopic prey, such as algae, detritus, and saprobic microbiota. Large numbers of eggs are typically produced at one time in order to assure that at least a few survive to adulthood. Such insects are often important links in the food web between microbiota and fishes.

To avoid excessive predation on the adults during the brief reproductive period, two additional factors are necessary. The first is a seasonal cue to stimulate all larvae or pupae in a water body to begin metamorphosis at a time when breeding will begin under ideal weather conditions. Details of the signals from the environment prompting the onset of metamorphosis and ripening of the gonads are known for few species. One of these few is *Clunio marinus*, a chironomid with larvae that develop among seaweeds in the sublittoral zone of the North Sea. Its biology is discussed in more detail in Chapter 18 of this book.

The fact that entire populations of larval *Clunio marinus* is prompted to begin the reproductive process by a combination decreasing day length following the summer solstice, the phase of the moon, and the occurrence of calm weather conditions that will last for at least two to three days demonstrates that the signals the larvae receive from the environment are quite complex. If the weather is stormy at the time the other conditions are detected, the breeding process will begin at the same phase of the moon one month later.

The second factor giving species with this reproduction strategy the chance for maximum reproductive success is an optimal chance for males and females to quickly and successfully mate. This is achieved by the habit of the adults to congregate in swarms so that the success of each insect to find a member of the opposite sex of the same species in no more than a few minutes is almost certain. In the case of the chironomids, swarms of some species seem to consist almost exclusively of males. Females approaching such swarms usually alight on nearby littoral plants, and a male descends from the swarm to carry her on a flight during which mating occurs. The fertilized eggs are deposited shortly later. The alternative to this mating ritual is for the female to fly up into the swarm of males and unite with one of them in the air before they return to land on a plant along the shores of the water body. Obviously, the quicker the eggs can be laid after mating, the better will be the chance that the fertilized eggs will survive long enough to hatch.

The phenomena related to the formation of swarms are assumed to be the result of pheromone production by the female. Whether other stimuli play a role still remains to be determined. Male chironomids are recognized by their densely feathered antennae. This is presumed to be the organ capable of detecting minute amounts of the pheromone produced by the species in the air. The insect is then able to navigate its way to the female or females producing the substance. With such an efficient way of finding a mate and spawning, the species is subjected to predation by flying predators and subsequent reduction in the size of the next generation for only a minimum period to time.

In spite of the opportunity for success provided by such a technically sophisticated system, there is still the need for another safeguard to reduce the danger of unsuccessful mating. Water bodies conducive to insect development often harbor a dozen or more populations of different chironomid species. Even though the reproductive period of each is relatively short, the probability is high that the swarms of two or more species will appear simultaneously, especially during periods of calm weather. Pheromones can lead a male to a swarm, but if the swarm consists of two or more species, there is a chance that a male might select a female of a different species. Female chironomids typically mate only once and deposit all eggs as quickly as possible. If individuals of two species mate by mistake, either the eggs will not hatch or the offspring will be infertile. Such an event would hurt the survival chances of both species.

Such futile interspecific mating does not seem to occur. There is a well-supported hypothesis that it is impossible for individuals of different species to copulate because the complex structures of the male and female genitalia function like those of a key and a lock. The female is able to receive sperm only from a male of the same species. If an interspecific mating is attempted, no sperm can be transferred because it is not possible for the genital openings to come together because they are blocked by intricately constructed appendages on the male or female genitalia. The failure to complete the mating process permits the female to continue the search for a suitable male of her own species. If there had been a successful mating with a male of a different species, the female would presumably have been unable to mate with any other male and be unable ever to produce a new, fertile generation of its own species.

For identifying chironomid species, the male genital structures are definitive for each species. Theoretically, the male of every species has structures that are distinctive enough from all others to be recognized after cursory examination. It is also considered likely that most insects have such safeguards against interspecific mating, accounting for the importance of the male genitalia in insect taxonomy. For insects which must mate and deposit their eggs rapidly, this is an important safeguard to assure the survival of the species from year to year. It is easy to understand that adults unable to consume food would quickly exhaust their nutrient reserves and be unable to reproduce at all if they had to undertake exhaustive searches for a mate.

The descriptions in this chapter are general ones, based on careful studies of a few species. Because the ecology and natural history of the vast majority of species in the family Chironomidae have not been carefully studied, it cannot be ruled out that some species deviate significantly in their behavior and reproductive habits from those discussed in this chapter. The same is true for species of Ephemeroptera, although there is more information available about the mayflies of the Holarctic Zone than there is about most chironomids.

SECTION 3

Diptera – Culicidae

The survival strategy of the species of Culicidae, the mosquitoes, is much more specialized for getting the maximum benefit from the available food supplies than that of the other insect taxa discussed in this chapter. Like most other species in the order

Diptera, the mosquitoes depend upon a high rate of reproduction to offset the high rate of mortality. The species in this family exploit the ability of females to receive and store sperm from one mating to fertilize all of the eggs that she will be able to produce in a lifetime. While the female of several mosquito species can attack and prey on much larger animals and use their protein-rich blood to produce a second or third clutch of eggs in a year, the male has usually made his entire contribution to the survival of the species within a few days, or even hours, after completing metamorphosis from the pupa to the adult. Whether there are species that employ variations to this basic strategy is unknown, since there are many species of mosquito in the world that have not been sufficiently studied. It is known that climatic conditions impose several restrictions on the basic reproductive schedules. For example, the female mosquitoes in the genus *Aedes* that live in Arctic regions usually lay eggs only once per year, and these eggs remain to hatch during the following year. The female *Culex pipiens,* on the other hand, continues to lay eggs during the warmer seasons until the arrival of winter ends her activities. This species does not thrive in Arctic climatic zones (Frohne, 1954).

Blood is an extremely nutritious food for the female mosquitoes. It contains sufficient quantities of proteins, carbohydrates, vitamins, and mineral elements to provide a female mosquito with all she needs to produce a whole group of eggs. However, the donor of this blood might be infected with any number of parasitic invertebrates, protozoans, bacteria, or viruses, which infect the mosquito just as they do the vertebrate host. Many of these organisms permanently infest the mosquito, forcing it to provide the nourishment and shelter for the production of infectious stages to be carried to and injected into the next suitable host. More details about this will be provided in Chapter 25.

The continued survival of the males is somewhat counterproductive. If they consumed food, the supply of edible material for the females would be correspondingly reduced. If they are preyed upon by another insect species, that would substantially contribute to the population of that predator, making life more dangerous for mosquitoes in general.

As long as there are female adults that have not yet mated, it is beneficial for the species to have adult males near the water, which have yet to mate. Females being preyed upon while seeking a mate could be considered unnecessary mortalities. Therefore, as soon as a male of most species has mated, it becomes only a liability for the survival of the species. This situation makes it clear that there is no reason for a male mosquito to survive after mating, and therefore it requires no food to eat after its final metamorphosis, although males of some species may be able to consume nectar or other fluids from plants.

Briefly stated, once the male has mated, he has no more useful purpose. In the case of some insects, including many members of the Culicidae, the male does not need to consume any nourishment during the adult stage and dies shortly after mating, while the female can feed after laying one clutch of eggs and continue to produce additional eggs, which are fertilized by stored sperm. How many groups of eggs can be produced during the life of one female remains to be discovered for many species. However, because some species transmit diseases by consuming the blood of an animal infested with a pathogen or parasite and then injecting the stages into a second, uninfected host, it is clear that at least two and probably several clutches of eggs can be produced by each female, wherever the climate permits.

Similarly, the life cycles of the males of many species have not been investigated sufficiently to know whether all of them die shortly after mating or whether some are capable of either mating again with another female or remaining alive for a relatively long period of time by subsisting on energy-rich foods, such as the nectar produced by flowers, prior to mating for the first time. The males of some culicids are known to feed on juices of plants, while the females take meals of blood, which provides a rich source of proteins utilized for the production of eggs. Which species have males that feed and survive appreciable periods of time and which have males that do not feed at all is still not known with certainty in the cases of many species, world-wide. A few aquatic insects are believed to be parthenogenic and produce no males at all, while others are parthenogenic seasonally but produce males at least once per year.

16

Life in Water Without Oxygen

Industrialization within the past 200 years produced large quantities of many kinds of wastes, which were most conveniently disposed of in the nearest stream or river. The effects of the widest varieties of toxic substances and pathogens were learned only through the stern school of experience, but not before much damage was done. While natural water bodies containing very little dissolved oxygen had existed long before the Industrial Revolution and the rise of densely crowded cities, problems caused by organic pollution of natural water bodies were first recognized as serious dangers to human life after the relationships between microorganisms and diseases came to be recognized during the late 19th century. Not long thereafter, the danger of organic pollution to the flora and fauna of streams and rivers came to be recognized, and demands for the improvement of water quality called attention to the quantities and varieties of organic compounds being thoughtlessly dumped into the nearest streams, rivers, and lakes.

It was not long before it was noted that some of the most polluted water bodies contained thriving populations of a few insect species, which had the advantages of inhabiting waters where many other insects that should be able inhabit them were completely excluded by the long-term lack of dissolved oxygen, and occasionally by highly toxic chemicals. How some of these insects managed to survive became very clear after a cursory examination of their aquatic larvae. The survival strategies of others required considerable effort to discover.

Of course, the lack of oxygen in the water is a problem that has to be overcome by any aquatic insect that must exist in water containing none at all, even only temporarily. However, this is not the only problem that they have to overcome. Much more complex problems involve the presence of highly toxic substances in some of the water bodies in which the insect larvae develop. Compounds of sulfur, such as H_2S, are frequently produced when large amounts of organic material, especially proteins, decompose in a small volume of water. H_2S is toxic to most animals, but its strong odor of "rotten eggs" repels most animals from water bodies that contain it. However, some insects have already found ways of protecting themselves from its effects and even seem to feed on some of the bacteria that produce it.

From prehistoric times, anaerobic water containing a variety of toxic substances produced by various microorganisms has existed near human settlements. The water was produced by human activities, including animal husbandry, food processing and preparation, leather making, cloth dying, and many other local arts and crafts. Small water bodies near human habitations were almost always surrounded by small assemblages of mud-like mixtures of water and excrement, rotting food remains, dyes, and substances used for producing items to be used in the village. The larvae of the species of Diptera currently found in the most polluted water undoubtedly altered their habits to utilize some of the wastes produced by human activities for their shelter and nutrition during the past few thousand years. Today, a few of these species are considered pests near almost every human habitation anywhere they might be present in the world, except where extreme climates exclude them.

SECTION 1

Diptera: Family Syrphidae

Apparently, many species of Diptera developed very ingenious adaptations to overcome the problems of developing as larvae on or in polluted, oxygen-free water. Many of them belong to terrestrial families and live in various kinds of waste produced on farms and ranches, some of the waste almost dry and others drenched with water and fluid. Among the most remarkable of these belong to the family Syrphidae, some of which are aquatic as larvae and are not considered pests. The adults of these species often resemble bees and wasps, relying on their mimicry of stinging insects to scare off predators, which would otherwise consume them. They make themselves useful within their biotic communities by pollinating flowers. While the larvae of species in the subfamily Syrphinae are terrestrial and consume herbivorous insects, such as aphids, those in the subfamily Eristalinae are aquatic and live in habitats characterized as being among the most polluted with organic compounds that can be found.

One group of eristaline larvae are known as "rat-tailed larvae" because of the long, narrow tubes arising from the posterior ends of their abdomens. While the larvae remain among the foul-smelling materials suspended in polluted water, the long breathing tubes are extended to the surface, where gases are exchanged directly with the atmosphere.

Among the ditches dug to drain rainwater from orchards near the Elbe Estuary in Germany, those left permanently without water exchange for long periods of time become so heavily contaminated with organic pollutants that they remain completely anaerobic throughout most of the year, except for short periods after particularly heavy rainfall. These habitats were found to be characterized by the development of massive populations of *Chromatium* sp., commonly called purple sulfur bacteria (Caspers and Heckman, 1982). These bacteria can engage in photosynthesis using H_2S rather than H_2O, and they produce elementary sulfur as the waste product instead of O_2.

This would seem to call for investigations to determine whether species of *Eristalis, Eristalinus,* and other eristaline genera have the capability of utilizing the sulfur compounds produced by the photosynthetic sulfur bacteria for their own metabolism. However, because of their access to oxygen through their breathing tubes, such a modification of their diet would not be necessary for them to survive. Although the possibility that the eristaline larvae can utilize elemental sulfur as a substitute for oxygen in their metabolism has never been conclusively demonstrated, this additional source of an oxidizing agent would not be necessary for their survival as long as the long, narrow breathing snorkel, often called the siphon, remains functional. This permits normal respiration utilizing atmospheric oxygen. Therefore, the only other necessary morphological modification to permit their survival in an extremely hostile aquatic environment would be a cuticle impervious enough to prevent a wide variety of toxic substances from passing through from the outside into the body of the larva. Presumably, pesticides and other harmful chemicals are also excluded by the impervious cuticle because obviously healthy larvae were frequently collected from ditches into which a wide variety of agricultural chemicals were sprayed on a regular basis (Caspers and Heckman, 1981).

A second group of eristaline larvae is characterized by short, spine-like breathing tubes, which are inserted into the stems or roots of aquatic plants. The plants inhabiting such polluted water bodies supply their roots with atmospheric oxygen through special plant tissue called aerenchyme, which conducts air into the root system deeply submerged in anaerobic sediments. The syrphid larvae are able to utilize some of this air by inserting their breathing tubes into the aerenchyme. This relieves the larva of the necessity of having any direct contact with the atmosphere through the surface layer of the water. The larva itself can then remain in the anaerobic mud at the bottom of the highly polluted water, keeping it in complete safety from typical aquatic predators.

Like the rat-tailed larvae of the first group of eristaline species, the larvae in the second group, which utilize the aerenchyme to obtain oxygen, must be able to protect themselves from the toxic effects of many different substances present in the most polluted freshwater sediments. Their bodies are protected by a similar cuticle that seems to be impervious to most caustic and weakly acid substances. One aquatic larva was reportedly swallowed by a person and successfully developed while passing through that person's digestive system. Its pupa was collected after it had completed passage through the stomach and intestines, and it metamorphosed into a healthy adult fly (van der Goot, 1981).

Eristaline species piercing the aerenchyme tissues of plants rooted in anaerobic sediments can also be found in anaerobic water colonized by both purple and green sulfur bacteria, which, as already stated, are capable of photosynthesis to produce hydrocarbons starting with CO_2 and H_2S and releasing elemental sulfur as a waste product. Therefore, where oxygen is always unavailable, the purple and green sulfur bacteria displace the typical saprophytic bacteria and substitute sulfur for oxygen in their metabolic activities to break down organic substances. In a milieu so rich in detritus, microorganisms, and simple organic compounds, it certainly would be of value to know exactly which of the food sources the individual syrphid species utilize for their growth and development. As adults, most of the species seem to utilize mainly the sugar-rich nectar from flowers and probably pollen, as well.

In the cases of eristaline larvae, it appears as if the insect does not deal physiologically with the many harmful pollutants it encounters in the water, it would fail to survive. However, the integument surrounding the larvae seems to be impervious to a great variety of potentially toxic substances it encounters in its habitat, so the inner organs of the insect never have to come into contact with the pollutants, at all. The long, thin siphon at the posterior end of its body, which can telescope to become longer or shorter, permits the larvae of some species to feed well below the surface of the water while maintaining continuous contact with the atmosphere. The shorter organ for piercing the aerenchyme tissue of plants permits the larva to descend even deeper into the anaerobic milieu. Aside from particles of food that the larva consumes below that water level, it is separated from its aquatic environment and all its toxic chemicals by its impervious integument.

Many species of Diptera are capable of surviving in water bodies that become anaerobic periodically, either for a few days at a time or during certain periods of individual diurnal cycles. The larvae of many such species contain respiratory pigments capable of storing oxygen for hours or even days at a time. In many eutrophic water bodies, photosynthesis results in the release of enough oxygen to make the water supersaturated during the daytime. At night, the oxygen is quickly consumed, and the dissolved oxygen concentration reaches almost 0% before dawn. In such cases, the respiratory pigments are sufficient to assure the larva a supply of oxygen throughout the diurnal cycle for long periods of time. Whenever the lack of oxygen persists for several days, the larvae must migrate to the surface of the water in order to renew the oxygen supply. The habitat of the syrphid larvae differs fundamentally from that in which the larvae depending upon pigments to store oxygen are found. The difference is that oxygen must be present temporarily long enough for the pigments to renew the oxygen supply held by the pigment at regular intervals. If this is not the case, then gases can only be exchanged with the atmosphere, and only members of the family Syrphidae are known to do this in most waters dominated by the sulfur bacteria. Under the most extreme conditions, the larvae of several families classified in the taxon Cyclorrapha develop in liquid manures and raw sewage, which they immerse themselves in, obviously relying of an impervious integument to keep their bodies free from toxic substances. A few syrphids also seek out this kind of habitat, for which a short siphon that makes direct contact with the atmosphere is sufficient (Table 16.1).

During the study of small standing water bodies in agricultural regions along the Elbe Estuary, several species of eristaline syrphid were very common during the warmer seasons (Caspers and Heckman, 1981, 1982). Many apples and cherries were grown in the orchards that were drained of excess water through many small ditches beneath the trees. At the blind end of such ditches, dead leaves, rotten apples, and other organic detritus accumulated, producing an ideal, oxygen-free habitat for syrphid larvae. Their "rat-tail larvae" were encountered in the most polluted water bodies, in which aquatic predatory insects were excluded by the lack of oxygen. Aquatic grasses, such as *Phragmites australis* (Cavanilles) Trinius and *Glyceria maxima* (Hartman) Holmberg, as well as *Typha latifolia* Linnaeus, *Typha angustifolia* Linnaeus, and species in the family Cyperaceae provide roots containing aerenchyme tissue, into which eristaline syrphids with the "short-tailed" larvae can insert their spike-like breathing tubes. Other larvae with short breathing siphons live in masses of wet detritus, compost heaps, and dung. Only one of the species observed is known from moist soils, somewhat drier than the other habitats. Although no larvae with short breathing tubes were observed during the study, their presence was confirmed by the continual visits by adults to flowers along the shores of the drainage ditches. This

Table 16.1 The aquatic members of the family Syrphidae observed in the most eutrophic ditches draining fruit orchards near Cranz, located in the State of Hamburg, Germany. During the warmer seasons of the year, the water lacked dissolved oxygen except immediately after rain storms, and it usually contained appreciable amounts of H_2S. The water appeared purple due to the presence of many colonies of the purple sulfur bacteria, *Chromatium* sp. (Caspers and Heckman, 1982). The long, telescoping, "rat tail" breathing tube is used by the syrphid larvae inhabiting such water, while those with short breathing tubes usually live in rotting compost or dung, or tap into the aerenchyme tissue of aquatic plant roots. Some information for this table was obtained from publications by Doležil (1972), Arnett (2000), and Miranda et al. (2013).

Species	Tribe	Breathing tube of larvae
Eristalinus sepulcralis (Linnaeus, 1758)	Eristalini	Long, "rat tail" breathing tube
Eristalis intricarius (Linnaeus, 1758)	Eristalini	Long, "rat tail" breathing tube
Eristalis arbustorum (Linnaeus, 1758)	Eristalini	Long, "rat tail" breathing tube
Eristalis pertinax (Scopoli, 1753)	Eristalini	Long, "rat tail" breathing tube
Eristalis nemorum (Linnaeus, 1758)	Eristalini	Long, "rat tail" breathing tube
Eristalis horticola (De Geer, 1776)	Eristalini	Long, "rat tail" breathing tube
Helophilus trivittatus (Fabricius, 1805)	Eristalini	Long, "rat tail" breathing tube
Helophilus hybridus (Loew, 1846)	Eristalini	Long, "rat tail" breathing tube
Helophilus pendulus (Linnaeus, 1758)	Eristalini	Long, "rat tail" breathing tube
Helophilus lineatus (Fabricius, 1787)	Eristalini	Long, "rat tail" breathing tube
Helophilus transfugus (Linnaeus, 1758)	Eristalini	Long, "rat tail" breathing tube
Helophilus interpunctus (Harris, 1776)	Eristalini	Long, "rat tail" breathing tube
Myathropa florea (Linnaeus, 1758)	Eristalini	Long, "rat tail" breathing tube
Chrysogaster macquarti Loew, 1843	Brachyopini	Short, for tapping into aerenchyme
Chrysogaster viduata (Linnaeus, 1758)	Brachyopini	Short, for tapping into aerenchyme
Neoascia podagrica (Fabricius, 1775)	Brachyopini	Short, for extending from soils
Neoascia tenur (Harris, 1780)	Brachyopini	Short, for tapping into aerenchyme
Neoascia meticulosa (Scopoli, 1763)	Brachyopini	Short, for tapping into aerenchyme
Rhingia campestris Meigen, 1822	Rhingiini	Short, larvae in cow dung
Syritta pipiens (Linnaeus, 1758)	Milesiini	Short, in rotting organic matter, dung

attests to the fact that larvae with the short tubes are better able to conceal themselves than those that must establish contact with the air through long, extendable tubes. If plants with aerenchyme tissue are not present in the water body, then only the rat-tailed larvae can survive.

For creating such an ideal habitat for them, the adult syrphids pay back the farmers by pollinating flowers of many of the plants they raise, such as the apple and cherry trees, which require insects for pollination. Adult syrphids usually feed on nectar from the flowers and readily transfer pollen. This service may become vital for continued agriculture of many crops if the disappearance of honey bees continues unabated.

Adults are strong flyers and travel far from water to seek flowers. Most closely resemble honey or bumble bees, but a few appear to be wasps, keeping them safe from many predators, which are fooled by their mimicry. If approached, they often make no attempt to escape, because many predators avoid attempting to capture them out of fear of being stung. When captured in the hand, the syrphids even make buzzing sounds resembling those made by bumble or honey bees when molested. When the flowers are approached by most species of insectivorous bird, they can be observed leaving the syrphids alone, showing how well the mimicry works. The very realistic buzzing sound is sufficient to make many people and most predators immediately release the insect. Of course, to avoid making a mistake, it is only necessary to determine whether one or two pairs of wings are present or just to observe the wing vein pattern of the insect. Such patterns of syrphids are distinctive. Stinging insects always bear two pairs of wings, while those that do not sting bear only one.

a. Syrphids of northern Germany

From research on the flora and fauna in ditches maintained for draining water from orchards in the *Altes Land*, located on the original floodplain of the Elbe Estuary near Hamburg, Germany, the physical, chemical, and biological conditions present in the ditches were characterized by their aquatic communities. They progressed through a sere from what were locally considered to be pristine conditions, shortly after the ditches were constructed or restored, to the terrestrial habitats into which they developed after undergoing a natural process of eutrophication. The process entailed an ongoing accumulation of material produced by plants in and along the ditches, increases in the organic and inorganic plant nutrients in the water, and a continual filling in of the ditches with material washed into them during rainstorms until they cease to be permanent water bodies. If the ditches were not periodically excavated by the farmers, they would no longer function to collect excess rainwater and drain it into the larger canals, which empty into the Elbe Estuary (Caspers and Heckman, 1980, 1982).

The final aquatic stage immediately preceding the complete transition to a terrestrial phase proved to contain oxygen-free water both during the day and at night. The water had the odor of H_2S and contained enough purple sulfur bacteria to be clouded by the colonies of *Chromatium* sp. Its surface was usually completely covered by species of *Lemna*. The predominant emergent plant at the edge of this ditch was the grass, *Glyceria maxima*. Occasionally, species of Collembola ventured out onto the surface of the floating beds of *Lemna* spp., and, at times, caterpillars of *Cataclysta lemnata* (Linnaeus, 1758), a moth in the family Crambidae in their portable silk cases incorporating whole *Lemna minor* plants, were observed moving slowly at the surface.

The conditions in this ditch during the summer were not conducive to the survival of most of the insect species found abundantly in parts of the ditches in which mesotrophic conditions prevailed. During winter days when the temperature dipped somewhat below freezing, an ice layer would develop on the surfaces of the ditches containing oxygen-rich water, but those smelling like H_2S and containing the sulfur bacteria did not freeze over.

During periods of heavy rainfall, the anaerobic water was found to contain measurable amounts of oxygen, which remained detectable for several hours to several days after the rainfall ceased. Otherwise, the water remained anaerobic continuously.

The physical and chemical conditions in such anaerobic water bodies appear to be hostile to most aquatic insects. Other observations confirmed that they actually inhibited colonization of these sections of the ditch system by most aquatic species. However, several species of insects were clearly adapted to such conditions. The species are abundant in northern Germany, and these perform important functions in the terrestrial biotic communities. As mentioned above, prominent among these are members of the family Syrphidae, subfamily Eristalinae. As adults, they feed on nectar from flowers and are increasing in importance with the decline of the honey bee. As already stated, their defense against natural enemies as adults is mimicry of various species of honey bee, bumble bee, or wasp, which the adult flies resemble, both in appearance and in behavior.

It is the larvae of these flies that develop in the anaerobic ditches draining the German fruit orchards. Adults were frequently observed on flowers along the edges of the ditches from late spring through autumn. The following species in the family Syrphidae were observed during the warm parts of the year in a small part of the system of ditches in which anaerobic conditions prevailed during the years from 1978 through 1981 (Table 16.1).

b. How do eristaline larvae survive?

Larvae in the subfamily Eristalinae are obviously suited to survive where few other insects are able to for many reasons. Their integument appears thick and impervious to liquids of all kinds, and this is assumed to provide an important defense against the influence of toxic chemicals of many kinds. However, the nature of this integument has not been investigated sufficiently to provide a cogent explanation of how it protects the insect in an environment that would prove fatal to all insects except for certain other members of the Diptera distributed among many families. Only a few families in the suborder Cyclorrhapha, such as Muscidae and Sarcophagidae, rival the subfamily Eristalinae in tolerance of rotting, putrid wastes. However, their larvae usually develop in habitats not considered aquatic. Only a few of the syrphids, including *Neoascia podagrica,* have larvae that are well suited for living in soils that are not waterlogged or submerged under water.

All eristaline larvae do not develop in water bodies so eutrophic that they are dominated by sulfur bacteria. Typically, they are found in water that is filled with many kinds of detritus and bacteria in the process of breaking it down. Some oxygen might be present at times, but usually the rat-tailed larvae prefer habitats that are anaerobic or nearly so. This eliminates most of their competitors and the predators that might consume them as food. The larvae which breathe by using their short siphons to pierce aerenchyme tissue must develop in soft sediments around the roots of the plants that have this kind of tissue. However, the larvae that successfully develop in the most extreme habitats would be the most interesting subjects of studies.

The first necessary function of the integument would be to separate the internal organs of the larva completely from the outside medium, which contains considerable amounts of H_2S dissolved in the water to form hydrosulfurous acid. An indication that the integument of the larva is actually impervious to this and similar chemicals is supported by the fact that such larvae have passed through the human digestive system without being killed by any of the acid or alkaline secretions of such digestive systems or the digestive enzymes.

The larvae of eristaline syrphids appear plump and cylindrical with a thick integument, tiny eyespots, and minute mouthparts. If the integument is as impervious as it appears to be, there are few places other than the mouthparts through which toxic substances could pass. This would mean that the secret of their ability to survive in habitats that would prove lethal to almost all other insects is their nearly complete isolation from the aquatic medium surrounding them. The only points of chemical exchange between the larva and the outside world are the mouth and the opening of the breathing tube. Through the mouth passes the material that the larva eats, while atmospheric air passes through the breathing tube, either directly from the atmosphere above the water or from air passed down toward the roots of aquatic plants through their aerenchyme tissue.

The alternative to the apparent protection of the internal organs by an impervious integument would be a complex set of biochemical defenses against the H_2S and other acidic, caustic, and toxic substances in the water. Studies to determine the composition of the integument and to find possible enzymes to eliminate various toxic substances will be needed before the ways in which syrphid larvae deal with the extremely hostile environments in which they develop can be determined with certainty.

c. Syrphids of Brazilian Amazonia

To compare the state of entomological progress in the region of the world where the systematic study of insect families began during the 18th century with that in the region which is probably the overall richest in insect species, Table 16.2 is provided showing a recent list of species known from the eastern Amazon Basin in Brazil (Miranda, 2017). The region surveyed encompasses the Brazilian states of Amazonas, Pará, Maranhão, and Roraima, which cover a vast forested area. In addition, recent attempts to introduce agriculture into the region have resulted in destruction of large areas of forest by cutting the large trees and turning the rest and by the construction of large dams. This has produced new kinds of habitat into which insects from similar habitats in other parts of the country can migrate.

The size of this region dwarfs the small agricultural region along the southern shore of the Elbe Estuary, yet the list of species in the subfamily Eristalinae in Amazonia is notably shorter than the number known from the research area in northern Germany. This underscores the matchless opportunities for entomologists working in the Neotropical Region and other parts of the world that have not yet been subjected to intensive surveys of the entomofauna.

Table 16.2 A list of species in the family Syrphidae, subfamily Eristalinae, reported from the Amazon Basin of Brazil in the comprehensive list provided by Miranda (2017). Insects not described or identified to species are omitted. For the details of his survey and a key to the syrphid genera, refer to Miranda's publication.

Species	Author reporting the species
Cepa margarita Thompson, 1999	Thompson (2007)
Meromacris milesia Hull, 1942	Hull (1942), Blatch et al. (2003)
Meromacris pachypus (Wiedemann, 1830)	Thompson (1981)
Nausigaster bonariensis Lynch-Arribalzaga, 1892	Carrera et al. (1947)
Palpada aemula (Williston, 1891)	Mengual and Thompson (2008)
Palpada agrorum (Fabricius, 1787)	Lagrange (1992)
Palpada fasciata (Wiedemann, 1819)	Morales and Marinoni (2009)
Polybiomyia bigotii (Williston, 1888)	Curran (1941)
Quichuana longicauda Ricarte and Hancock, 2012	Ricarte et al. (2012)

Ecological reports about the species listed in Table 16.2 are also scarce, and almost nothing has been reported about the habitat preferences and diets of the larvae of most species. Even in Europe, where the number of species present is presumably not as great as in Amazonia and where the research on the habits of these insects started more than 100 years earlier, there are gaps in the knowledge about many of the species. This is a further incentive for a diligent entomologist to focus on the Neotropics as a frontier of the natural sciences.

SECTION 2

Diptera: Other Families

Other insects survive in temporarily to semi-permanently anaerobic water in other ways than storing oxygen, which is bound to respiratory pigments. Most of these belong to families encompassing species with a variety of developmental histories, mainly in terrestrial habitats. A few species, however, live in water bodies containing a considerable amount of organic detritus and are often poor in oxygen. There is little information about the respiration of most of them. Many of the species of dipterans from different parts of the world are known only from the imago, and it is not known where and how their larvae develop.

In almost every case, the larvae of such midges and flies breathe in direct contact with the atmosphere. Because almost all have no more than a short extension of the abdomen through which to take in air and release waste gases, they typically remain among detritus floating on the top of the water, relying on their color and resemblance to odd-shaped fragments of leaves and branches to escape the attention of predators.

Although the number of aquatic species with larvae that develop in oxygen poor water is relatively low, they belong to a fairly large number of Diptera families. A summary of the families which include such species is provided in Table 16.3.

Among the families listed in Table 16.3, three in the subfamily Nematocera encompass predominantly aquatic larvae: Psychodidae, Culicidae, and Dixidae. Those in the family Psychodidae are the most typical of those that develop in moderately

Table 16.3 Families of Diptera encountered in heavily polluted water, mainly in the Holarctic Biogeographical Region. Except as noted, the great majority of species in these families are terrestrial.

Family	Habitat of most species	Habitat in polluted water
Subfamily Nematocera		
Tipulidae	Soils	Waterlogged soils, littoral
Limoniidae	Semi-aquatic, marshes	Submerged soils and plants
Psychodidae	Wastewater, sewage	Hypertrophic water, sewage
Culicidae	Water with dissolved oxygen	Heavily polluted water
Dixidae	At the surface of marshes	Marshes occasionally anaerobic
Subfamily Brachycera		
Ephydridae	Extremely moist habitats	Water with much detritus
Sphaeroceridae	Wetlands and littoral	Water with much detritus
Empididae	Terrestrial soils	Surface of forest pools
Tabanidae	Small water bodies	Wastewater from farms
Subfamily Cyclorrapha		
Muscidae	Dead animals and fecal matter	Heavily polluted wastewater

to heavily polluted water. The mosquito larvae in the family Culicidae are usually found in cleaner water, but with sufficient bacteria and other microscopic organisms to supply them with sufficient nutrition. Relatively few prefer heavily polluted water, even though they breath using atmospheric air obtained at the surface. Larvae in the family Dixidae are usually encountered in marshes with emergent vegetation, often with much detritus and a film of bacteria at the surface. The oxygen concentration often drops to near zero during the night. The larvae orient their bodies horizontally beneath the surface tension layer, and assume a typical U-shape, which makes them immediately identifiable as dixids.

The family Chironomidae were not included in the table because none are known to live in water that is anaerobic over the long term. There is also an absolute minimum oxygen concentration that the larvae of chironomids can endure because their larvae must exchange oxygen with the water rather than utilize the oxygen in the atmosphere, as the families listed in Table 16.3 can. They are also omitted because excessive amounts of pollutants will eliminate all of the known species. Even though other chironomid species have adapted to some extreme habitats for insects, including the ocean, their capacity for obtaining oxygen through their larval system of gill filaments has limits that other families of Diptera have overcome by keeping contact with the atmosphere. It would still not be surprising if a chironomid species were discovered living in anaerobic water somewhere in the world using a novel method of obtaining oxygen, but that is only speculation.

Those insects that developed as larvae in the sections of the ditches that were studied in the fruit orchards along the freshwater section of the Elbe Estuary, which were in the final transition to terrestrial habitats, had to be adapted to anaerobic water containing a particularly low number of insect species. Larvae of Diptera species belonging to several other families have the capability of surviving and even thriving in anaerobic waters, but most of them are known to keep some physical contact with the atmosphere at the surface. Larvae of species in the family Ephydridae are among the most adaptable species to water containing a wide variety of unusual pollutants. Larvae of aquatic species in the families Tipulidae and Limoniidae generally live in detritus-rich soils, muds, and water bodies. The aquatic species seem to depend upon the atmosphere for gas exchange and limit themselves to accumulations of detritus in shallow water so that they can place their short breathing tubes directly in contact with the surface tension layer.

Ditches and ponds containing water that remains anaerobic for relatively long periods of time are suitable habitats for a few species in the family Ephydridae. These species are not as exquisitely specialized for living in water free of oxygen and containing substances that could be toxic to most insects. However, a few species of adult insects were often observed on the surface tension layer of water that was usually free or nearly free of dissolved oxygen. The larvae of these small flies lack long breathing tubes. They are sometimes found among detritus on the underside of the surface tension layer. This keeps them in contact with the atmosphere just above the water.

The most abundant ephydrid inhabiting the ditches was *Hilara manicata* Meigen, 1822, which was observed at times on both the anaerobic ditches and those with water containing sufficient oxygen to support many species of aquatic insect.

Species in the family Ephydridae are not only tolerant of water containing little dissolved oxygen, certain species in that family are also able to live on the surface of water containing high salt concentrations, large amounts of organic detritus, crude oil, and other kinds of unusual chemical contamination. Some of these species will be discussed in subsequent chapters.

Many members of the suborder Cyclorrapha show great tolerance for muds and liquid manures with consistencies ranging from liquid wastes to waterlogged soils, such as those frequently encountered in places used for the intensive raising of farm animals. Others develop in the carcasses of dead animals. Their larvae are often encountered in large numbers and are not selective for suitable substrate according to the amount of water it contains. Although they live in liquefied substrates, they are not usually classified as aquatic. Little is known about the respiration of the larvae of many species, although they have been observed to be very tolerant of what appears to be extremely polluted water. The tolerance seems to be considerably different from species to species. Those in the family Muscidae, for example, may survive in liquid manures, while related species avoid such substrates. Those that are capable of surviving in anaerobic waters heavily contaminated by manures have very active larvae, suggesting that they may obtain oxygen by frequently coming into contact with the atmosphere to renew their oxygen supplies.

As a general rule, it can be said that insect larvae that exchange gases with the water through gills are most affected by shortages of oxygen. Those species that can exchange gases directly with the atmosphere, on the other hand, are seldom limited by problems caused by limitations of access to oxygen. However, those that need to keep contact with the atmosphere in order to breath are limited in their movements into a water body. They are usually prevented from going too deep into the water and from living in submerged habitats with impediments keeping them from access to the surface. To become an aquatic insect, a species either gains complete access to the aquatic environment by eliminated all need for coming into contact with the atmosphere, or it can venture only so far into the water that it can still reach the atmosphere before its air supply is depleted.

17

Insects on the High Seas

Although aquatic insects make up a relatively small portion of the hexapod fauna throughout the world, many of the aquatic species are generally known, even to the casual observer. The oceans and other saline water bodies, however, are not generally known to support insects. In fact, to many people, the seashores constitute geographical limits to the distribution of members of the classes Collembola and Insecta. The exceptions to this rule are poorly known, but their habits are extremely interesting.

The best known of the saline water species are seen occasionally in large numbers on the surface tension layers of inland saline pools, sheltered bays, and mangrove swamps. These will be discussed in the next chapter. What few entomologists have observed are the insects inhabiting the surface tension layers on the high seas, often more than a thousand miles from the nearest land. These pelagic water striders have been subjected to few systematic studies. From those observations reported, however, it is known that the pelagic species can form relatively dense populations, which move about on the surface of the open oceans. According to Ikawa et al. (1998), only five of the more than a million insect species already described inhabit the pelagic zones of the oceans.

A somewhat similar colonization pattern by insects invading the sea is shown by marine species of midges in the dipteran family Chironomidae. Members of this family appear to be poor candidates for invading the oceans, with their turbulence and exposure to severe weather. However, they are hardier than they appear at first glance, and they require no food as adults. Most marine species have been found in shallow coastal waters, where their larvae develop. However, one species is known to be pelagic and almost cosmopolitan, apparently absent only from the colder parts of the oceans, as discussed in the second part of this chapter, based on the recent studies of Huang and Cheng (2012) and Huang et al. (2014).

SECTION 1

The Pelagic Water Striders

a. The genus Halobates

The species of truly pelagic water striders belong to the genus *Halobates* in the family Gerridae. Five species in this genus are presently recognized: *Halobates micans* Eschscholtz, 1822, from the Atlantic Ocean and Gulf of Mexico; *Halobates sericeus* Eschscholtz, 1822, from the tropical Pacific Ocean; *Halobates germanus* White, 1883, from the tropical Indian and western Pacific Ocean; *Halobates sobrinus* White, 1883, from the tropical eastern Pacific off the coasts of North, Central, and the northern part of South America; *Halobates splendens* Witliczal, 1886, from the tropical eastern Pacific off the coast of South America.

Much is still to be learned about the habits of these marine water striders, especially about the ways they are able to maintain their contact with the atmosphere during periods of stormy weather and rough seas. None of these species have functional wings as adults, so they cannot take to the air to seek calmer seas. The parts of the oceans they inhabit may provide some clues to the apparent success of these insects. Seas close to the equator are noted for being becalmed for long periods, which would minimize the turbulence encountered at the surface of the water. To the north and south of the calm equatorial zone are belts of steady easterly winds, often called trade winds, a name left over from the time of sailing ships. Where such winds prevail, the seas become extremely turbulent only when tropical storms develop.

Ocean currents also play an important part in creating a generally stable environment for the pelagic species of water strider. There are pelagic regions throughout the world in which the surface currents move along large circular paths surrounding areas in which little water movement occurs. The calm waters in the middle of these roughly circular systems of surface currents typically accumulate large amounts of flotsam, floating material of both natural and artificial origin which decomposes slowly. It may also be the habitat of floating species of marine algae, such as *Sargossum* spp., from which the Sargosso Sea near the middle of the Atlantic Ocean takes its name.

Oceanographic factors are therefore of primary importance in shaping the habitats in which the pelagic *Halobates* species spend their lives. They must also provide a habitat for marine invertebrates that form the food supply in the same areas where the insects congregate. However, little has been learned about the food supplies utilized by the pelagic water striders. It is obvious, however, that these marine gerrids rely on prey consisting of arthropods other than insects or other kinds of marine animals, upon which no other group of insects is known to subsist.

In addition to studies of the nutrition of the *Halobates* species, research will be required to determine their patterns of seasonal activity, reproduction, and migration, both passive and active.

As in the case of freshwater members of the family Gerridae, the survival of the pelagic water striders depends fundamentally on special morphological modifications permitting the insects to support themselves and move about on the surface tension of the water. The presence of relatively high concentrations of various salts in seawater produces differences in the physical nature of the surface tension layers of fresh and salt water. An obvious feature of *Halobates* morphology is the relative shortness of the legs. Species of *Gerris* and *Limnogonus*, which remain on the smooth surfaces of calm, freshwater ponds typically have much longer middle and hind legs, which spread the weight of the insect over fairly large areas of the surface tension layer. It can be assumed that on a choppy surface of the sea, the bodies of the insects would be forced into contact with the water or even briefly submerged more frequently than the freshwater species of their family. This would require the tiny setae providing the "waterproofing" of the surfaces of the abdomen on which the spiracles are located to be more effective in preventing the water from blocking gas from entering the trachea than those on most freshwater gerrids.

Studies of the integument covering the thoracic and abdominal surfaces of *Halobates* species will be required to elucidate the mechanisms protecting their spiracles. Electron micrographs of the surfaces of freshwater gerrid species have revealed that a double layer of setae protects the integument from contact with the water (Heckman, 1983). The outer layer of setae is rather sparse and consists of relatively long setae, which are often slightly curved near their apices. This is sufficient to prevent water from reaching the main protective layer of setae and probably to inform the insect that its body is about to come into contact with the water.

The setae in the inner layer are short and fine. The setae in this layer are much denser than those in the outer layer. When forced against the water surface, this layer can retain a layer of air sufficient to keep the integument below it completely dry, up to a point. If the insect is forced under water, it can remain there for a while, usually long enough to permit the insect to survive until whatever danger that caused the submergence has passed. However, it does not provide long-term protection from drowning like the "plastron" of heteropterans that live beneath the surface does. Once the inner layer of setae becomes saturated with water, the insect usually drowns (Heckman, 1983). A few observations on *Halobates* forced below the surface of the water indicated that they also drown in the same manner (Herring, 1961).

Halobates species would very likely have difficulty surviving stormy periods on the surface of the ocean if they did not have a more impervious layer of setae protecting the integument than that on freshwater gerrids. Research on the layers of setae protecting the pelagic gerrids from becoming wet would therefore be welcome.

Another question that should be answered concerns the method by which *Halobates* species obtain their fresh water. Marine vertebrates often accomplish this by ingesting seawater and excreting concentrated salt solutions to eliminate the excess salts. Other possibilities include the intake of rainwater as it falls on the surface of the ocean or obtaining fluids from foods containing body fluids of lower salinity.

Finally, how the *Halobates* species obtain their nourishment must certainly differ fundamentally from how other gerrids feed. Freshwater species capture and feed primarily on freshwater insects, usually captured alive or found shortly after death on the surface of the water. The actual ingestion of nutrients occurs after external digestion by enzymes injected by the gerrid into the prey or produced by bacteria. The facts that pelagic gerrids would not be expected to encounter any other insect species at all and that the materials ingested would be predigested and contain no identifiable chitinous material precludes conventional studies of their diets. Under such circumstances, direct observation of the feeding of these species in the field would provide the most reliable results. However, transportation to the habitats of these insects in the middle of major oceans constitutes an impediment to these studies. Grant applications for more conventional studies of marine life generally could be expected to receive priority.

Those few observations of *Halobates* species feeding in the ocean include those reported by Herring (1961) near Hawaii. He reported observing a pelagic *Halobates* species feeding on species of Cnidaria. This constitutes a surprising addition to the foods known to be acceptable to water striders. It may be that enzymes found in no other members of the family Gerridae are produced by these species, or perhaps the material constituting the jelly of the cnidarians is easily reduced to a fluid mass by the enzymes already among those produced by most species of water strider.

A final question concerning the biochemical capabilities of *Halobates* species is whether or not they are capable of producing the noxious chemicals similar to those produced by other aquatic heteropterans, and if so, are these substances as effective in repelling marine predators as they are in making the water striders unpalatable to freshwater fishes.

Halobates species on the high seas reportedly complete five larval stages prior to their metamorphosis to adults. The populations of *Halobates* on the open Atlantic and Pacific Oceans are not well known, but in waters in the regions of the Banda Sea off Indonesia and among the islands of Micronesia, each larval stage requires 7 to 14 days to complete, depending upon the temperature (Chang, 1981; Chang et al., 1990).

b. Occupying the Niche

Life on the surface of the open ocean obviously requires a large number of morphological, physiological, chemical, behavioral, and ecological modifications of basic insect biology. The difficulties insects have had in colonizing the marine pelagic habitat is reflected by the fact that so few species have been able to do so.

The family Gerridae is characterized by a large number of extreme modifications just to permit many of its species to elegantly pass their lives on the surface of calm, freshwater habitats. These freshwater insects have two pairs of legs that can support the weight of the insect on the surface tension of the water and propel the insect rapidly forward by one or a series of rapid springs. They also have a pair of fore-legs modified for capturing prey, a coat of setae to hinder contact between the surface of the abdomen and the water, a dense coat of shorter setae that bind a layer of air to the abdominal surface in case the insect is forced below the surface, and a sensory system capable of detecting vibrations passing across the surface tension layer for detecting prey that fall into the water and apparently also for receiving "messages" from conspecific insects. The eyes seem to be effective for detecting and avoiding potential predators and helping the insect navigate to struggling terrestrial insects that fall into the water. The mouthparts are modified for stabbing such insects and injecting venom and digestive juices so that the prey can be liquefied and then drawn into the digestive system of the water strider. Like most other species of Heteroptera, water striders inhabiting ponds and lakes have chemical repellants against predators and the ability to stab and inject venom into large animals that attempt to capture them.

Pelagic water striders have different requirements for surviving in their respective geographic niches. Unfortunately, detailed studies of the adaptations the insects have for their habitat have not been completed, and their habits and biological adaptations remain to be elucidated. The surface tension layer of seawater has different properties than that on fresh water bodies, and the salinity of the ocean is usually greater than that of the body fluids of the terrestrial and freshwater animals inhabiting it. For this reason, many researchers have speculated that the salinity of sea water was significantly lower when life first appeared in the oceans than it is today.

Although the salinity of seawater has obviously proven a formidable obstacle to its colonization by insects, far more marine species are known from shallow coastal waters than from the open ocean. Obviously, exposure to storms and dietary factors would explain why so few insects can survive in the pelagic regions of the world's oceans. However, less obvious problems could also contribute to excluding insects from life on the high seas. One of these is navigation by the insects to unite with breeding populations of the same species on vast areas of open ocean. Another is finding the locations of significant populations of suitable prey.

In addition, the turbulence on the open ocean would interfere with the kinds of movement typical of freshwater gerrids. The relative lengths of the middle and hind legs of *Halobates* species are quite different from those of most other gerrids, indicating that the modes of locomotion are different. Unfortunately, little is known about the differences in the movements themselves. Significant modifications in the movements would be expected so that the insects could navigate in choppy seas and avoid being forced under the surface of the water by turbulence.

It would not be expected that vibrations travelling through the surface tension layer would facilitate detection of prey or communication with other insects. Information about the capture and digestion of prey is still unavailable, but it is certain that the supply of terrestrial insects blown from distant islands onto the surface of the sea is insufficient to nourish a large population of a pelagic *Halobates* species. This would tend to support the hypothesis that marine animals, such as cnidarians, form the basis for the nutrition of the pelagic water striders.

The greatest open questions about the survival of *Halobates* species on the open ocean concern the methods of protecting the respiratory system from the input of water forced into it by turbulence. In addition, it is still not known whether the insects navigate the seas in some seasonal pattern or are drifted together by surface ocean currents. Needless to say, almost nothing is known about the capabilities of the insects to digest various kinds of food they encounter in the ocean. The role of vision in hunting prey and escaping predators is unknown. If the insects feed predominantly on marine invertebrates just below the surface of the water, as it seems they must, then vision above the surface could hardly be of much help in finding food. How do they find suitable prey?

Because so little is known about the feeding habits of pelagic water striders, it is not even possible to say whether or not all of the species and the populations in different oceans have the same diets. Seasonal differences and specific preferences for food remain unknown. Because most pelagic species are confined to one of the oceans, it seems likely that there are significant differences in the diets, and seasonal differences in the composition of the plankton would also seem to suggest that seasonal diet changes would be necessary to assure the survival of each species. However, that still remains speculation.

SECTION 2

The Brachypterous Midge on the High Seas

a. The taxonomy of and distribution of Pontomyia

Midges assigned to the genus *Pontomyia* belong to the Diptera and clearly belong to the family Chironomidae, a family long known to encompass marine species. However, they differ from the other known marine species in that family by inhabiting the open ocean, far from land. They are remarkable for the short life spans of the adults, their brachypterous wings being used more as oars than for flying, and their world-wide distribution (Fig. 17.1).

Recent studies employing comparisons of DNA markers and morphological traits have shown that one of the pelagic species, *Pontomyia natans* Edwards, 1926, is conspecific with *Pontomyia cottoni* Womersly, 1937, and with a previously unidentified larva found in coastal waters near Puerto Rico. After the revision of the genus by Huang et al. (2014), it still encompassed two other nominal species, formerly considered to have different ranges in the Pacific Ocean: *Pontomyia pacifica* Tokunaga, 1932, from sites ranging from the Pacific Ocean near Japan as far south as the waters near Australia, and *Pontomyia oceana* Tokunaga, 1964, from coastal tidal pools near Palau and Taiwan (Soong et al., 1999).

Postponing speculation on the statuses of *Pontomyia pacifica* and *P. oceana* until more information becomes available, we are left with a single species of a fully marine, pelagic midge, *Pontomyia natans*, which is found in parts of all three major oceans of the world, the Atlantic, Pacific, and Indian, which remain unfrozen throughout the year. This means that *P. natans* is a species sometimes found in the pelagic zone of all three major oceans located in parts of the world within the tropical, sub-tropical, and

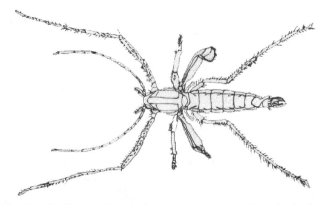

Figure 17.1 The short-lived imago of a male *Pontomyia natans,* a species in the family Chironomidae, which lives entirely in the pelagic regions of the Atlantic, Pacific, and Indian Oceans. The male does not fly but rather skims across the surface of the ocean seeking a female, which is wingless. The figure was redrawn and modified using the example provided by Huang et al. (2014).

temperate climatic zones. The almost cosmopolitan distribution cannot be attributed to the migration of the adults for reasons that are clear from its activities during the imago stage. Ocean currents, wind, and weather have the capability of transporting the adults and larvae far more quickly than the vestigial wings of the adults are capable of doing.

Because they are pelagic, there have been relatively few reports by observers of their reproductive behavior. Nevertheless, from those few reports of its activities, we know that the species is remarkable for several reasons. The males have wings considered to be brachypterous, although enough of them are present to propel the males rapidly over the surface of the sea. They appear only to have been distorted due to contact with seawater while unfolding during metamorphosis. Females lack wings, so the males propel themselves with the vestiges of their own wings rapidly across the surface of the water to find them. Both males and females have extremely short life spans, during which mating and spawning occur rapidly. The search for a female by a male can continue for up to about three hours before the male dies. The female resembles a worm, which does not move around. As soon as it has mated, it lays its eggs and then dies. It has been suggested that these are the imagoes with the shortest life expectancies of any insect species in the world.

An especially remarkable accomplishment of this tiny midge is that with little more than vestiges of wings in the male and no wings on the female, it has become the first insect to colonize the warmer parts of the entire World Ocean. It has reached both the pelagic and neritic zones of the oceans, apparently more by passive than active migration, and it has apparently been able to find suitable food everywhere it occurs.

In contrast to the extremely short life of the imagoes, the larvae take about 45 days to develop, and it has been speculated that they can be carried to other parts of the ocean during that time. It has been suggested that they can attach to the shells of sea turtles or the hulls of ships to reach the distant parts of the oceans in which they are known to occur (Huang et al., 2014). However, this is not necessary because the ocean currents alone could account for its long migrations.

b. Occupying the niche

The fact that *Pontomyia natans* is the only member of the Diptera and one of only three nominal species of insect known to have mastered life beneath the surface on the high seas demonstrates that insects have extremely great difficulties to overcome before they can take over an ecological niche in the open ocean. Obviously, life in the ocean close to the shore is not quite so difficult, as a considerable number of species have established niches there. While it has been suggested that this insect has established colonies in the three major ocean systems of the world by riding on sea turtles (Schärer and Epler, 2007), this would be difficult to prove. In any case, they have obviously populated remote parts of the world already, making it even more difficult to prove whether their world-wide migration was actually accomplished with the help of sea turtles or that they could have travelled using one of many means of transportation available to the midges. Certainly, development of the larvae among the aufwuchs on the hulls of ships could also account for the distribution of the species, and there are many more ships of all sizes plying the seas today than there are sea turtles.

Once a group of midges had succeeded in traversing several thousand kilometers of ocean to reach a new habitat in another part of the world, it would have already demonstrated that it could survive in a seawater medium with all of the disadvantages to insects that have kept almost all other species out. It would also have demonstrated that it could vary its diet sufficiently to survive in water with a continually changing menu of foods. Finally, it would have shown that it had defenses sufficient to keep from being devoured by any of the many small marine predators that abound in the oceans. The facts that its adults are short-lived and do not consume food would suggest that passive migration during the larval stage followed by one or two days of intensive reproduction could explain the appearance of the species in new locations throughout the world.

The other requirements for occupying a pelagic niche are the general ones that all insects must overcome to live in the sea, including adjustment of the salt content of the body fluids, finding of one or more reliable food sources and learning how to exploit them in the habitat, developing a way of avoiding or warding off natural enemies, and establishing a reproductive pattern to assure that mating is regulated temporally and gives the species a reliable way to propagate itself without losing too many of the new generation.

There are relatively many species that have adopted strategies to utilize special morphological and physiological modifications for surviving along parts of the sea bordering on continents or islands. The fact that so few of them have been able to establish populations far from seacoasts strongly suggests that breaking all ties with the land presents a barrier in itself that is not very easy to overcome. Simple physiological processes for things like adjusting the salt content of the body, avoiding the harmful effects of ultraviolet light, regulating the body temperature without the ability to produce body heat, and similar necessities for life in the sea have all been developed by the marine insects that live near shore. Once out of sight of all land, however, certain aids for insect survival are lost.

First, entering the pelagic zone removes the insects from fixed locations that can give them orientation in space and time needed to coordinate the beginning and end of mutual activities, especially reproduction, that the individual insects must participate in to achieve maximal success in seeing to it that their population survives. For example, coordination of the time and place of metamorphosis to the adult stage, mating, and spawning are usually required so that all members of one generation of the population can complete these activities at the same time. This will be explained in the next chapter in the discussion of the chironomids living in shallow coastal waters of Europe. It is sufficient to say that coordinating a simultaneous emergence of the imagoes of a single population is highly complex but necessary to assure the survival of an insect species from one generation to the next.

First, the emergence will take place on a specific date, usually when a predetermined phase of the moon is reached. Second, for a species surviving as an adult for only a few hours, the time of day of the metamorphosis must be scheduled according to the position of the sun. Finally, the weather shortly before the scheduled time for metamorphosis must be determined so that the entire population has time to postpone its emergence as adults, usually until the same phase of the moon, one month later. A single thunderstorm during the emergence could easily spell doom for an entire population.

Just how the complex and carefully calculated planning for the emergence is done and communicated to the whole population defies the imagination. Whatever is done must be based on inherited processes, however, and all members of the population must be able to perform the same complex calculations if the reproduction is to be successful. Acquired knowledge cannot be passed on because the parents die before their eggs have hatched. It is not easy to understand how tiny insects which will only survive in the adult stage for no more than a few hours and all correctly calculate the exact day and time that they should undergo metamorphosis and to cancel the plans at the last moment if a storm is approaching. The alternative to making such calculations is for the males to emerge at different times and find no females to mate with, and for the females to emerge and fail to find a mate.

Another unanswered question is how the males find the females. Male freshwater chironomids are known to use pheromones released by the females to start their search for them. Their feathery antennae can filter and sense the tiniest amounts of these chemicals and lead the male to the location of an unmated female. But what about the marine species? Does the presence of seawater droplets in the air inactivate such pheromones? If pheromones are not used, then how does the male find the wingless female in the middle of a vast ocean? These and more questions must be answered before we can begin to understand just how, against all odds, these few species have managed to find a home in the middle of vast oceans.

18

Marine Littoral Hexapods

The number of hexapod species known from marine waters near land is considerably greater than the very few known pelagic species. One reason for this is obviously the sheltering effect of the nearby land, which greatly reduces stress from the turbulence of storms and rough seas. There are also sources of freshwater in most neritic habitats, rich communities of benthic and periphytic microorganisms, and easy access to terrestrial and freshwater insects belonging to most major orders, which can serve as suitable prey for the littoral predators. These insects are the kind of prey most frequently captured by aquatic predators, and they serve suitable littoral predators with food, just as well. Abundant terrestrial insects from adjacent habitats fall onto the water that moves back and forth across the beaches, die, and provide food for aquatic scavengers. Even herbivorous insects are usually able to find plants near the water's edge sufficient to provide them with suitable foods, while others are capable of subsisting on seaweeds washed up on the beaches. Among the dying seaweeds at the high tide mark of the day, there are often many individuals belonging to tiny insect species in the classes Collembola and Insecta.

The littoral zone is not a single habitat but rather a series of distinct habitats forming distinctive zones roughly parallel to the line separating the ocean from the land. The exact location of the line separating sea from land is constantly moving back and forth with the tides. Before we examine the hexapod species inhabiting these zones, we should make a very general classification of the distinctive habitats.

The habitats are located in distinctive kinds of coastline, the most common of which are sandy beaches, coasts with cliffs or large rocks, and salt marshes. Along coastlines, there are also river mouths forming estuaries wherever tidal influence of the water level can be detected. In the estuaries, there is an ecotone from freshwater flowing seaward out of the river to salt water with the full salinity of seawater. The estuary may be limited to a relatively small area at the river mouth, or it can be wide and long, extending many kilometers inland from the mouth.

Moving landward from the sea, the first littoral habitat encountered is located in the sublittoral zone, which is relatively shallow seawater with a benthic community that is rarely or never exposed directly to the atmosphere. At low tide, the water may be very shallow during spring tides, but its benthic flora and fauna are always protected from desiccation. The benthos may be dominated by animals that burrow into sand, live among large rocks, or live on extended muddy ground. In stretches of the coast relatively far from outlets of estuaries, the water has the full salinity of seawater. Only heavy rain showers can temporarily dilute the seawater. At the mouths of large estuaries, diluted seawater can be detected far seaward from the coast, especially during ebb tide.

Another kind of habitat encountered along almost all seacoasts throughout the world is the intertidal zone. The habitat varies according to the nature of the topography of the land bordering on the sea. Rocky seacoasts, sandy beaches, extended mudflats, and river deltas all have their distinctive biotic communities, most harboring at least a few insect species. Only parts of the coast along the Arctic Ocean and much of the coastline of Antarctica do not have an identifiable intertidal habitat because a layer of ice separates the water from the land for the entire year.

The Intertidal Zone is a region of continually changing transition between sea and land. Its lowest lying parts are exposed to the atmosphere only briefly during each tidal cycle. Its uppermost margin is the highest location reached briefly by the seawater during any tidal cycle, except during storm tides. It is generally marked by a long line of marine detritus washed up during high tide. It borders on a terrestrial habitat that is influenced directly by the sea only during storm tides. As a general rule, the terrestrial hexapod species that live on the landward side of the seacoasts are distinctive for these kinds of habitat. They will not be discussed in detail in this book.

There are specific kinds of biotic community along seashores, which are noted for their insect populations. Mangrove swamps are among the seawater habitats richest in insect species. This is due to their locations in tropical and subtropical regions and the variety of food sources found in them. Other seawater habitats supporting communities comprising fewer insect species include beaches in the temperate zones, where decomposing marine algae accumulate along the high water line. As the material from the sea rots, it provides a rich source of nutrients and microorganisms, which meet the demands of insect larvae, as long as they can survive great variations in salinity, periodic dryness, and high or low temperatures, according to the season.

Clearly, there is plenty of food, as well as several mild microclimates to choose from according to their positions in the intertidal and subtidal zones. There is also a minimum number of predatory species that can feed on the insects. These factors combine to create an ideal habitat for insects with suitable adaptations, accounting for the vast populations of a few insect species that have succeeded in colonizing one of the neritic or intertidal habitats that exist along the world's seacoasts.

It can be surmised that the only factor limiting the survival of most insect species in coastal habitats is the overabundance of salts, especially NaCl, sodium chloride. It is the tolerance to salinity that permits all marine species to survive in the ocean, in spite of the fact that the body fluids of many vertebrates and possibly certain other organisms belonging to other higher taxa must be maintained physiologically at a lower salinity than that of undiluted seawater.

In some marine invertebrates, the seawater is simply ingested and incorporated into the body fluids as it is. Other species are able to concentrate and eliminate some of the salt to keep the salinity inside of the organisms at a lower level than that of seawater in the open ocean. However, in coastal waters, a whole series of complications are encountered, making physiological studies of neritic species unusually difficult. The greatest complications include the variability of the salinity in neritic waters, which can range from that of freshwater to almost full salinity within estuaries, and which can range from full salinity to much higher than that found in the open ocean in shallow tidal pools and man-made salt pans along the coasts. Organisms living in the intertidal zones are also subject to dehydration, which can concentrate the salt in the body to the point where some of the organs can no longer function. Added to this is the problem of extremely high temperatures in shallow tidal pools during sunny summer days and freezing temperatures in such pools during low tide on winter days.

To date, only a few of the physiological adaptations to subtidal and intertidal habitats by insects have been subjected to detailed studies. In this chapter, the spectrum of habitats in coastal waters that have been colonized by insects are briefly reviewed, and some of the unanswered questions about how the insects have managed to adapt to a highly stressful environment are enumerated. Undoubtedly, research to answer these questions will provide some surprising results and will also raise some new questions. Adaptations to hypersaline water will be discussed further in Chapter 19.

A general entomological inventory of the most familiar kinds of coastal habitat is provided in this chapter. Rare or unusual habitats are discussed in separate chapters.

SECTION 1

Insects of Mangrove Swamps

Mangroves are encountered along tropical and subtropical seashores, where they form intertidal forests. These marine littoral habitats add considerably to the stability of the coastline and protect breeding areas for various species of fishes and invertebrates. Among the best known insect species in such habitats are marine water striders in the family Gerridae. In his report on the genus *Halobates,* Herring (1961) distinguished between pelagic species, which are found near shores only after heavy storms, and neritic species, which are frequently encountered near the shore and seldom, if ever, found on the open seas. The pelagic species were already discussed in Chapter 17.

Mangroves are trees especially adapted to thrive on tropical mudflats along seacoasts where other trees and nearly all other tracheophyte plants cannot survive. Sodium chloride acts as a poison to most aquatic and terrestrial plants. In addition, the roots of most trees are suffocated if submerged continually in water. Furthermore, seawater contains a variety of other salts, which exhibit various degrees of toxicity to plants. As if these dangers to the existence of the trees were not enough, the soft consistency of the sediment on flooded coastal mudflats does not give the roots of most trees sufficient support to prevent them from being blown over during strong windstorms.

The mangroves are superbly adapted to survive on coastal mudflats, and mangrove swamps were once common in the tropics throughout the world. These characteristic habitats acted to stabilize parts of the seacoast along which soft, fine-grained sediments were deposited. The mangroves can tolerate high salt concentrations. To prevent the roots from suffocating, they have specially modified roots that grow upward and extend above the surface of the water, even during high tide. They conduct air directly into the submerged roots.

To increase their hold in the soft sediments, mangroves are able to grow supplemental trunks from above downward into the sediment. The trunks of most are also able to grow moderately large, supplemental roots that grow downward diagonally from points on the trunks above the usual high water level of the swamp. These help to prop up the tree trunks during periods of high wind or rough seas. Eventually, the spreading roots and new trunks form an anastomosing network, which gives the trees a firm hold in sediment spreading over a large area. As the mangrove swamp develops, it increases its hold on ever growing areas of the sediment. This protects the trees from being dislodged by typhoons or tsunamis. Obviously, cutting away some of the tree trunks or otherwise damaging some of their root system has the effect of weakening the whole stand of trees. Hence, removing some of the trees can easily cause the loss of a larger area the mangrove swamp than intended. Some experts fear that removal of mangroves can result in a rapid erosion of the seacoast. Therefore, before a mangrove swamp can regenerate from careless cutting of the trees, an entire section of the coastline may be eroded away. After the damage to coastal areas of Thailand caused by removal of mangroves became evident, that country launched an effort to replant the trees in order to re-stabilize the coast (Fig. 18.1).

Figure 18.1 A small plantation of trees and other plants typical of a Southeast Asian mangrove swamp along the shore of the Chao Phrya Estuary in Bangkok. These attempts were made after it was realized that erosion increases greatly when mangroves are removed. The photo was taken at high tide.

Surveys of the tree species that form the mangrove swamps in various parts of the tropics are typically limited to relatively small geographical regions. A study of the mangroves in the estuaries along the West Coast of India near the city of Goa yielded the following list of plant species growing in the coastal forests (Untawale et al., 1982): *Avicennia officinalis* Lamarck, *Avicennia alba* Blume, *Avicennia marina* (Forsk) Vierh., *Rhizophora mucronata* Lamarck, *Rhizophora conjugata* Linnaeus, *Rhizophora apiculata* Blume, *Acanthus tlicifolius* Linnaeus, *Sonneratia alba* J. Smith in Rees, *Sonneratia caseolaris* (Linnaeus) Engl., *Excoecaria agallocha* Linnaeus, *Aegicerus corniculata* (Linnaeus) Blume, *Kandelia rheedii* Wight & Arn, *Acrostichum aureum* Linnaeus (a fern), *Ceriops tagal* (Perr.) C.B. Rob., *Derris heterophylla* (Willd.) K. Heyne, *Bruguiera parviflora* (Roxb) Wight et Arn ex Griffith, and *Bruguiera gymnorhiza* (L.) Lamarck ex Savigny.

Studies of mangrove swamps in Colombia provided a cursory list of a few species of mangrove encountered in the coastal waters there (West, 1956). The following mangroves were identified to species: *Avicennia nitida* Jacq., *Rhizophora brevistyla* Salvosa, *Rhizophora samoensis* (Hochr.) Salvosa, *Laguncularia racemosa* (L.) C.F. Gaertn., *Conocarpus erectus* Linnaeus, *Peliciera rhizophora* Planch and Triana, and *Acrostichum aureum* Linnaeus.

Many of the predominant species in mangrove swamps around the world are congeneric with others, and some of the species are cosmopolitan or almost so. South American mangrove swamps have been disappearing at an alarming rate, and there are attempts being made to preserve them. However, they take a long time developing, and when they are damaged, they continue to deteriorate before they begin to recover. Within these habitats, fishes, crustaceans, and a few insect species interact more than in almost any other habitat. The insects present range from unmodified terrestrial species in the upper branches of the trees to typically marine species, such as several in the gerrid genus *Halobates*, which stay close to the shore. The differences between the pelagic *Halobates* species and those that stay in or near the intertidal zone seem to be based mainly on behavior. Whether there are special morphological or physiological adaptations that improve the chances of the pelagic species to better survive on the open ocean with its occasional storms and periods of rough water is still not known.

A study of the mangroves and mangrove-like species along the coasts of the Seychelles Islands (Frith, 1979) did not report the mangrove or insect species present. However, it revealed the higher taxa of the local insects that were attracted to light traps. Species belonging to typically aquatic families, as well as members of terrestrial ones, were present near stands of mangroves and at other localities on small islands and atolls. Surprisingly, the best represented order near the outer coasts was Lepidoptera, while Heteroptera, Coleoptera, Diptera, and Hymenoptera were also represented. Along the shores of atolls facing a lagoon, some less common taxa were represented, including Odonata, Isoptera, Dictyoptera, Embioptera, Orthoptera, and Neuroptera, as well as the more abundant Coleoptera and Diptera. This study indicated that even small islands isolated by the ocean from larger pieces of land can develop an endemic insect fauna encompassing many species largely adapted to life on the landward side of the seashore. Obviously, a detailed study of the native species along with their likely zoogeography could not fail to yield interesting results.

SECTION 2

Hexapods of Frigid and Temperate Zone Beaches

Beaches with substrates of fine sand, coarse gravel, or other sediments are familiar throughout the world. Although most hexapod species avoid them, there are a few that are characteristic inhabitants of this kind of intertidal habitat between the tropics and subarctic or subantarctic regions. The hexapods most frequently encountered belong to the classes Collembola and Insecta. The insect orders best represented are Coleoptera, Diptera, and Hymenoptera. They share their habitat with amphipods and other small crustaceans that have come to inhabit the hypersaline fringe of the ocean beyond the limits of most of their marine relatives.

Because both decaying plant material originating from the adjacent terrestrial habitats and marine seaweed beds, as well as carcasses of small animals, are present together in the detritus deposited near the high water mark during each tidal cycle, a variety of scavengers thrive in the littoral zone on beaches and rocky coastlines. Generally, the larger scavengers are crustaceans, and most hexapods are among the smallest ones. The necessary modifications for life in this zone include tolerance to high salinity and temperature variations over short periods of time and a method of surviving periodic inundation beneath salt water. In the temperate zones, however, is added the need for surviving the winter, which, in some localities, can challenge species otherwise adapted to seashore conditions by irregular and sometimes severe physical damage done by the formation and movement of large masses of ice.

Unlike the seashores in the tropics, those in the frigid zones are usually considerably warmer than inland areas during the winters. Warm ocean currents bring water long distances from subtropical areas throughout the winter, and they warm the seashores they flow close to. As a general rule, the oceanic currents carry warm water along the West Coasts of the continents, while cold currents flow along the East Coasts. However, the exact paths of the individual currents are relatively complex.

While terrestrial insects living farther inland from the coast are spending the winter at extremely cold temperatures in an inactive state, many adult aquatic insects simply move to higher elevations on the beach during the flood tide and return to the deposits of detritus during the ebb. The warm Alaska currents and the Gulf Stream in northern Europe are able to keep the adjacent coasts in the Pacific Northwest of North America and the North Sea coasts in Europe warm enough for limited activity by insects even during the winters of some years.

The temporarily hypersaline fringe of sandy seashores is inhabited mainly by insects from predominantly terrestrial families. Most of these insects are relatively small. Some are too small to examine properly without a dissecting microscope.

a. Collembola

During the brief summers, there are periods of intense activity by hexapods along the seashores. The most abundant species inhabiting the decomposing material at the high tide line on many beaches are species of Collembola. These hexapods often escape notice because they are wingless and cannot annoy people visiting the beach, as small flies do. They become very abundant at times, but only a few species can tolerate the extreme conditions encountered at the upper end of the intertidal zone, especially during hot summers and cold winters.

Common collembolans encountered along the seashores of Europe include *Anurida maritima* (Guérin-Méneville, 1836). In Europe, it sometimes occurs in massive numbers along the line of dead seaweeds and other kinds of detritus at the level of the last high tide along the beach. Frequently, they are accompanied by tiny dipterans, the species depending upon the geographical locality. The species of Collembola themselves are more likely to be cosmopolitan in distribution than the species of any insect order. It is sometimes speculated that the distribution of common collembolan species had reached all of the continental land masses before the continents drifted apart, but the fact that members of this order are also capable of surviving trans-oceanic migrations on large pieces of wood and other floating objects also demonstrates that they could have migrated to all parts of the world in relatively recent times.

In an experiment to determine whether terrestrial species of Collembola can survive on or in seawater, six such species were selected and tested experimentally. Many species of springtail are cosmopolitan in distribution, or nearly so, so the experiment was designed to determine how many of the six species could survive a long journey across an ocean. It was learned that five of the species without special adaptation for life on or in seawater could survive an oceanic voyage of more than 14 days, although one of the species had a survival rate of only 27% (Coulson et al., 2002).

In an additional experiment, the species *Tetracanthella arctica* Cassagnau, 1959, was found to be capable of surviving more than 14 days completely submerged in seawater, although only 12% managed to survive. The other five species tested were held in small containers, in which the insects could remain on the surface of the water in contact with air. The containers were agitated continuously.

The results showed that 100% of the *Onychiurus arcticus* (Tullberg, 1876) and *Hypogastrura tullbergi* (Schäffer, 1900) were able to survive at least 16 days under these conditions when the water temperature was maintained between 6° and 10°C. The following survival rates were obtained for the other three species: *Tetracanthella arctica*, > 80%; *Onychiurus groenlandicus* (Tullberg, 1876), > 80%; *Folsomia quadrioculata* (Tullberg, 1871), 27% or 29%. Unfortunately, there seem to be some discrepancies in the data reported in various parts of the paper, although the differences are small (Coulson et al., 2002).

The results of this study indicate that even unspecialized terrestrial members of the class Collembola are capable of surviving at sea for long periods of time. That indicates that surviving in seawater for periods exceeding 14 days is not difficult, even for unspecialized species of Collembola. The association of certain species with the intertidal zone along the seashore is therefore the result of the species preferring that habitat rather than the failure of other species to be capable of surviving there, at least under the conditions that prevail in Northern Europe. In view of the great variety of detritus from plants, cadavers of marine and terrestrial animals, algae, fungi, and bacteria swept together by the tides, it is not surprising that so many species prefer it.

b. Insecta: Coleoptera

Specialists in the exquisitely adapted species of aquatic beetles familiar in freshwater habitats have little to interest them along the seashores. Species in such families as Dytiscidae and Hydrophilidae have little to look for at the edges of the oceans. Few species of these and the other families of the well-known swimming beetles have been discovered anywhere near seashores. It is rather the members of overwhelmingly terrestrial families that have found their ecological niches along the seacoasts of the world.

Typical examples of the beetles encountered in the intertidal zones of the ocean include members of the large, mainly terrestrial family, Staphylinidae. Along the Pacific Coast of North America from Alaska to Mexico, the line of detritus at the high water mark is the habitat for such staphylinid rove beetles as *Thinopinus pictus* Leconte, 1852. This species remains concealed in detritus above the high tide line during the day and hunts in the intertidal zone at night (Bouchard, 2012).

A member of the Ptiliidae, *Motschulskium sinuaticolle* Matthews, 1877, inhabits the same beaches from British Columbia to Baja California, but it prefers the intertidal zone, into which few insects venture voluntarily while it is flooded. In this habitat, it must survive regular periods of submergence on sand or pebbles, often made unstable by breaking waves.

Seashores can also be inhabited by insects in the Frigid Zone, especially where warm ocean currents protect the habitats from long periods of extremely cold weather. A rare member of a relatively new family, Agyrtidae, inhabits seashores of some islands in the Aleutian Chain. Little is known about this species, and its known habitat is limited to tiny areas on small islands, which are hard to reach.

While the casual observer might mistake the tiny beetles concealed among the rows of detritus at the high water mark of the intertidal zone for terrestrial species that fell into the sea and were washed ashore, an entomologist should be able to recognize how easy it is to make discoveries of undescribed species of terrestrial families that have adapted themselves to life in the marine littoral habitat.

c. Insecta: Diptera

While beetles are merely represented among the seashore fauna, the greatest number of species in the habitat belong to the Diptera. Thanks to the abundance of decaying plants and animals from both the sea and from adjacent terrestrial habitats, scavengers and predators find an ample food supply. There are a great many larvae of Diptera species, which live primarily on decaying vegetation or on animal carcasses. Those species that have the ability to survive in a temporarily hypersaline habitat subjected to periodic desiccation according to the movements of the tides are able to take advantage of this food supply. However, some predatory insect species also have the capability to adjust to the high salinity and great diurnal temperature changes during the summers and cold temperatures during winter. The scavengers are therefore subjected to intense predation in some of the seashore habitats.

Among the seashore scavengers, larvae of a dipteran species, *Coelopa vanduzeei* Cresson, 1914, in the family Coelopidae, develop as larvae in decaying seaweeds, notably kelp (Hardy, 1956). It is widespread along the Pacific Coast of North America. Because of their association with seaweed on beaches, these flies have been given the name, kelp flies.

Many species of shore-flies found along beaches throughout most of the world belong to the family Ephydridae. On public beaches, they are regarded as pests because they frequently form dense swarms of flying imagoes, usually concentrated near the strip of decaying seaweeds marking the upper limit of the last high tide on the beach. However, the adults can appear almost anywhere in large numbers for various reasons, perhaps the most common of which is the formation of swarms engaged in mating and producing eggs to deposit in places where the developing larvae can consume the nutritional foods in the material washed ashore by the waves during the flood tides.

In a recent revision of the ephydrid genus *Mosillus* Latreille, Mathis et al. (1993) provided information on the great range of the six species now recognized as belonging to that genus. The species inhabit freshwater, seawater, hypersaline water, alkali, and waters polluted by man. The range maps provided by Mathis et al. (1973) suggest that the spread of the species deep into the continents of North America and Asia proceeded through river systems, indicating that some of the species of *Mosillus* have spread far inland along lotic water courses, and that the species penetrated into mountainous regions following the water without regard for its salinity or pollution. In spite of their great abundance and vast ranges throughout the world, there is still almost no information on the habitats of their larvae or their ecological requirements.

The species recognized by Mathis et al. (1993) include the following: *Mosillus bidentatus* (Cresson, 1926), *Mosillus bracteatus* Schiner, 1868, *Mosillus stegmaieri* Wirth (1969), *Mosillus subsaltans* (Fabricius, 1794), *Mosillus tibialis* Cresson, 1916, and *Mosillus asiaticus* Mathis, Zatwarnicki and Krivoscheina, 1993. The genus has a world-wide distribution, except for New Zealand and nearby islands. Only one of the species shows any strong preference for water with the salinity of sea water. The rest can live along the seashore or move inland following freshwater courses. A strong habitat preference for the sea has been suggested for only one species, *Mosillus stegmaieri*, the adults of which were collected among grasses near the seashore. Its known range includes mainly coastal areas of North and Central America surrounding the Caribbean Sea and the seashores of several of the West Indies. A few have been found along the Atlantic Coast of North America, where winter frosts seldom occur, and along the Pacific Coast from Mexico to Peru. It has not been reported far from the sea.

The first species described in 1794, *Mosillus subsaltans*, lives mainly in the temperate zones of Europe and Asia, all the way to the Far East. It has been found in North Africa near the coast, and it seems to be fairly abundant north of the high mountain ranges in Central Asia. It has not been reported from the northernmost regions of Europe and Asia nor from the tropics.

Two of the species are confined to the Americas, including the Hawaiian Islands, where they are found on seacoasts and far from the sea, usually along river valleys. One of them, *Mosillus tibialis,* has a range that demonstrates its strong preference for seacoasts, as well as its ability to migrate to sites near the center of continental North America along the major river systems. However, it seems to be most abundant along the Atlantic and Pacific coasts of the United States. It has reached the Hawaiian Islands and ranges south to Baha California, the border region between Mexico and Central America, and sites along the Pacific Coast of South America. Like other species in the genus *Mosillus,* it avoids cold weather and is found only rarely along the southern border of Canada. It has not been encountered farther north.

The second species, *Mosillus bidentatus*, is confined entirely to North America. It is known from British Columbia and Baja California. Otherwise, it has been found only in the United States, ranging from the West Coast eastward as far as Michigan. It is found along the Pacific Coast and eastward along major river systems, becoming gradually scarcer towards the East.

The final two species are confined to the Old World. They apparently have very limited ranges, although it is not certain whether they simply live in regions that are overlooked by dipterologists and have not been noticed. *Mosillus bracteatus* appears to be confined to the seacoast of South Africa, while *Mosillus asiaticus* has been found only in Mongolia and the region of China very close to the Mongolian border.

Another genus of shore-fly that has recently been reviewed and revised is *Parydra* Stenhammar, 1844. While the revision is limited to the species inhabiting the State of Paraná and neighboring territory in the States of Saõ Paulo and Santa Catarina, Brazil, it includes information on the species of this genus as far north as Florida and California (Mathis and Marinoni, 2011).

Adults belonging to this genus live on muddy sediments at the edge of lentic and lotic water bodies and feed on algae (Bischof and Deonier, 1985; Foote, 1995). Most of the species are Nearctic in distribution, known only from the Frigid and Temperate Zones (Mathis and Marinoni, 2011). Only a few are confined to the Neotropical Region. However, several of the Nearctic species have ranges that extend into the Neotropics. These include *Parydra humilis* Williston, 1897, which is known from Northern Mexico, Cuba, the Virgin Islands, throughout Central America, and as far south as southern Brazil, as well as *Parydra ochropus* Thomson, 1868, which ranges from California in the north, through Mexico and Central America, and southward along the Atlantic Coast to Uruguay (Mathis and Marinoni, 2011).

A few of the flies associated with beaches in North America belong to families not usually associated with the oceans. For example, the family Tabanidae is represented on the beaches in Florida by the species *Stenotabanus psammophila* (Osten Sacken, 1876). This species had been frequently noted flying over the water along Florida beaches and coming to rest from time to time on the sand near the water's edge, as reported by Blickle (1958), who used the synonym, *Aegialomyia psammophila.*

Other shore flies are found throughout Europe and Asia, but many of them are clearly terrestrial and are not further discussed in this book.

What the general ranges of the shore flies indicate is that most of the species tolerate salinities over a wide range from very low to much greater than seawater in evaporating pools of seawater. This adaptation permits them to colonize the intertidal zones of the seashore, where the salinity changes from only slightly above rainwater after periods of heavy rainfall to well above seawater in the upper intertidal zone during hot spells in the summer. For example, the range map of the Nearctic shore fly, *Dimecoenia spinosa* (Loew, 1864), shows that it lives beside water bodies all the way across the United States and Canada from east to west and from the Boreal Zone as far north as New Brunswick and south to Cuba, Jamaica, and southern Mexico. It appears sporadically along inland water bodies, but many specimens have been collected along the Atlantic seacoast, and it is well represented in collections from the Pacific shores as far north as the San Francisco Bay area and south along the coast of Baja California. The greater number of specimens found along the Atlantic Coast may be explained by the great number of bays and harbors offering suitable habitats for the species, or it may also be due to fewer entomologists seeking ephydrids there. A review of the genus *Dimecoenia* was provided by Mello and de Oliveira (1992).

What the pattern of distribution shows is the ease with which shore flies are able to spread along seacoasts. Few other insects can colonize the intertidal zones of the oceans, apparently giving the ephydrid species an open avenue for migrating as far as the temperature limitations of the species permit. Obviously, migrations through river systems is also possible, but the fact that stream and river water usually has relatively low salinity makes it hospitable for a great many other aquatic species, which could prey on or compete with the shore flies.

The great toleration of shore flies to high salinity, alkalinity, organic enrichment, pollution, and other forms of contamination opens the way to ecological niches in a habitat that is poorly utilized and occupies a narrow but contiguous strip at the edges of every continent and island in the world. Tolerances of members of the family Ephydridae to other kinds of extreme conditions are also described in several other chapters of this book.

SECTION 3

Insects with Marine Benthic Larvae

Although the intertidal zone is the habitat of most insects adapted to life along the seashores of the world, a few insect species have larvae that develop in the subtidal zone. Most of these seem to use various kinds of seaweed as a substrate for their larvae. None are yet known from loose, sandy substrates.

a. Trichoptera

Caddis flies are very familiar members of the various communities of freshwater organisms. However, a few species of Trichoptera develop in shallow coastal waters of the sea. All known species of the family Chathamiidae are marine. The family is found only from the South Temperate Zone, where the species develop in the coastal waters of Australia, New Zealand, and Catham Island, after which the family was named. They build their shelters out of pieces of coral.

The species studied the most thoroughly to date is *Philanisus plebeius* Walker, 1852, a caddis fly that deposits its eggs in the body cavity of *Patiriella regularis* (Verrill, 1867), a starfish in the family Asterinidae (Winterbourne and Anderson, 1980).

One main characteristic of caddis flies that can live in the ocean is certainly a physiological method of controlling salinity in its body. The larvae are obviously capable of living in undiluted sea water. A study by Leader (1972) elucidated some of the ways in which the marine caddis flies conduct osmoregulation in seawater.

b. Diptera

Most known subtidal insect larvae belong to several families of Diptera. The best-studied species is *Clunio marinus* Haliday, 1855, a member of the family Chironomidae inhabiting relatively shallow marine habitats along the European coasts of the Atlantic Ocean, from Iceland to Spain. A population inhabiting the subtidal zone surrounding the island of Helgoland is known to breed and lay eggs according to a lunar cycle during each summer. As in the cases of most other chrionomid species, emergence of all adults occurs almost simultaneously to maximize the chances of successful mating and deposition of eggs. However, in the case of *Clunio marinus,* the emergence of the adults occurs strictly according a lunar cycle governing the movements of the tides. Through sensory signals not yet understood, if the weather is stormy at the time in the lunar cycle during which emergence of adults and mating should occur, the larvae will postpone metamorphosis until the next suitable period in the lunar cycle, when it will occur hopefully during calmer weather.

Marine benthic larvae like *Cluneo marinus* are relatively rare, but all such species may not yet have been noticed or described. Its larvae inhabit shallow coastal waters along the North Sea coasts. A population off the island of Helgoland has been well studied because of its proximity to a marine biological laboratory (Caspers, 1951).

Most of the life of this insect is spent as a larva living among marine algae. There are mass emergences of the adults at certain phases of the moon during summer. Metamorphosis and mating may be postponed for one month in case of bad weather. Like other chironomids, the adults eat nothing and do not survive more than a day or two. They simply mate, deposit their eggs, and die.

Other nominal species of *Clunio* inhabit shallow marine habitats along other coasts in the Palearctic Region. However, they have not been studied as thoroughly as *Clunio marinus*, and additional taxonomic studies will be necessary to determine how many species should be included in the genus. These species also develop as larvae among seaweeds in coastal waters.

Meanwhile, marine biologists working in other parts of the world began to discover the larvae of other chironomids living in shallow marine habitats. For example, in relatively shallow water off the coast of Brazil, chironomids were found with larvae that inhabit seawater. They were described under the name *Thalassomya gutae* DeOliveira, da Silva, and Trivinho-Strixino, 2013.

In widely separated parts of the world, species of *Pontomyia* have been described. Their larvae were found in the sublittoral zones, where the water temperatures were usually higher than those in the waters in which *Clunio marinus* develops. *Pontomyia natans* Edwards, 1926, was discussed in Chapter 17 as a pelagic insect, the first species of Diptera to have been found in the open ocean, far from all continents. However, some populations in this genus have been found not too far from islands. The taxonomy of the genus has recently been revised, and it is best to wait until it can be determined whether there are both pelagic and neritic species, as in the case of the genus *Halobates*. In contrast to *Cluneo*, the adult males of *Pontomyia* species have only vestigial wings and do not fly, but rather skim across the surface of the water. Females are completely wingless (Huang and Cheng, 2011).

The question of whether the spread of this species throughout the world has occurred because of oceanographic processes or whether international shipping has contributed significantly to the dissemination of marine chironomids has yet to be answered. Obviously, the study of these insects requires much travel, extended periods of time for the research, and special equipment, and there are no economic problems believed to be caused by these insects. As a result, the secrets that thorough studies of these marine insects would reveal may remain concealed for the foreseeable future.

Some of these secrets include the diets of the various species of marine insects. Do the species subsist mainly on microscopic algae, planktonic organisms, or macro-algae? Do all of the species have similar diets, or are there specialists that consume and digest only one group of microorganisms? In the case of the chironomids, studies are made difficult because of the great variety of organisms, which are available to the insect larvae within their shallow marine habitats. Do all larval instars feed on the same group of oganisms, or does the diet change somewhat with each shedding of the integument?

Other questions are related to the role of salt water in the metabolism of the insects. In some, and perhaps all insects inhabiting marine seacoasts, fresh water is available each time it rains. However, is it also possible for an insect inhabiting the intertidal zone to utilize seawater to supply its need for moisture?

In the cases of most insects inhabiting the intertidal zone, it will await special studies to answer this question. It is likely, however, that many differences among the insects present will be encountered. Furthermore, different salt tolerances will probably be encountered in larvae and adults of the same species. This question is particularly important to answer when dealing with the subtidal species. Obviously, insects find it hard to adjust to life in the sea because so few of them have done so. For most insects, adjustment to the relatively high salinity in standard seawater can be assumed to be a major impediment to life in the sea.

Insects can adapt to sea water by simply using it in small amounts to meet their physiological needs, by taking in the seawater and excreting highly concentrated salt solutions, waiting for rain or an inflow of fresh water to renew their physiological water supply, flying inland to find water, or by obtaining fresh water from their food. If the insect develops in the subtidal zone, only two of these possibilities could solve its problem. Either it uses seawater at full strength to meet its physiological needs, or it concentrates and excretes the excess salts. There are a few species of terrestrial insects which inhabit arid regions capable of surviving without ingesting any water in liquid form during their entire lives, so it is possible for at least a few insects to adjust the salinity of the water in their bodies. It would certainly be worth the time and effort to carefully perform experiments to determine how each of the insect species obtains water to meet its physiological needs.

It is hard to estimate the effects of the absence of almost all other insects from the environment. For example, the larvae of *Cluneo marinus* live on submerged seaweeds in water apparently free of all other insect species. Even the adults would not be likely to encounter another insect species during their short life spans. Naturally, the absence of predatory insects would favor the survival of the maximum number of their species, but it would still be subject to predation by fishes, crustaceans, and other invertebrates. However, predators from other taxa would seldom contact any insect in the sea, and which of them would be able to hunt and find the larvae would be questionable.

19

Living in Hypersaline Water

Two categories of hypersaline habitats are commonly colonized by a few insect species with special adaptations for living in them. Some hypersaline water bodies are permanent and show little variation in salinity over the course of a year, while others have a salinity that varies with the season or even during single tidal cycles. Each present special difficulties for insects to overcome in order to take advantage of the benefits such habitats offer to insects that can survive in them, especially the paucity of predators that are able to live in hypersaline water. Those few insect species that have suitable adaptations to survive in or on the surface of such water bodies sometimes appear in massive numbers. They benefit from nearly limitless nutrient supplies in the form of detritus and sometimes microorganisms thanks to a lack of competitors and safety from most aquatic, insectivorous animals.

There are also chemical categories of hypersaline water bodies. One is characterized by a great similarity to seawater, usually due to periodic water exchange with the sea. Other water bodies contain specific salts in proportions clearly different from those in sea water. Such seas are also different from one another because there are differences in the salts present and their concentrations relative to one another. The unique mixtures of salts in the water of such lakes or seas almost always result from the geological peculiarities of their respective watersheds. In recent times, however, the activities of the human population in each watershed have played an increasing role in the nature of the substances found in the hypersaline water bodies. Agricultural activities generally bring about contamination with other substances, including a variety of pesticides, as well as increases in the amounts of organic substances, nitrates, nitrites, and phosphates. It is therefore important to distinguish between ancient hypersaline environments, including their parts that still exist in a pristine state, and artificial ones resulting from recent human activity.

As a general rule, the more a saline water body deviates from pure seawater, qualitatively or quantitatively, the fewer species it will support (Boix et al., 2008).

SECTION 1

Permanent, Naturally Hypersaline Water Bodies

The largest hypersaline water bodies are well known geographical features of the regions in which they occur. The Dead Sea, in western Asia, is located at the deepest point on the earth below sea level that is not at the bottom of a sea. The Great Salt Lake in North America is somewhat less saline but more variable in salinity due to periodic, long-term variations in precipitation within the watershed. Other such water bodies, such as the Aral Sea in Central Asia, have become much more saline in recent years due to human activities (Aladin, 1995). After much water was used irrigating fields for short-term agricultural projects, the rest of the Aral Sea began to dry up completely. So much water has evaporated from this sea that the area it covers decreased to only a fraction of its size a few decades ago. What was formerly occupied by a large inland sea turned to desert. Only after the use of its water was discontinued did it begin to slowly refill.

As the salinity of the water increases, the fish fauna becomes increasingly impoverished. Finally, all fish species disappear, and the aquatic fauna becomes dominated by a few species of invertebrate that can survive in water too salty for almost all other animals. The greatest successes by species of macrofauna in adapting to high salinity have been accomplished by crustaceans. Aquatic insects have not been as successful, although a few can survive after all vertebrates have disappeared from the water. In the end, however, it was of no avail to any of the sea life. When the sea completely dried up, all of the aquatic organisms disappeared with it.

One of the best species adapted to spend its entire life in water too salty to permit any potential predator to survive is the crustacean, *Artemia salina* Linnaeus, 1758. It belongs to an ancient order of Crustacea, Anopleura, which is not able to survive in water bodies with salinities low enough to make them hospitable to predatory animals. The brine shrimp, as they are called, have little or no natural defenses against organisms seeking to devour them, and they quickly disappear from habitats with a fauna showing any degree of diversity. It was noted by the first naturalists to visit the Great Salt Lake in Utah that a species of *Artemia* was one of the few animals of any kind that would tolerate the saline waters, and it was abundant (Vorhies, 1917).

Most insects occurring near hypersaline water bodies tend to live only along the shores or on the surface tension layer. Those insect species that can tolerate such extreme habitats have more complex ways of adapting than by employing simple physiological adaptations. Almost all insects successful in colonizing extremely hypersaline water bodies belong to the order Diptera. For such insects, rainwater could easily provide a source of freshwater, permitting them to utilize the hypersaline water only as a source of food organisms. The fact that most of the hypersaline water bodies in the world are located in regions of very little rainfall precludes most insects from adopting this survival strategy. With few exceptions, the insects are forced to adapt to using saline water or avoid such habitats.

The studies of many water bodies on the Crimean Peninsula by Belmonte et al. (2012) confirmed the relationship between the salinity of the water and the diversity of the species encountered. They showed that insects have not been very successful in finding adaptations to high concentrations of salt, and that only a few crustaceans have become characteristic species of the most hypersaline habitats. The insects that are the most notable exceptions are brachyceran flies in the family Ephydridae, perhaps the most successful family of insects in adapting to physical and chemical features of aequeous media that would be deadly to almost all other insects. Although none were reported from the hypersaline water bodies on the Crimean Peninsula, the larvae of a member of the family Chironomidae and an unidentified beetle larva were collected in other hypersaline lakes (Belmonte et al., 2012). Unfortunately, the species of these two salt-tolerant insects were not identified. Similarly, larvae of an unidentified insect, "*Tendipus group salinarius*", were reported as the only insects found in saline lakes in northern Tibet by Zhao et al. (2005). The reference is probably to a species in the *Chironomus salinarius* Kieffer, 1915, group, because *Tendipes salinarius* is its invalid synonym. The species *Chironomus salinarius* itself was originally described from European specimens, but midges found in seawater off South Korea have been redescribed under the name *Chironomus salinarius* Kieffer, 1921, by Ree and Yum (2006).

This illustrates the problem in recognizing the extraordinary adaptability of larvae in the family Chironomidae to extreme conditions and occurrence in places where insects would not be expected. Unlike the species in the family Ephydridae, the other champions of adaptability among the insects, adult chironomids from stressful environments, usually have very short life spans. They are easily overlooked, and only a small percentage of the native species have been described in most parts of the world.

Adult flies and midges are generally encountered along the edges of hypersaline water bodies. Ephydrids rest on the surface tension layer, while chironomid males fly in swarms hovering in the air, where females seek them when ready for mating. The larvae of aquatic ephydrids develop in the water, usually near the surface, or sometimes in fine, water-soaked sediments at the water's edge. Chironomid larvae usually remain submerged, living in tubes they spin among debris on the epibenthos or on surfaces of leaves or inanimate objects.

SECTION 2

Physiological Ways of Eliminating Excess Salt

Most vertebrates that can live in hypersaline water bodies seem to maintain a lower salinity in their body fluids than that in the ambient water by removing excess salts and excreting them through special glands, which expel highly concentrated salt solutions from the body. Except for species deemed to be primitive, which lack strict control of the salinity of their body fluids, marine vertebrates typically have body fluids with lower salt concentrations than ordinary sea water. These marine species have to excrete excess salts constantly to maintain the salinity of the blood and other body fluids at a physiologically optimal level for the species. For example, sea turtles appear to be shedding tears when on land. Glands near the eyes secrete these "tears" constantly, but they only become visible when the turtles are on land. They contain very high concentrations of salts from the sea water in order to reduce the salt concentrations inside the body.

That the need to maintain an optimal salinity in body fluids is physiologically challenging is demonstrated by the fact that vertebrates are among the first organisms to disappear from habitats with increasing salt concentrations. Salt tolerance, on the other hand, seems to be highest among certain species of Crustacea. As already mentioned, the most tolerant aquatic animal of hypersaline water seems to be the so-called brine shrimp, which are members of the genus *Artemia*.

In the case of the insects inhabiting hypersaline waters, the problem of coping with hypersalinity is more difficult to solve because the ambient water is even more saline than seawater, and most of the insects inhabiting it are relatively small and lack the body mass to store appreciable quantities of desalinized water. Therefore, it must be assumed that the insects present have a greater toleration of high salt concentrations than related species inhabiting fresh or slightly salty water bodies. Although the adults can fly to nearby terrestrial habitats and obtain rainwater or freshwater from other sources, it is obvious that their aquatic larvae can depend only on the hypersaline water in which they develop to meet their needs from hatching through metamorphosis.

Studying the methods of adaptation to high concentrations of salt would be a fruitful field of research because there are very likely many different methods adopted by the insects belonging to all of the various higher taxa that encompass species with different solutions to disposing of excess salt, or, alternatively, avoiding methods of consuming it in the first place.

For example, taking the two families of Diptera that are apparently the least sensitive to unusual concentrations of salts in their habitats, Chironomidae and Ephydridae, the fundamental differences in their relationship with the ambient water would suggest that their methods of controlling the effects of high salt concentrations in the ambient water must be fundamentally different. Chironomid larvae remain in complete contact with the ambient water for respiration. Ephydrid larvae live immersed in the water but almost always remain just below the surface, keeping their breathing tubes in close proximity to the atmosphere to permit gas exchange at all times.

SECTION 3

Natural Water Bodies that Periodically Become Hypersaline

In some parts of the world, coastal flood plains with restricted access to the sea become flooded with seawater during spring tides or due to heavy wind storms. After being flooded, the water remains until it evaporates. Without regular periods of flooding, the salinity continues to rise as the water evaporates. The water then goes through a succession of species involving the progressive disappearance of the animals with relatively low tolerance to salt. Often, this is accompanied by increases in the sizes of the populations of species tolerant to high salinity. The death of the members of species killed by the salt increases the food supply for insect larvae that feed on the carcasses of dead animals.

A common strategy for insect species to profit from the conditions in such natural salt ponds or marshes is for them to live in the hypersaline water bodies as larvae, and after metamorphosis, to remain along the shores waiting for rainfall, wind storms, or spring tides to replace the water that evaporated and lower the salinity to tolerable levels for new generations of the insects. When conditions are right for spawning, a stimulus of some kind must prompt the adults to deposit their eggs and produce a new generation of larvae.

Such habitats are generally small and located near the sea, most commonly in the tropics. Few of them have been studied to determine their fauna, and the popularity of seacoast areas for recreation and tourism have undoubtedly eliminated many of them before the species inhabiting their water could be identified. Coastal mangrove forests are among the locations at which the temporary development of shallow, hypersaline water bodies have been found. Regions affected by monsoonal or similar weather patterns producing wet and dry seasons will typically be the sites of periodically hypersaline water bodies.

In desert regions, insects adapted to the local conditions must be capable of getting water from any source available without being too selective. Williams (1998) investigated the fauna of lakes filled with water containing different relative contents of specific salts in Australia and concluded that salinity has less impact on the community structure than previously assumed. Unfortunately, he did not fully explain what he considered to have been previously assumed. Generally, there is a considerable tolerance to water with very low salinity to values less than half that of sea water. In a desert region with long periods between rainfalls, such as that in Australia, insects certainly had the chance to adapt to a wide range of salinity, more so than in regions where the climate is characterized by regular rainfall. Therefore, deviations in salinity of inland water bodies have less of an impact on the native fauna than it would in most Holarctic or Equatorial water bodies.

A good example for showing the impact of salinity on the aquatic biotic community was provided by personal experience while performing research along the Elbe Estuary and helping others with research on the North Sea. The Estuary extends for more than 100 km from an artificial barrier constructed inland from Hamburg to part of the North Sea receiving the outflowing water during successive ebb tides. The water at the inland end is fresh water flowing into the estuary from upstream. The tides cause this water to alternately flow seaward and then change direction to bring seawater from the North Sea into the estuary. The middle sections of the estuary contain brackish water produced by the mixing of the freshwater and seawater. The total number of animal species observed decreases slowly from the upper freshwater section toward the middle of the estuary, and then it increases again from the middle to the North Sea. The only species found both in the freshwater section of the estuary and the North Sea are anadromic and catadromic species, which live as adults in one during part of their lives and then migrate to the other to spawn.

Breaking the fauna down into freshwater elements and seawater elements, the dynamics within the biotic community becomes clear. As the salinity increases from fresh to brackish water, the number of freshwater elements decreases continually, with some species disappearing as soon as the water becomes slightly brackish, while others persist until the salinity almost reaches that of seawater. At the same time, euryhaline species first appear in the brackish water section, and the number of salt water species increases steadily as the water becomes more saline. The lowest total number of species is encountered somewhere near the middle of the brackish water section of the estuary.

During this transition, the greatest number of insects is in the freshwater section of the estuary, while in pure seawater, none of the freshwater insect species are encountered. Those insects that might be found in the sea, such as *Clunio marinus* Haliday, 1855, do not occur at all in the freshwater section. A curve plotting the total number of animal species against the distance from the uppermost part of the estuary to the sea will show a continual decrease in the number of species present until reaching a point near the mid-length of the estuary. The curve will then show a continual increase in the number of species present until the water reaches the sea. The exact stage of the tidal cycle must be noted because the mid-length of the estuary continues to change its position as the water body moves back and forth with the tides. The curve simply shows a numerical depiction of the freshwater species being eliminated and replaced by marine species as the water moves seaward.

SECTION 4

Artificial, Temporary, Hypersaline Water Bodies

In addition to landlocked seas, suitable habitats for insects with tolerance to high salt concentrations include temporary, man-made hypersaline habitats in salt pans, in which seawater is allowed to evaporate to commercially obtain the salt, as well as small, natural water bodies along seacoasts, in which seawater can enter periodically and evaporate, leaving a hypersaline water body. Such natural salt lakes and lagoons are most common where deserts border on the sea. Most of these are permanently hypersaline, although the salinity may occasionally decrease if the region is subject to periodic rain storms.

Among the insects most typically associated with hypersaline water bodies are species in the family Ephydridae, which are flies in the suborder Brachycera. This family is very rich in species, many of which have been able to adapt to extreme conditions of various kinds and thereby develop massive populations in places avoided by most other animals.

The least adaptation to hypersaline habitats is required for species that live in marine littoral marshes, which become hypersaline only for relatively short periods of separation from the sea, during which evaporation causes an elevation of the salinity. Shore flies, sometimes called salt flies, are commonly found in such habitats. Along the Atlantic and Pacific coasts of North America, flies belonging to the subfamily Ephydrinae develop as larvae in marshes flooded with seawater. Examples of such flies in North America include species of *Dimecoenia* Cresson, 1916. Such species were already discussed in previous chapters, but relevant only to contact with seawater and conditions promoting its evaporation along the seashore during the tidal cycles occurring twice during periods of about 25 hours.

In many coastal regions, seawater floods areas inland behind natural barriers during stormy seasons. During long summer periods of calm weather, the trapped seawater evaporates and creates temporary pools of hypersaline waters. In such seasonal habitats, shore flies and other insects with tolerance to high salt concentrations are able to reproduce in an environment with little competition for food and safety from most predators.

Dimecoenia fuscifemur Steyskal, 1970, and *Dimecoenia spinosa* (Loew, 1864) are two species of their genus, which show distribution patterns clearly associating them exclusively with salt marshes along the seacoasts. *D. fuscifemur* is associated only with Atlantic Coastal mud flats and marshes of North America from Delaware to Florida. *D. spinosa* inhabits salt marshes along both the Atlantic and Pacific Coasts. It is much more commonly collected, and it is apparently more abundant at all latitudes along the Atlantic Coast where both species occur. Adults of *D. spinosa* were observed occasionally in freshwater habitats, but no evidence was found that they could survive in the water there as larvae (Mathis and Simpson, 1981).

A second genus with a comparable distribution is *Cirrula* Cresson, 1915. Of the two species in this genus that were studied by Mathis and Simpson (1981), one was found only along the Atlantic Coast between about 41°N and 49°N latitude, while the other inhabits saline habitats along both the Atlantic, including the Gulf of Mexico, and Pacific Coasts. The species found only along the Atlantic Coast, *Cirrula gigantea* Cresson, 1917, is restricted to relatively cold climatic zones. The more widespread species, *Cirrula austrina* (Coquillette, 1900), occurs only south of 40°N latitude (Mathis and Simpson, 1981). Unless there are subsequent reports of sympatric populations of these two species, it appears that their ranges do not overlap.

With populations already adapted to surviving or even thriving in natural hypersaline pools, it is only a matter of a minor adaptation for an insect to move into similar habitats deliberately constructed by man.

a. Commercial salt pans

Artificial habitats along seacoasts in various parts of the world would seem to offer many insect species an ever increasing, physiological stress in their environment, which is not present in natural coastal marshes. Coastal regions have been increasingly modified for construction of houses, businesses, and recreational facilities, and this eliminates many of the habitats which offer niches to suitable wetland insects. Commercial salt pans have been built for the production of sea salt at suitable locations around the world since prehistoric times. When natural saline lagoons and salt marshes are replaced by salt pans, conditions are artificially regulated to produce a controlled rate of salinization leading to the salt being fully separated from the water by precipitation. Salt pans are generally kept separated from the sea until salt crystals are produced and harvested. That creates a temporal halocline, starting with the salinity of the local sea water and ending as fields covered with salt crystals. After initial evaporation of the excess water, the saturation of the common salts in the water reaches 100% and remains there as excess amounts of each salt form crystals and precipitate out of solution. Eventually, all of the free water evaporates, and only salt crystals remain. During this process, one species of salt-tolerant insect will be eliminated after another until aquatic insects will no longer be found.

Studies of salt pans and their insect fauna are similar in many ways to laboratory experiments. Natural coastal habitats along seacoasts do not normally evaporate so rapidly or completely that 100% salt saturation is reached. There is normally some exchange of water with the sea, which dilutes the hypersaline water trapped in depressions temporarily cut off from tidal influences. These have higher salinity values than normal seawater, but the values may become quite high during dry periods in summer. Salt pans, on the other hand, are artificially prevented from being periodically exposed to dilution with normal sea water, and it is likely that all insect species would be eliminated before the end point of complete evaporation would be reached. If not, further physiological investigations on that species would almost certainly be advisable to substantially increase the knowledge of how insects can regulate the salt content of their bodies in hypersaline waters.

Salt pans along the coast of Portugal were studied by Talveres et al. (2009), who found that the diversity of species inhabiting them was lower than in unmodified stretches of the seashore. As in other stressed ecosystems, the number of individuals belonging to each species present is greater than in unstressed, natural environments.

The authors described the structure of the biotic communities present in the various habitats of the salt pans. The families noted for species that have adapted themselves to the most extreme habitats were well represented. These included the families of Diptera: Ephydridae, Chironomidae, Culicidae, and Limoniidae; the families of Coleoptera: Hydrophilidae and Dytiscidae; as well as one species in the family Corixidae and one in Notonectidae of the order Hemiptera.

The results of this study provided no surprises in view of the available information about insects that can adapt to extreme conditions of various kinds. The family showing the most adaptability to the conditions in the salt pans after the measures imposed by commercial management were relaxed was Chironomidae. The species in this family were the most successful. It was also predictable that Epydridae would wind up in second place among the Diptera. A smaller number of species present belonged to each of the families Sphaeroceridae, Dolichopodidae, Limoniidae, and Culicidae. Each of these families is often represented among the fauna living in habitats stressed by various factors. In addition, a small number of aquatic species from the order Hemiptera belonged to the families Corixidae and Notonectidae. Finally, beetles in the family Hydrophilidae were fairly well represented as both larvae and adults, while at least one species in the family Dytiscidae was also found. Unfortunately, the species were identified only to genus.

The results of the study supported the hypothesis that habitats under stress show a reduction of species present, while the number of individuals in each species tends to increase. This conclusion was also not unexpected. Hypersalinity is only one stressful factor, which can be compensated for by many insects in various ways. It is combinations of stressful factors which are especially harmful to the insect fauna, making some of the natural habitats exposed to extraordinary combinations of harmful influences less conducive to the development of all insects.

20

Petroleum Flies

SECTION 1

An Insect with Special Abilities

Petroleum is a mixture of hydrocarbon compounds, many of which are toxic or otherwise harmful to most living organisms. Near oil wells, it is not uncommon to see open pools of petroleum, beneath which rainwater collects. Usually, the owners of the oil wells clean up or cover up these unsightly pools after the prospecting and drilling operations have ceased. While they exist, the pools of petroleum on the oil fields act as fatal traps for almost any insect that comes into contact with them. The stickiness of the petroleum almost immediately immobilizes insects and blocks their spiracles, causing a quick death. Those few aquatic insects capable of surviving the physical challenges to life in sticky fluids succumb to the chemical toxicity of one or more of the many compounds of which the petroleum consists.

The lethal conditions for most insects in the petroleum pools makes it particularly surprising to encounter many small adult flies on the surface of the petroleum and to find the larvae of these flies in or beneath the sticky black mass of crude oils, water, and soil. The best known species of these flies are appropriately known as petroleum flies, members of the family Ephydridae, which is also the family of the commonest species noted on the surfaces of other inhospitable shallow water habitats, such as the hypersaline lakes and lagoons discussed in Chapter 19.

The adult petroleum fly, *Helaeomyia petrolei* (Coquillett, 1899), originally described under the name *Psilopa petrolei* Coquillett, 1899, is about 2 mm long. The adults often congregate in large numbers on the surface of tar-covered pools on oil fields and on asphalts, where natural seepage of petroleum occurs (Thorpe, 1934). They are best known from California, although they have also been said to occur throughout the world where suitable conditions prevail. The reports about the extent of the range of this species are often contradictory. The flies were first discovered near oil wells located at the La Brea Tar Pools, a location at which many animals became trapped in the asphalt left after the lighter fractions of crude oil seeping from the ground had evaporated. Since prehistoric times, animals of all kinds had met their doom by breaking through the asphalt and sinking into the deep deposits of tars, and their remains were preserved to some extent by the heavy fractions of the crude oil. Ongoing research has resulted in a great many bones of all kinds being unearthed and identified.

Most of the crude oil in California and most other parts of the world is located deeper under the ground, and there is no seepage detectable at the surface, as there has been at La Brea for thousands of years. The light fractions of the oil cannot evaporate until the oil is brought to the surface by drilling and pumping. Before petroleum was recognized to be a precious and limited resource, whenever a strike was made, the oil was allowed to gush out of the well for a while to clear out extraneous material introduced during the drilling. As a result, oil fields were very heavily contaminated with crude oils, substances that *Helaeomyia petrolei* had come to depend on for making its own living. Therefore, reports of its appearance wherever oil had been discovered are credible.

Schelkle et al. (2012) investigated water bodies containing asphalts and other residues of petroleum that have seeped to the surface from underground pools. They enumerated the major habitats of this kind throughout the world, listing three in California: the La Brea tar pools, the McKittrick tar pits, and the Carpinteria tar pits. A fourth large tar pit, Lake Guanoco, is located in Venezuela. The fifth of these habitats, which the group investigated, is Pitch Lake, on southwestern Trinidad. It is the largest such habitat in the world (Schelkle et al., 2012).

Petroleum from different parts of the world contains distinctive mixtures of hydrocarbons, and the waters that fill the different lakes also have different salinities and other distinctive chemical characteristics. If oil continually or periodically seeps into a lake, the light hydrocarbon fractions evaporate, leaving behind a tar or asphalt containing less active and usually less toxic fractions. As a result, it can be assumed that the water chemistry in each water body contaminated with petroleum is distinctive.

Helaeomyia petrolei has successfully colonized all of the large petroleum lakes in the New World which have been thoroughly investigated by entomologists. Whether the flies reached these locations during prehistoric times or were introduced by the petroleum workers remains to be determined.

Petroleum flies have developed a series of remarkable physiological adaptations for life on and in petroleum and its component substances. Those who have examined the habits of *H. petrolei* agree that its main food is the remains of arthropods of many species, which fall into the petroleum and perish. The dead insects are usually protected from rapid decomposition by the petroleum, which is presumably toxic to most abundant scavengers and saprobic microorganisms. The remains of the carcasses may break apart, but their rate of decomposition while they are surrounded by the mineral oils is presumed to be greatly slowed, leaving a sufficient accumulation of food for the larvae of the petroleum flies.

These larvae have been found to have their digestive tracts filled with petroleum, which is not thought to be an important source of nutrients for them. Although specialists who have studied the habits of the larvae agree that the body parts of dead insects trapped and killed by the oil residues serve as the main food sources on which the larvae subsist, definitive information about of the digestion of the foods and protection against the specific toxic substances in the oil and water is still lacking.

Naturally, this raises a considerable number of questions concerning the biochemical reactions which decompose the food items in the digestive systems of the larvae. Nearly all known digestive processes of arthropods involve water and chemical reactions that occur in aqueous solutions. Is this also true of the digestion of insect remains contained in nearly pure petroleum residues? The chemical pathways involved in the digestion and assimilation of the food items of petroleum fly larvae certainly must have some significant modifications from those of insects from more conventional habitats.

Studies on the microbiota in the digestive tracts of *Helaeomyia petrolei* have shown that bacteria were 100 to 1000 times more abundant in the guts of the larvae than they were in the free petroleum of the habitat. Interestingly, the pH of the gut contents of the larvae was found to be only 6.3 to 6.5, meaning that the medium in which the food was being digested was only slightly more acid than neutral (Kadavy et al., 1999). This speaks against the insect producing its own special digestive juices to digest insect carcasses imbedded in the crude oil or asphalt that it consumes. It would rather suggest that the larvae have a symbiotic relationship with certain bacteria that break down chitin for their own benefit and also for the benefit of the host.

The oil fields in and near Los Angeles, California, are distinctive because they have been open on the surface at least for millennia. The La Brea Tar Pools have been subjected to intensive scientific studies for many years. They are remarkable as sources of preserved carcasses of many kinds of animals, which entered the pools covered with water at the surface and were trapped and killed as they sank slowly into the tar. This gave the flies plenty of time to colonize the tar pools from similar habitats that exist or once existed in the vicinity or to have reached the open petroleum pools from more remote habitats. Where the species first adapted themselves to life in natural pools of petroleum is unknown, but the existence of such flies in many parts of the world suggests that their mode of existence is not relatively new.

By adapting to this mode of life, the ephydrids gained a considerable number of ecological advantages. They benefit from a rich food supply. Not only do a vast number of insects perish in the oil pools each year, the petroleum not only directly interferes with the most common decomposition processes in uncontaminated water bodies, it also slows the action of saprobic microorganisms by reducing the oxygen supply required for rapid decomposition of organic substances. Similar to food for human beings that is vacuum packed, the remains of the insects which perished in the oils probably remain in a good state of preservation for many years. The availability of large amounts of this kind of food could explain how such large populations of adult petroleum flies could be supported on relatively small pools. The second great advantage to the fly larvae that can survive in mixtures of water and petroleum components is the elimination of almost all predators that would otherwise prey on the small ephydrids and their larvae in any more common and less toxic kind of aquatic habitat. In most cases, pools of petroleum also contain abundant water, either on top of the tar or beneath the oil, creating multiple combinations of potentially toxic mixtures.

There are exceptions to the exclusion of predators that would be able to prey on the larvae of the petroleum flies. In Pitch Lake, there is a population of guppies, *Poecilia reticulata* Peters, 1859, which appear to suffer no ill effects from the toxicity of the habitat. Members of the same species brought in from less contaminated water bodies are also able to survive in the lake, and even seem to be freed of gyrodactylid parasites by substances in the water (Schelkle et al., 2012).

The adults and larvae of petroleum flies have distinctly different habitats. The adults usually do not fly frequently or far from the pools. Instead, they walk on the surface of the mixture of water and oils in which they apparently deposit their eggs. The larvae live below the surface of the water, where they consume mixtures of the petroleum residues and the bodies of insects that fell into it. It is assumed that they subsist on the remains of the dead insects, but it remains to be determined whether any of the components of the petroleum in the gut are assimilated and used as a nutrient source for the insects.

The most obvious questions that remain to be addressed pertain to the behavioral and biochemical processes which protect the petroleum flies from the properties of the substances in which their larvae live. The mixtures of such substances quickly kill almost every other kind of insect that come into contact with them. Several possible processes could be responsible for protecting petroleum flies from being killed by toxic substances in their habitat. The integument covering their bodies could exclude all foreign substances which might be toxic to them. Larval syrphids inhabiting small pools of highly polluted waters seem to have similar impervious integuments that protect them from all kinds of toxic pollutants.

Such an integument would also have to line the digestive tract because the petroleum residues are consumed and remain in the gut for a period of time. This method could protect the insects, although there would have to be locations along the digestive system through which certain necessary nutrients could be assimilated into the body. An alternative would be for the insect larvae to secrete substances that would block or neutralize the most harmful components of the petroleum residues. Still another would be a biochemical pathway for breaking down petroleum itself to obtain energy for the larvae to utilize. The problem is that all of these processes would have to deviate considerably from those known from other insect species, making any such physiological process or biochemical pathway for permitting the larvae to survive in petroleum residues seem very improbable.

SECTION 2

Finding an Ecological Niche in an Oil Field

It appears that the habitat required by *Helaeomyia petrolei* is a body of water on the surface of which there is a combination of fresh water and petroleum residues. Due to a relatively rapid evaporation of the lighter fractions of the petroleum, the residues consist mainly of asphalts. However, there is a constant, but usually slow, seepage of newly emerging crude oil, which contains light as well as heavy fractions of the petroleum. The pools of these oils on the surface of the water are death traps for countless insects, the carcasses of which are considered to be the main food sources for the petroleum fly larvae. The petroleum fly was first considered to be a member of the Diptera family Psilopidae. Only after being subjected to closer examination was it discovered to be a member of the Ephydridae, the family with the most tolerance of several common environmental pollutants. Psilopidae was reduced to a subfamily of Ephydridae: Psilopinae.

The adaptation of the petroleum flies to life in the remains of crude oil that seeped through natural processes from the ground probably occurred during relatively recent times, on a geological scale. The bones found in the tar are predominantly from birds and mammals known to have existed during the Pleistocene. Of course, it could be possible that the petroleum flies appeared on earth much earlier at sites where petroleum once seeped from the ground but no longer does and spread from there to the sites at which they are presently known to live.

Major adaptations must have occurred to permit the tiny flies to settle in tar pools and other petroleum seepage areas, which appear from time to time due to shifting of tectonic plates and similar geological activity.

A fundamental question concerning the decisive biological change that took place in the original ancestor of the petroleum flies that enabled their larvae to take advantage of a trophic resource that has remained unavailable to other insects is how they overcame the potentially fatal inability to exchange carbon dioxide for oxygen across membranes that would normally be subject to blockage by many fractions of the petroleum. Once this problem was solved, an environment rich in food resources was opened to them.

The second fundamental question is how the larvae solved the problem of digesting the carcasses of dead insects while these were entrapped within a protective casing of heavy fractions of crude oil. The answer that suggests itself here is the cultivation of bacterial species in the digestive tract with the biochemical capability of doing this. However, this is not the only possible solution to the problem. If such a bacteria were isolated from the larvae, it may well be the answer to cleaning up petroleum spills and quickly ridding many places of petroleum contamination.

Once the fundamental capabilities necessary for surviving in crude oil were acquired, the petroleum flies gained undisputed possession of a valuable habitat and food supply. It can also be assumed that the crude oil traces in their bodies made them unpalatable or even poisonous to their former natural enemies. For survival of their species, this has certainly been an inestimable gain of a secure place to live without either competitors or natural enemies to contend with.

Petroleum flies are small and easily overlooked. In spite of their extremely interesting capabilities that make their way of life possible, they have maintained their secrets of existence in otherwise hostile environments for over a century after being discovered. It remains to be determined whether there are other species of petroleum flies living in other parts of the world where oil seepages or surface pollution of oil fields exploited by man exist. These questions remain to be answered.

21
Larvae that Live in Sponges

The Planipennia is considered to be a suborder of Neuroptera by some systematic biologists, while it is considered to belong to the order Neuroptera itself within the superorder Neuropterida by others. In any case, almost all families of this taxon are known to be overwhelmingly terrestrial without any affinity for water as a habitat, except for two. One of them, Osmylidae, includes species known mainly from littoral habitats, while the other includes a genus with larvae known to develop well below the surface of freshwater bodies inside freshwater sponges in the family Spongillidae.

SECTION 1

The Species and Their Habitat

This small family of insects, Sisyridae, includes about five aquatic genera, including *Sisyra* Banks, 1905. The family is almost cosmopolitan in distribution, but it has not yet been reported from Australia or New Zealand. Its other New World genus, *Climacia* McLachlan, 1869, is absent from the Old World (Bowles, 2006). The other species belong in three additional genera found only in the Old World. The modes of development of all of these insects which have already been reported are remarkable. All known larvae of *Sisyra* species develop inside of freshwater sponges and possibly also among colonies in the phylum Ectoprocta (Heckman, 2017). Little is known with certainty about the larvae of *Climacia* species, other than that they develop in or near water (Flint, 2006).

In the Old World, Sisyridae is also represented by the small genera *Sisyborina* Monserrat, 1982; *Sisyrella* Banks, 1913, and *Sysyrina* Banks, 1939. These genera, if they turn out to be valid, are found in the Palaearctic and Oriental Regions. Other generic names found in the literature are now considered by most to be invalid synonyms.

The common name for species of *Sisyra* is spongillafly. What they do during their development in the sponges remains disputed. It is therefore not known with certainty whether they should be considered predators, parasites, commensals, or mutual symbionts, a relationship that is beneficial to themselves and to their hosts. The larvae of one species has also been found on of colonies of an unidentified ectoproct species. However, whether this observation indicates that at least one species of spongillafly can also develop in colonies of species in the phylum Ectoprocta or whether this was just a chance occurrence remains uncertain.

What is known is that the larvae of *Sisyra* species live inside of freshwater sponges in the genera *Spongilla* Lamarck, 1816, and *Meyenia* (Penny, 1981). The sponges do not seem to show any permanent damage from the relationship, although the *Sisyra* larvae have often been referred to as parasites. This is based on the assumption that the insects feed on the cells of the sponges.

The family Spongillidae encompasses freshwater sponges, which are supported by spicules of silica. They generally prefer clean, unpolluted water and feed by creating currents of water to pass through their flagellated chambers, in which microscopic organisms are filtered out and consumed by their collar cells.

The mouthparts of species of *Sisyra* appear to be narrowed into sharply pointed probosces, which were assumed to be used for piercing and sucking. This stands out among members of the Neuroptera, which normally have larvae with jaws suitable for grasping and chewing. For this reason, early researchers assumed that the larvae of spongillaflies could be considered either predators or parasites. More recently, however, it has been noted that the mouthparts of the larvae are also suited for pipetting and consuming microorganisms within the chambers of the sponges. This means that the relationship between the insect and the sponge could also be considered commensalism, a simple form of symbiosis. Whether this kind of feeding could provide the sponge certain advantages, such as removal of harmful microorganisms, is still speculative and will remain so until more detailed studies of the habits of the larvae of each species are undertaken.

The genus *Sisyra* is known from most parts of the world with suitable habitats for freshwater sponges. It occurs in the tropics and warmer parts of the temperate zones. In the New World, this genus in found in water clean enough for freshwater sponges to thrive. The adult spongillaflies are small and inconspicuous, encountered near the water bodies in which their larvae develop. Their fore-wing length is seldom longer than 4.0 mm (Parfin and Gurney, 1956).

Obviously, there is much to be learned about what the insect larvae are doing in the sponges and whether or not there are significant differences in how the individual species behave inside of the sponges. Because nobody has yet found anything of economic importance in the relationship between the insect and the sponge, there have not yet been any detailed studies undertaken on this topic. Furthermore, research on the insect would not be easy. The sponges themselves are sensitive to water quality, and the insects live deep inside the sponges. Therefore, investigations in a laboratory would require a variety of special facilities and development of new research methods.

At least in the temperate zones, and probably in the tropics, as well, many species of sponges have seasonal periods of activity and dormancy. Each sponge produces small resting bodies called gemmulae, which are left after the sponge dies and disintegrates, either during autumn or at the end of the rainy season. In spring or at the beginning of the rainy season, these gemmulae each begin to develop into new sponges. Therefore, in many habitats, there is a season during which active sponges are not present. Observations and insights will be required to demonstrate where the spongillaflies spend the time during which there are no active sponges in the local water bodies, how they find sponges in which their larvae can successfully develop, and how the females deposit their eggs so that the larvae can successfully enter a host.

All of the freshwater sponges do not display seasonal periods of activity. Some remain active the year round in large oligotrophic lakes, where they can reach very large sizes. It is therefore likely that individual species of *Sisyra* have markedly different life cycles, depending upon the bionomics of each and those of its hosts.

SECTION 2

The Nature of the Ecological Niche

The lack of research on the spongillaflies is evident when the nature of the relationship between the sponge and the insect is considered. Unlike those of most members of the Neuroptera, the mouthparts of larval spongillids appear to be modified for piercing and sucking or for pipetting microorganisms, as mentioned in Section 1. The first question that arises concerns what the insect larvae are doing in the sponges. A second question arises from the first. Do all species of *Sisyra* have the same kind of relationship with its host sponge?

Because typical spongillids produce only several kinds of cells, which are arranged along a dense matrix of interlocking silica spicules, true organs and tissues cannot be distinguished. Therefore, it must be possible that the mouthparts of the spongillaflies must be used for pipetting microorganisms rather than piercing and sucking. However, the larvae have also been found in a colony of ectoprocts, which do have tissues and could be pierced by properly shaped mouthparts. Some specialists have attributed the presence of larval spongillaflies on ectoproct colonies to be the presumed fact that the larvae feed on the colonies. Others doubt that the larvae are capable of feeding on ectoprocts and attribute the presence of the insects to the artifacts of the sampling methods. In large, clear lakes or ponds, both sponges and ectoproct colonies can be expected to develop in close proximity to each other, and a larva could easily stray from a sponge to an ectoproct while being hauled up from the bottom of the water body.

Early reports on spongillaflies suggested that the larvae could be feeding on the whole cells of the sponge, making the relationship one between parasite and host. However, it might also be one involving commensalism if tiny invertebrates inside of the coanocyte chamber of the sponge were being eaten by the insect larvae.

Typically, a spongillid sponge contains chambers lined with coanocytes, also called collar cells, which create steady currents of water through the sponge by movements of their flagella. A sisyrid larva inside or near a coanocyte chamber could benefit greatly from the presence of a large food supply in the form of microorganisms drawn in from the ambient water.

If the insect larva fed upon microorganisms that could attack and harm the cells of the sponge or block the water flow inside of the chambers, the relationship would be one of true symbiosis, which would be mutually beneficial for the insect and the sponge. Certainly, larvae that feed on their hosts could well make their own survival difficult by killing their host before their own metamorphosis. Truly symbiotic relationships, however, are not at all rare in the animal kingdom. The research methods to discover what is happening inside of the sponges are not complicated in principle, but they would be very tedious and require much patience, as well as a system to keep both sponges and the insects healthy.

Elucidating the nature of the relationship between one species of *Sisyra* and its host sponge would by no means provide answers that would apply to all species in its genus. The larvae of many species of spongillafly have never been identified, and there is little information even about the habits of those species that have been found and studied. It is likely that there are many variations in the ecological niches occupied by these still unknown larvae and that their host species are different.

What would be particularly interesting to know is how the adult sisyrid finds suitable host sponges and deposits its eggs in a way that the larvae can get into the host. Freshwater sponges do not always appear at the same locations from year to year, especially in water bodies near human habitations. The adults of the various spongillafly species are very small, often less than 5 mm, and they have not been reported to swim or fly underwater, as species in the other Hymenoptera have been known to do. Therefore, there must be a great deal more to learn about the survival strategies employed by these inconspicuous insects.

While the life cycles of the *Sisyra* species are not at all well understood, those of other members of the family Sisyridae are scarcely known at all. While the larvae of *Climacia* species are presumed to be aquatic or semi-aquatic because their pupae have been found at the edge of streams, the basic places and methods of larval development remain speculative. It is assumed that they do not develop in sponges, but there is no certainty about this assumption.

22

Parasitoid Larvae in Aquatic Insects

During research on the shallow ditches draining apple orchards southeast of Hamburg, Germany, tiny black flying insects were occasionally captured by sweeps of a net. Many casual observers, including students, assumed that these were just small terrestrial flies that fell into the water while the net was being swept. This has apparently been a common mistake, which resulted in one of the largest insect taxa being overlooked until the past few decades. It appeared to the untrained eyes of casual observers that these insects were tiny black flies, but a glance through the magnifying glass revealed two pairs of wings rather than one. Although the body looks small and frail, it turns out to be covered by a strong and thick exoskeleton. In addition, the tiny wings turn out to be strong and durable. Just a glance through a magnifying glass makes it evident that these species belong to the order Hymenoptera. Furthermore, a particularly careful observer can discern that the insect did not fall into the water but is rather exiting the water quickly to escape capture.

Because the larvae of these small insects develop in the bodies of other insects, it was not realized until fairly recently how many different species of these parasitoids exist. Those with aquatic insects as hosts can be considered aquatic. The great majority of them are terrestrial, however. The adults of the aquatic species must usually seek out the host underwater in the kind of water body inhabited by insects suitable as hosts.

SECTION 1

Life Cycle of Parasitoids

The small black brachionids and slender ichneumonids that are occasionally observed on or underwater are never seen during their larval development except under unusual circumstances. Their larvae develop as parasitoids inside insects from other orders, and they emerge to form pupae on their dying hosts. As adults, they must seek a suitable host, usually a larva. Terrestrial species have developed very unusual morphological modifications and behavior patterns to achieve their goal of depositing one or more eggs on a suitable host. Some species of Ichneumonidae have extremely long ovipositors capable of penetrating wood. Some of these species have been reported to be parasites of plants, but these reports are questionable. Trees with larvae of beetles eating burrows through the wood may be the ones through which the ovipositors of the ichneumonids are meant to bore. An egg of the parasitoid placed on the beetle larva deep in the wood of a tree trunk provides a fairly secure guarantee that the next generation of the parasitoid species will survive. Some of the aquatic species accomplish feats that are even more amazing.

The most common aquatic species found personally during studies of water bodies on the floodplain of the Elbe Estuary in northern Germany were members of the family Braconidae, but species belonging to Ichneumonidae were represented, as well (Caspers and Heckman, 1982; Heckman, 1984). In addition, aquatic hymenopteran parasitoids belonging to other nominal families were reported from northern Germany and other parts of Europe (Wesenberg-Lund, 1943), although the classification of some of these species has been changed to accommodate hundreds of species more recently described. In 1971, a catalog of species in the family Braconidae, subfamily Opiinae, was published (Fischer, 1971), but since then, there has been a considerable expansion in the number of described species and in the knowledge of the biology of some of the more newly discovered ones.

As already mentioned, parasitoids differ from parasites by killing their hosts at the end of their development. Parasites usually lack life stages that live independently of the host. There are two superfamilies of parasitoid insects presently assigned to the insect order Hymenoptera in Europe. The families encompassing the most aquatic species in the watersheds of the North and Baltic Seas seem to be Braconidae and Ichneumonidae, both of which belong to the superfamily Ichneumonoidea. A substantial majority of species in both of these families have terrestrial arthropods as hosts and do not seem to seek suitable hosts underwater.

In addition to the species in the superfamily Ichneumonoidea, aquatic insects in the superfamily Chalcidoidea have also been found in Europe (Baquero and Jordana, 1999). These species seem to be less selective in the choice of their hosts than members of the Ichneumonoidea, but it is premature to come to this conclusion when the habits of so many species remain unknown, worldwide.

Many parasitoids were long overlooked, mistaken for terrestrial species, or treated as curiosities. In fact, although most parasitoid species of Hymenoptera develop only in terrestrial arthropods, those few that develop in aquatic insects turned out to be superbly modified to reach their hosts underwater. Their behavior is so well oriented toward hosts living underwater that it would be hard to imagine them also accepting terrestrial species as hosts. In spite of the fact that most or all of their aquatic hosts remain submerged, the imagoes of the parasitoids successfully find them so that they can deposit one or more of their eggs in or on the body of each host. Larvae are the stage most frequently selected for the initial infestation by the adult parasitoids, although a few species seem to select adults for the hosts of their offspring. Apparently, the pupal stages of hosts are avoided by the parasitoids. After an egg is deposited on or just beneath the integument, it hatches, and the larva begins to feed on the host tissues, avoiding any of the vital organs. Because the hosts remain healthy enough to go about their normal activities of swimming or creeping around to look for food, they show no recognizable signs that they are infested with a parasitoid until shortly before they die. This makes it difficult for entomologists to find and study the development of the larvae.

The adults of many of these species fly directly into the water body without hesitation and are capable of motion resembling flight using the wings beneath the surface of the water. Some species land on emergent parts of aquatic plants and creep down the stem into the water (Hirayama et al., 2014). The ease of braconid wasps at breaking through the surface tension layer in each direction has been personally observed. When a water sample is scooped out of a pond, occasionally a tiny insect breaks through the surface layer and continues flying away, apparently making little change in the movement of its wings as it transitions from the water to the air (Wesenberg-Lund, 1943). The aquatic parasitoids in the water belong to those species that have specialized in seeking out aquatic hosts, presumed to belong to only one or a few closely related species. As the larva grows, it eats its way through the soft body parts of the host, still feeding on tissues that are not vital for the host's survival. After the larva has completely developed, it consumes vital organs, killing the host, just before the parasitoid emerges for pupation. In some cases, one pupa of a parasitoid appears, while in others, groups of several may emerge at about the same time, either from different parts of the host's body or in small clusters. After the adult parasitoids emerge from their cacoons, often attached to the body of their dying host, they seek new hosts for their own progeny, presumably after mating.

What appeared to be a rare or unusual behavior of insects that were only occasionally observed has turned out to be one of the most common activities in the insect world. The insect order with the most known species in Great Britain today is Hymenoptera, and this is due to the great species diversity of parasitoids belonging to the superfamily Ichneumonoidea, which, as noted above, encompasses the huge families Ichneumonidae and Braconidae. A few families of Ichneumonoidea formerly recognized in Europe have been demoted to subfamilies of one of these two families. By 1991, there were roughly 3,200 species in this superfamily known from Great Britain out of about 13,000 species described throughout the world (Shaw and Huddleston, 1991). Estimated numbers of all species, described and undescribed worldwide, exceeds 40,000 (van Achterberg, 1984), with some estimates much higher (Shaw and Huddleston, 1991). Of course, these figures are obsolete today. As in Europe, the vast majority of these species infest terrestrial insects and other non-aquatic arthropods. However, it is the aquatic species that display some of the most interesting deviations from typical insect behavior, such as flying through water. Of course, it is the small window of opportunity for actually observing these interesting insects that has delayed appreciation of their importance. They are only observable free in the environment between the time their larvae emerge from the host and pupate and the time their imagoes mate and deposit their eggs on or in a new generation of hosts. The rest of the time, they cannot be detected by any known means without dissecting an infected host.

Many parasitoid species can only be studied after elaborate preparations of facilities for raising the hosts. If such facilities were available, methods of recognizing hosts infested with a parasite might successfully be developed. Even if characteristics of infestation were known, the development of the parasitoid remains concealed inside the host throughout most of its development. Most vexing of all is the large number of species of such parasitoids that have been described already, while extremely few of them are recognizable as larvae. The simplest study methods seem to entail obtaining adults emerging from pupae attached to hosts, which permits the host of the parasitoid to be identified with a reasonable degree of certainty. The next step is to place the adult together with suitable hosts in an aquarium to observe when and how the parasitoid initiates infestation of its host. This, of course, is time consuming and requires spacious laboratories and expensive equipment. However, such research might well be profitable when the host turns out to be an insect that damages crops or domestic animals, or, alternatively, transmits serious diseases.

When research concentrates on finding parasitoids to control insects harmful to crops in the tropics, more attention may be given to parasitoids that kill aquatic hosts. For example, there is a long list of tropical plants cultured for food production, which might be protected by species of Hymenoptera that develop in aquatic insect pests living on important cultured plants, such as rice, taro, water chestnut, lotus, yam, sweet potato, or one of the many green vegetables that grow in water.

One of the remarkable features of the parasitoids is an extreme robustness of the tiny adults. Most braconids are roughly 5 mm or less in length, but they have firm, armored bodies and are not easily crushed. Their wings are also unlikely to be torn by their movements underwater. In order to place eggs beneath the integument of suitable insect hosts, the tiny wasps must be swift and strong. They must overcome the efforts of the host to evade and avoid being infected by a parasitoid.

Egg laying does not seem to be dangerous for the tiny parasitoids. They are so small that they apparently fail to cause any alarm or defensive reactions by the intended host of its larvae, and they are also strong and robust enough to ward off any attacks by other insects. The aquatic insects are fast and agile underwater, and the deposition of eggs on the host selected is completed quickly.

SECTION 2

Taxonomic Studies of Aquatic Parasitoids

Obviously, the study of parasitoids in the superfamily Ichneumonoidea will be overwhelmingly directed toward the many terrestrial species. Such studies require much less preparation than those of aquatic species because the deposition of the parasitoid eggs on or in their hosts do not require aquaria to observe. They can be watched directly, even during field studies.

Adult parasitoids appear occasionally in nets that have been swept through water bodies in which suitable hosts are abundant. Braconids are usually very small, with a length of less than about 5 mm and often under 3 mm. Some of the ichneumonoids are larger, however. Apparently, some aquatic species deposit only a single egg on each host, but the larvae of few species have been carefully studied, and it is not known in most cases whether laying only one egg is a general rule for most species. It is still not known which host is acceptable to any given species of imago, so it cannot always be determined whether or not a species of parasitoid should be considered aquatic based on one observation. Wesenberg-Lund (1943) did not include parasitoid hymenopterans among the aquatic species if they develop in insects that live on the emergent parts of aquatic plants because their habits were no different from those of terrestrial parasitoids, and their adults did not have to enter the water to find hosts.

Most of the recent taxonomic publications on parasitoid hymenopterans have included thorough anatomical descriptions of new species, thorough enough to easily distinguish the imagoes. However, there are usually few notes on the kind of habitat in which the specimens were found and the host or hosts in which the larvae can develop. Exceptions to this will be found in Section 3 of this chapter.

The taxonomy of the parasitoid species of Hymenoptera is one of the last pioneering research fields for specialists in taxonomy and systematic. To the hundreds of species of parasitoid that have recently been named and described are hundreds of others still waiting to be described throughout the world. Apparently, most of those that have recently been discovered were collected as adults or as pupae attached to the dead or dying host. This leaves the larvae of hundreds of species about which little information of any kind has ever been reported.

When the hymenopteran parasitoid is "flying" underwater, it is probably in search of a suitable host for its young. Therefore, the species can be assumed to be aquatic and have aquatic hosts if the adults are searching for one underwater. Apparently, each parasite is limited to one species of host or to a few that are closely related phylogenetically and probably in the same family. This assumption conflicts with earlier reports indicating that aquatic species seek out one of a fairly large number of potential hosts. Which assumption is correct cannot be confirmed yet, and it may be that some species are relatively unselective while others are highly specialized in the choice of a host. Obviously, there is little likelihood that most species of parasitoid seek to deposit their eggs in both aquatic and terrestrial hosts. Both locomotion underwater resembling flight and actual flight to find a suitable host undertaken by one and the same insect have not yet been observed. It seems likely that no parasitoid would accept either a fully terrestrial species or an aquatic one as a suitable choice as a host for its larvae. However, the hosts of most terrestrial parasitoids are still unknown, and it would be premature to speculate on the ways in which the parasitoids choose the hosts for their offspring. It seems likely, however, that the adult female of any given species of parasitoid undertakes an instinctive journey in search of individuals of one or a few hosts, which usually share common habitats, and on this journey it deposits only one egg on each suitable host that it detects. Because several pupae have been observed attached to a single host, it also seems that one rule does not hold for all species.

The parasitoid hymenopterans which develop within the bodies of other arthropod species had apparently not previously been distinguished with accuracy. For example, Wesenberg-Lund (1943) reported that the aquatic Hymenoptera native to Central Europe belonged to five families: Agriotrypidae, Braconidae, Chalcididae, Ichneumonidae, and Proctotrupidae. Of these, three have been promoted to super-families, which also encompass smaller families in their groups. Obviously, the number of aquatic species present was underestimated on a massive scale. Only in recent decades was the great size of two super-families appreciated, and the vast majority of the species were relatively recently described.

Wesenberg-Lund (1943) also assumed that single parasitoid species were capable of developing in hosts of many species, when it is now known that each species of

Braconidae develops in only one or a few closely related species of host. This is because only a few species had already been described at the time the book by Wesenberg-Lund (1943) was published, and many of the aquatic species were simply not distinguished.

The two families of Ichneumonoidea known to include aquatic species in Europe are large and represented in most parts of the world. At the present state of knowledge, only adults can be identified with reasonable certainty in Central Europe, and it is hard to even locate larvae in living hosts. There is much less knowledge of these parasites outside of Europe, but the taxonomy and systematics of the group is presently advancing at a rapid pace. It is premature to speculate how many families will eventually be recognized worldwide. There is still little knowledge about the hosts of those species that have been described. Furthermore, there are relatively few research institutes throughout the world with the facilities for keeping the hosts alive until the emergence of the parasitoid imagoes. Discovering the hosts for more than 40,000 species is a daunting undertaking, although the percentage of aquatic species in this superfamily is probably rather small. Because the promise of using these parasitoids to control economically important insect pests that cause economic losses in agriculture is so great, the costs of research on their habits will probably be repaid many times over. However, a negligibly small proportion of the research is concentrating on aquatic species.

At the present time, a tentative classification of the accepted families encompassed by each of the superfamilies can be presented (Table 22.1). However, within such a large and active field of taxonomy, there is no guarantee that the systematics of what are regarded as the two main families of parasitoid hymenopterans will still be valid tomorrow.

SECTION 3

Ecological Studies of Aquatic Parasitoids

Once a species has been determined to be aquatic and its host or hosts are known, studies of its ecological requirements, mating behavior, and search for suitable hosts can be undertaken. Although many such studies are in progress in many parts of the world, very little is known about the ecology of most aquatic species, and it is still not even known whether or not some of them have aquatic or terrestrial hosts.

The fact that the larvae develop as parasitoids in one or a limited number of aquatic insect species somewhat simplifies studies of their ecology. However, it makes it more difficult to observe them. They are active outside of the host for only short periods of time, and their appearance outside of the hosts is likely to occur simultaneously or in conjunction with one phase in the developmental cycle of the host. In this way, the parasitoids can mate and lay eggs with the best chances of success.

A few of the things that must be determined include which life phase of the host is initially infected, whether only one host or a group of alternative hosts are sought, how long it takes before pupation occurs, and what seasons the various events in the life cycle of the parasitoid and its host are likely to occur. When only a few aquatic species were known, it was thought that pupae were never actively infected by the egg-bearing female parasitoid, and that the most parasitoids infected only larvae. It

Table 22.1 The families belonging to the two major superfamilies of parasitoid Hymenoptera and two minor ones. Based on Goulet and Huber (1993), including information on numbers and distribution of nominal species known as of 1993. Most species are not aquatic, but a few aquatic species apparently belong to each superfamily, and the habits of some remain unknown.

Superfamily	Ichneumonoidea	Notes
Families	Ichneumonidae	By 1993, over 60,000 species, cosmopolitan
	Braconidae	By 1993, over 40,000 species, cosmopolitan
Superfamily	**Chalcidoidea**	
	Agaonidae	More than 650 species living only in fig trees
	Aphelinidae	About 1120 species, cosmopolitan, used for pest control
	Chalcididae	More than 1875 species, cosmopolitan
	Elasmidae	c. 260, cosmopolitan, but most Paleotropical
	Encyrtidae	About 3825 species, cosmopolitan parasitoids
	Eucharitidae	About 380 species, most tropical
	Eulophidae	More than 3900 nominal species, control leaf miners
	Eupelmidae	Routhly 900 species, cosmopolitan but more tropical
	Eurotomidae	About 1425 species, cosmopolitan
	Leucospidae	About 240 tropical or subtropical species
	Mymaridae	About 1400 species, cosmopolitan egg parasitoids
	Ormyridae	Over 100 species, many Old World, few New World
	Perilampidae	Over 260 nominal species, cosmopolitan
	Pteromalidae	About 4115 nominal species, cosmopolitan
	Rotoitidae	One species in New Zealand, a second in Chile
	Signiphoridae	c. 80 species, used against aphids, agricultural pests
	Tanaostigmatidae	About 90 species, cosmopolitan gall wasps or parasites
	Tetracampidae	c. 50 species, parasitoids of beetles & leaf-mining flies
	Torymidae	At least 1150 species, ectoparasitoids of gall insects
	Trichogrammatidae	c. 675 nominal species, egg parasites of larger insects
Superfamily	**Mymarommatoidea**	
	Mymarommatidae	About 9 species in one genus, cosmopolitan
Superfamily	**Proctotrupoidea**	
	Austroniidae	One genus and three Australian species
	Diapriidae	c. 4300 species, cosmopolitan, many undescribed
	Heloridae	c. 7 known species, worldwide
	Monomachidae	20 species, Neotropical, Australia, and New Guinea
	Pelecinidae	One species, the Americas, Canada to Argentina
	Peradeniidae	One genus, two species in Australia and Tasmania
	Proctotrupidae	Estimated 1200 species, cosmopolitan
	Roproniidae	c. 18 species, Holarctic and Oriental Regions
	Vanhorniidae	Five species, Holarctic and Oriental Regions

is also not certain which species deposit only a single egg in each host and which deposit more than one egg. It is also important to know whether infestation by one parasitoid larva deters other imagoes from depositing additional eggs in a host. If not, do multiple infections interfere with the survival of the additional parasitoids developing within the host?

Insights into the relationships between the parasitoid and host could be gained by learning whether there are behavioral changes that occur after a host becomes infected. Do all species of braconids kill their host before leaving its body for metamorphosis to the pupa stage?

Finally, the activity of the aquatic ichneumonids and braconids while they are acting as free-living pupae and adults, should be better described. With so many species in these families, it is clear that differences will be found in how they seek suitable hosts for their eggs and larvae. These hosts have a great many ways of life in the aquatic community. Therefore, it is clear that the life cycles of the parasitoids must also be adjusted to those of the hosts. The discovery of braconids employing the larvae of flies that mine aquatic plants as hosts for their larvae show that the relationships between parasitoid and host can require very specialized interactions to be successful.

Feitosa et al. (2016) reported that many species of aquatic insects attach their eggs to the leaves of submerged aquatic tracheophytes. A parasitoid in the State of Amazonas, Brazil, places its eggs on aquatic macrophytes of the species also preferred by the larger insect hosts, allowing their own larvae to enter the eggs of several kinds of aquatic insect belonging to the orders Odonata, Hemiptera, and Lepidoptera. They are then able to develop by taking their nourishment from the eggs. The species which was reported to be the parasitoid of the eggs is *Anagrus amazonensis* Triapitsyn, Querino, and Feitosa, 2008, is a member of the family Mymaridae, in the superfamily Chalcidoidea.

The adult parasitoids in this superfamily are among the smallest of insects. The smallest of the imagoes among the species in Chalcidoidea already described averages between 0.1 and 0.2 mm long. As in the case of the superfamily Ichneumonoidea, the great majority of the known species develop in terrestrial hosts.

SECTION 4

Geographical Distribution of Aquatic Parasitoid Species

It appears to be fairly certain that the known zoogeographical ranges of the individual species of parasitoid hymenopterans will be greatly expanded as more studies of the group are undertaken. The numbers of species now known from each part of the world will substantially increase as more specialists undertake research on this group. A complication that researchers will certainly face is that ecological requirements and life histories of the parasitoids do not follow phylogenetic patterns. Their habits, reproduction, and search for nutrients depend almost entirely on the ecological characteristics of the hosts. One family of closely related species will include parasitoid species that must seek suitable insect hosts underwater, inside of the stems of both terrestrial and aquatic plants, inside of wood, high in trees, inside of mushrooms, and underground. They are found almost everywhere entomologists have looked, although sometimes patience is required in order to see one.

For example, in Japan, the host of the ichneumonid species, *Apsilops japanicus* Yoshida, Nagasaki, and Hirayama, 2011, is an aquatic moth: *Neoshoenobia teatacealis* (Hampson, 1900), which belongs to the lepidopteran family Crambidae. Some of the moths of this family develop from larvae, which mine the leaves and stems of plants. This particular moth mines the floating leaves of a yellow water lily, *Nuphar lutea* (Linnaeus) Sm., entering a part of a leaf above water. It later moves into the petiole of the leaf beneath the water, where it transforms into the pupa (Hirayama et al., 2014).

In most species of leaf-mining hosts for hymenopteran parasitoids, coping with life underwater would not be a requirement necessary for survival. The moths of other species of Crambidae do not mine aquatic plants, and the parasitoids of such species would not normally have to cope with life in plant tissues below the surface of a water body. A perusal of literature on the moth species would not immediately reveal that the larvae of species in the family Crambidae are typically found in aquatic plants, and *Apsilops japanicus* would appear to be an oddity with an aquatic plant for a host. Its habits would therefore differ considerably from those of closely related parasitoids.

Judging from the number of species being described in Europe, the numbers of undescribed parasitoids in the New World must be amazingly large. There is not yet enough information to make a reasonable estimate of the number, but those already described run into the tens of thousands. Many of those in the Nearctic Region can be assumed to invade aquatic insects (Kula and Zolnerowich, 2008; Kula et al., 2009), but the great majority of all known families are terrestrial. However, it is almost certain that in the Neotropical Region, the percentages of parasitoids of aquatic insects in the total number of hymenopteran species belonging to families of parasitoids would be much larger than elsewhere due to the extent of the wetlands in South America, in which potential hosts live. Pioneering studies on these insects have already yielded results that are supporting this assumption, in spite of the fact that obtaining specimens of Ichneumenoidea from living aquatic insects requires much more effort than collecting specimens of other aquatic species.

The only terrestrial regions of the world in which the number of endemic species in the superfamilies Ichneumenoidea and Chalcidoidea is not extremely great are those surrounding the poles. This is not surprising because the number of insects that could act as potential hosts is smaller in the taiga and tundra. However, given the large populations of those insect species that can survive in the coldest places in the world not covered by glaciers, any parasitoid that could also survive there would have no difficulties finding food sources for its larvae.

These parasitoids have already been found almost everywhere that people have looked for them. Naturally, those countries with the most entomological research institutes usually have the longest lists of hymenopteroid parasitoid species. Personally, I would not even venture a guess for the number of valid species that will eventually be described.

23

Flies and other Insects that Eat Snails

One family of flies, the Sciomyzidae, is noted for the development of the larvae of some of its species inside of the shells of living snails or occasionally other invertebrates, while the victim goes about its daily activities until the very end of the relationship. Discussions of whether the fly larva is a scavenger or a parasitoid almost always end favoring parasitoid (Wesenberg-Lund, 1943). Some of the approximately 500 species apparently produce larvae that develop in the dense mats of algae found in eutrophic water bodies, but the information about the development of all European species is not yet available. Like the species in the order Hymenoptera with larvae that develop inside of other insects, the known sciomyzid fly larvae must be classified as parasitoids rather than parasites because just before they become pupae, the host is killed.

Other aquatic insects attack snails and feed on them as predators. However, there are not many aquatic insects that can get past the formidable shells of the mollusks.

SECTION 1

Parasitoid Fly Larvae

Sciomyzidae is a family that is well represented in Europe, so the species present there have been better studied than those on other continents. Some of the flies are known to develop in only one species of snail, while others select two or more as hosts. The larvae of some species have been found to have a shape that fits into the whorls of the snail shells of the species it invades. Each species has its own special habits and preferred host or hosts. All but one African species that has been studied attacks or invades the shell of a mollusk. The exceptional species was raised in a laboratory entirely on annelids (Resch and Carde, 2003), but is that a typical diet?

All other species that have been studied feed on one or more species of snail, slug, or bivalve. The life history of each is distinctive. Apparently, all species that have been studied can be described as parasitoids or predators. No true parasites have yet been discovered among the larvae of these flies.

After the larval development is complete, at least some of the species molt and pass their pupal stage inside the shell of the host snail. Just prior to the molt, the larva feeds on the vital organs of the host, causing it to die, leaving the shell vacant for the pupa.

The family Sciomyzidae is large and distributed throughout the world. A book about this family alone supplies a great many details about its numerous species (Knutson and Vala, 2011).

SECTION 2

Snail-Hunting Water Bugs

At various times, aquatic snails were deliberately introduced to other continents. It was frequently noticed that freshwater snails were eaten by local farmers in various parts of the world and that they were good sources of protein (Heckman, 1979). Although the snails being eaten were native to the countries where they had become favorite foods, the dangerous practice of transporting species to distant lands was recommended by international organizations to improve local nutrition.

In South and Southeast Asia, at the time when rice was being grown almost in monoculture, snails in the family Pilidae that had been introduced from South America turned out to be extremely destructive to the rice. They were larger than the native snails in the same family, which were popular foods, and they lacked natural enemies. When they were released on their new continents and islands, they left behind all of the predators and pathogens that had controlled their populations back home. Their rapid growth rate was achieved by eating rice plants and other aquatic vegetables faster than the native species. They also had a rapid rate of reproduction, and they more than replenished the few snails that local farmers could consume. The result was another environmental emergency that required a remedy.

The search for a solution to the self-created emergency was the search for a method of biological control for the snails. One of the introduced snails was *Pomacea bridgesi* (Reeve, 1856), which allegedly needed to be controlled in India because of the damage it was doing to the taro crop. Through a search for an agent of biological control, the large predatory insect in the family Belostomatidae, *Sphaerodema rusticum* (Fabricius, 1781), was identified (Aditya and Raut, 2001). This same species was found by Aditya and Raut (2002) to prey on two other species of aquatic snail: *Lymnaea luteola* Lamarck, 1822, and *Physa acuta* Draparnaud, 1805. What is remarkable about its feeding on *Pomacea bridgesi* is that this snail species has a thick operculum which can tightly close the shell. *Lymnaea luteola* and *Physa acuta* have no such protection.

SECTION 3

Fireflies that Consume Snails and Making Decisions on Biological Control

In small water bodies and wetlands throughout the world, the larvae of species in the genus *Photinus* Laporte, 1833, members of the family Lampyridae, usually live in dense masses of submerged plants. They share the habitat with freshwater snails, which these carnivorous insect larvae find to be excellent sources of protein and easy to catch. Reportedly, they prefer very young snails with shells that are easy to crush.

Photinus is not the only genus in the family Lampyridae with aquatic species that feed on freshwater snails. A study of the species *Luciola leii* Fu and Ballantyne, 2006, in Hubei Province, China, by Fu et al. (2006), revealed that the preferred foods of this species are two species of snail in the family Lymnaeidae: *Lymnaea stagnalis* (Linnaeus, 1758), and *Lymnaea auricularia* (Linnaeus, 1758), one species in Planorbidae: *Gyraulus convexiusculus* (Hutton, 1849), and one species in Viviparidae: *Bellamya purificata* Heude, 1890. The larvae of many terrestrial lampyrids feed on snails, so it is not surprising that the species with aquatic larvae prefers aquatic snails.

The larvae of *Luciola leii* develop in small, shallow water bodies, including rice fields, so they could be considered beneficial agents of biological control of snails that feed on aquatic crops. Because *Gyraulus convexiusculus* is also the first intermediate host of the intestinal trematode, *Echinostoma lindoensis* (Sandground in Bonne, 1940) and the second intermediate host of *Echinostoma malayanum* (Leiper, 1911), the activity of the firefly larvae should help to protect the local human population from these debilitating parasites by reducing the number of intermediate hosts for both species.

In this situation, the importance of good taxonomic and systematic research should be emphasized as a basis for programs of biological control of species that can cause problems for public health or agriculture. The first problem was encountered when the ranges of the species involved in vectoring the parasites to human beings had to be determined. The first problem detected that was caused by one of these parasites was mentioned in a medical journal published about two years prior to the outbreak of World War II in Europe (Brug and Tesch, 1937). It reported a high incidence of infestation of local residents of villages located on The Celebes by a parasitic trematode tentatively reported to be *Echinostoma ilocanum* (Garrison, 1908). The first intermediate host was reported to be *Anisus (Gyraulus) sarasinorum* Bollinger, 1914, and the second intermediate hosts were supposedly any one of several mollusks, including the aquatic snail, *Vivipara javanica rudipellis*, and the bivalve, *Corbicula lindoensis* Bollinger, 1914. Among the information about extant species listed in various publications, the name *Vivipara* is sometimes spelled *Viviparus*, and deciding which of these two names should be valid has been the subject of much contention in the past.

Vivipara javanica and *Vivipara javanica rudipella* appear in various publications, but there are still some differences of opinion of what the valid name of the species in Lake Lindoe should be. Although these names are listed in more recent publications as second intermediate hosts of *Echinostoma lindoensis* in Lake Lindoe, the exact range of the species *Vivipara javanica* has never been precisely delimited. The second problem is determining the range of the nominal subspecies, *Echinostoma lindoensis rudipella,* and revising its status if necessary. Until this is done, the name will be copied from old publications without any real understanding of whether it should be considered a valid subspecies, an independent species, or just a minor variety of one species, the name of which has no taxonomic validity. An additional annoyance is that so many authors fail to follow the rules for introducing the name of a species. A full species name includes a generic name, a specific name, the name of the author who first described the species, and the year the publication appeared in which it was described. If the generic name has been changed since the description was first published, the name of the author and year of publication should be placed in parentheses. If this rule is not followed, confusion concerning the species referred to can occur in some cases.

Because neither the parasites nor most of the intermediate hosts had been identified in the first publication correctly, the paper was published by Sandground and Bonne (1940), which described the parasite as a new species: *Ecinostoma lindoensis* (Sandground in Bonne, 1940). The first intermediate host has been corrected to *Gyraulus convexiusculus* in recent publications, but without regard for the ranges of congeneric species in Southeast Asia.

Because a different nominal species with a reported range mainly in Malaysia was reportedly infesting a large and increasing proportion of the rural population in parts of The Philippines, the question to ask would seem to concern the similarities in the life cycles and intermediate hosts of the two parasite species. Are these species really distinct, and if so, is there a chance that both could share the same mollusks as intermediate hosts?

This other species, *E. malayanum,* was made the subject of a research publication, which showed it was having great success infesting large numbers of people in The Philippines (Belizario et al., 2007), just like *E. lindoensis* had been having in The Celebes 70 years earlier. The snail, *G. convexiusculus,* could have been serving as an intermediate host of both intestinal fluke species, at least in theory. But what were the geographical ranges of the snail species?

Perhaps a second hypothesis could be added to the first in order to find a solution to the rapid spread of the intestinal fluke species in the Philippines and elsewhere. *Gyraulus convexiusculus* is present over a large range in East and Southeast Asia. Studies in China suggest that the larvae of the newly described and named firefly, *Luciola leii,* show a special liking for the snail, *G. convexiusculus.* In this case, it should be determined how large the range of the firefly is or if a firefly species with similar feeding habits is already present on the Celebes. Could breeding and releasing such a firefly provide a means of controlling the snail populations, which are the chief intermediate hosts of *E. malayanum* and perhaps other species of intestinal fluke that have long been a health problem in insular Southeast Asia?

In this case, a dearth of information about the flukes, snails, and insects of the region certainly makes feasibility studies in many parts of the world more difficult than necessary. If a similar problem had to be solved in Europe, where a much more limited insect fauna had been under systematic study for the better part of 300 years, a feasibility study for implementing a solution would have been much easier to complete. However, there would still be difficulties in arriving at fully satisfactory conclusions.

In a case like the one just analyzed, improvements in public health and sanitation would be the best way to solve the problems associated with fluke infestations. Another consideration is that there would be complaints in some places that *Gyraulus convexiusculus* is on the Red List of Endangered Species, which would make reducing its local populations a strong cause for objection. Finally, if there were no native aquatic fireflies with a liking for snails in their diets, the introduction of species from other parts of the world would have unforeseen impacts on the local fauna and flora, which would probably turn out to be objectionable.

The rest of this hypothetical feasibility analysis speaks against the success of a project using a firefly to prevent the spread of the liver flukes to human beings. Even if it could theoretically reduce the number of metacercaria in the water by eliminating an intermediate host, the firefly is native to a place so far from The Celebes and The Philippines that climatic differences would make it doubtful whether it could survive and reproduce at all. The firefly was first described in 2006 from specimens taken in Hubei, China. There are no records showing how widespread the firefly actually is in China or elsewhere, and there are no reports on its habitat requirements.

An additional fact speaking against the success of the project would be that at the outset, it was assumed that only one snail species was serving as an intermediate host for both species of liver fluke, but other species of snail are also capable of doing this and are also present at each of the locations the flukes are invading human hosts.

Finally, if any of the reports cited here contained an error in identification of a species or the determination of which can serve as intermediate hosts for a specific parasite, then all of the above analyses could have led to false conclusions, and the likelihood of complete failure of any such project would be considerably increased.

What all of this discussion was meant illustrate is that predictions of environmental changes due to attempts at biological control projects are much easier and more accurate when data bases with ecological information about all species involved are available. Of course, the solution to the problem of infestation by intestinal flukes does not require biological engineering. Like pathogens, parasites and their eggs can be killed by simple boiling in water at sea level or cooking food in pressure cookers at higher elevations. Public education, rural electric power, setting standards for cooking food, treating wastewater, and frequent hand washing can go a long way toward prevention of parasites like liver flukes from establishing themselves in the bodies of ordinary people.

24

Flies that Give Wedding Presents

The superfamily Empidoidea belongs to the suborder Brachycera of the order Diptera. It encompasses a large number of species in the families Empididae, Dolichopodidae, Atelestidae, and Hybotidae (Meyer and Speth, 1995). The adults of aquatic species in these families are typically observed flying rapidly in small swarms back and forth just above the surface of water along slow-flowing streams and lentic water bodies. Within the swarms of these flies, which can sometimes combine to include many individuals, males and females form mating pairs so that the females can obtain enough sperm to last them throughout the rest of their lives. Typically, during courtship, the male will catch an insect suitable for food and present it as a gift to the female inside of a silken package. Acceptance of this gift by the female generally formalizes the bond between the two insects, and the male will provide her with the sperm that will be stored in her body to fertilize all eggs she will produce. Early observers described this courtship activity as giving "wedding presents," also called nuptial gifts (Chvála, 1994). There are indications that females of aquatic species may stay together with the same male in case more sperm is needed, but a contrary explanation that older females may mate with a different male in order to spawn for a second time was provided by Svensson et al. (1990). The species studied in the second case was *Empis borealis* (Linnaeus, 1758), a terrestrial species, however. Empididae is a large family, and the details of the reproduction of most are still not known. Judgement on the number and degree of differences among the individual species must wait until the behavior of most of the species is learned.

The habits of most species in the dipteran superfamily Empidoidea have not yet been carefully investigated in most parts of the world, but many observations suggest that the habit of giving wedding presents could be common to most or all European species in the superfamily. Unfortunately, reliable reports of this behavior by species in the family Dolichopodidae are lacking, and because the flies in this family do not usually form large mating swarms, the actual transfer of a nuptial gift would take a great deal of patience to observe.

Whether giving "wedding presents" is a characteristic behavior of the family Empididae in all parts of the world must still be confirmed through research on other continents. Most European species have already been classified as either aquatic or terrestrial, but even this is not known about many species that are not endemic to Europe and are known only from morphological descriptions of the adults.

It seems that most species in the superfamily Empidoidea are terrestrial during both the adult and the larval stage. In northern Germany, Meyer and Speth (1995) examined 4873 specimens of the native species in the superfamily and reported that out of 81 species present, 7 were aquatic. Those species with larvae that develop in water were classified as aquatic. The family Dolichopodidae was represented by 26 species, and the family Empididae, by 55. No species in the family Hybotidae were encountered among the specimens, all of which had been collected in emergence traps. All of the species reported to be aquatic belong to the family Empididae. The following empidid species were identified: *Chelifera diversicauda* Collin, 1927; *Chelifera precatoria* (Fallén, 1816); *Chelifera precabunda* Collin, 1931; *Chelifera stigmatica* (Schiner, 1862); *Clinocera stagnalis* (Haliday, 1833); *Hemerodromia unilineata* Zetterstedt, 1842; *Wiedemannia zetterstedti* (Fallén, 1826).

In a survey of the empidids in the state of Hessen, Germany, Wagner (1983) also reported only seven of the species he found to be aquatic.

Like other members of the superfamily Empidoidea, the family Empididae encompasses species that give wedding presents. It includes a large number of species, the adults of which are typically observed flying rapidly in swarms moving back and forth near the habitats in which their larvae develop. Many of these flies belong to species in the cosmopolitan genera *Empis* Linnaeus, 1758, and *Hilara* Meigen, 1822, neither of which accounted for any aquatic species in the area of northern Germany investigated (Meyer and Speth, 1995).

Few members of this family have been well studied in other parts of the world, so much of the information obtained about this taxon was obtained from studies of native European species. The habitats preferred by the individual species vary greatly, and most of them apparently develop in moist soils of forests, meadows, or locations along the margins of water bodies. Several species seem to be able to develop in more than one kind of habitat (Meyer and Speth, 1995; Niesiołowski, 2006).

As a group, the adults are thought to play an important role in the pollination of flowers (Niesiołowski, 2006). They seem to subsist mainly on the nectar, and as bee populations decline, their ecological and economic roles become more important. The larvae prey on small invertebrates, apparently mainly insects. The majority of the species about which information is available seem to be terrestrial during all phases. The larvae of most of them develop in terrestrial soils, especially those in forests or meadows. However, aquatic species of *Empis* are known to swarm in the littoral zones along mountain streams and above or near standing pools in wetlands, especially near the detritus-rich brownish waters of forest pools, moors, and bogs. The breeding of each species occurs during the same season each year, which maximizes breeding success and minimizes losses due to predation.

Swarms of aquatic empidids are often seen just above the surface of the water or over the shores and littoral plants along both fast and slow-flowing streams in northern Europe, as well as above or near lentic water bodies. Within the swarms of these small flies, which can sometimes include many individuals, males and females form mating pairs. Typically, during courtship, the male will catch a small insect suitable for food, wrap it up in a silken package, and present it as a gift to the female of his choice. Acceptance of this gift by the female generally formalizes a permanent bond between the male and the female that will produce a new generation. The terms, "wedding present" and "nuptial gift" were aptly given to the insect wrapped in a silken package by early observers, and it has been used ever since. Various interpretations of what was going on were proposed by observers from time to time (Chvála, 1994).

Mating often takes place during or just after the presentation of the small white package containing the gift, and as a general rule for members of the order Diptera, the females would be expected to store enough sperm from this mating to use for the fertilization of all eggs she will produce during her lifetime as an imago. Apparently, the wedding present provides the female with substances, such as amino acids, which are needed to produce the eggs. Since the nectar from flowers lacks these substances, the male not only supplies the sperm to fertilize the eggs, he provides the female with some of the nutrients which will be necessary to produce the eggs themselves. Niesiołowski (2006) reported some of the nuptial gifts that were given by the *Empis* species to be members of the hexapod class Collembola or the insect orders Ephemeroptera and Hemiptera, including members of the both suborders, Heteroptera and Homoptera, as well as other families of Diptera, often members of the suborder Nematocera.

There are other explanations for the purposes of the gifts, however. For example, prevention of cannibalism by the female, acquisition of a mate, and reduction of evasive reactions by the female (Svensson et al., 1990). The debate has not been settled, but all arguments seem to presume that there is only one possible strategy behind all giving of nuptial gifts. There are a great many species in the superfamily Empidoidea, and only a few have been studied. It may be that there are fundamental differences between the strategies involving "wedding presents" used by different species or groups of species, and this will not be conclusively demonstrated until this phenomenon is observed as practiced by many different species in different parts of the world.

The habits of most species of empidid fly have not been carefully investigated. However, the habit of giving wedding presents is common, and it may well be practiced by all species around the world. This cannot yet be confirmed or refuted, however, because the habits of so few of the species have already been described. According to recent estimates, there are presently about 4000 species of Empididae worldwide, and relatively few have been carefully observed or investigated.

The silk to wrap the wedding gift comes from special silk glands located in the basal tarsal segment of the male (Young and Merritt, 2003). Its presentation to the female seems to follow a species-specific ritual, which is necessary to complete prior to or concurrent with mating. The adults of the species observed seem to prey on other insects and also feed on nectar from flowers. That means that the females of the known species do not have to depend entirely on the wedding present to obtain the proteins needed to produce eggs. However, more studies will be necessary to determine differences in the diets of the individual species.

Some males of the species *Empis opaca* Fabricius, 1805, were reported to have attempted to trick females into mating using hairy willow seeds as presents. These seeds, of course, have no value as food for the females, and the males trying to give them as presents were usually, but not always, rejected by the females (Preston-Mafham, 1999).

The family Empididae is found throughout the world. It might well be that the temperate zones offer more conducive conditions to insects that form large mating swarms. This would conform to the hypothesis that moist tropical climatic zones are characterized by larger numbers of species and smaller numbers of individuals belonging to each of the species present than habitats in the temperate zones. In any case, the reproductive strategy of empidids in the temperate zones is obviously a successful one, as indicated by the large areas of distribution, the very great numbers of many of the species observed every year, and the many species belonging to this family throughout the world.

25

Aquatic Insects that Kill People

That a tiny insect could be a danger to the life of a human being is often forgotten. That an aquatic insect could kill large numbers of people seems even more unlikely. However, few people in the United States realize that stings from honey bees kill more Americans each year than all other animals in their country, combined.

World-wide, several species of mosquito in the family Culicidae outdo even bees as killers of human beings through their spread of contagious diseases caused by nematodes, protozoans, bacteria, and viruses. Outside of the tropics and a few lands surrounding the Mediterranean Sea, people are seldom concerned with diseases caused by insects with aquatic larvae. Mosquitoes are, at most, minor annoyances for a few months during the summer. However, in Africa south of the Sahara, the warmer parts of South America, and parts of Southeast Asia, many millions of people have died during the past two centuries of diseases transmitted by mosquito bites. Many others have survived the diseases but thereafter been chronically ill from the continued existence of the pathogens in their bodies. Such people then serve as reservoirs for the infectious stages of the potentially deadly protozoans or viruses, and whenever their blood is eaten by a mosquito of a species adapted to transmitting the respective disease, they can cause a small-scale epidemic of the disease in their own immediate environment or in their communities. This happens when the pathogens from their blood infect the mosquito, go through one or more metamorphoses in the mosquito's body, and their infectious stages are injected into the blood of a new human host along with the mosquito's saliva the first time it is injected into an uninfected person.

For many centuries, the city of Rome was effectively protected by mosquitoes. Invading armies during the Middle Ages successfully overcame the defenders of the Eternal City, only to be decimated by the malaria borne by the *Anopheles* species that developed as larvae in the Pontine Marshes. Native Romans were seldom afflicted because they left the lowlands and stayed in the mountains each year during the summer months, when the adult mosquitoes were active.

After a project was initiated in Brazil to resettle poor families from the big cities to farmland in Amazonia, the authorities recorded a considerable number of small outbreaks of malaria in the southern parts of the country, where it had never been recorded before. Each such outbreak was linked to a visit of one of the people who had settled in the Amazon region to friends or relatives in their former home towns.

Before this resettlement project, malaria also protected the vast rainforests of Amazonia from immigration and settlement for more than 400 years after the Portuguese began to settle in Brazil. During the second half of the 20th century, the attempts were begun to resettle thousands Brazilians from crowded cities and towns in the southern parts of the country into Amazonia. A long highway called the Transamazonica was constructed from the South to the North, and many dams were also built to generate electric power.

Ignoring the many other problems that beset the settlers, those caused by the three endemic kinds of malaria in the region and other less serious forms of the disease were responsible for much of the impoverishment and misery faced by the new settlers. One of the direct effects of exposure to the various forms of malaria endemic to Amazonia was the death of many settlers from the disease and some deaths in the southern states of Brazil that were recorded during each of the many local outbreaks of malaria in the southern part of the country, where the disease had never been known to occur. Presumably, mosquitoes susceptible to infection by one or more of the *Plasmodium* species had always been present throughout most of Brazil, but the parasites themselves had never been introduced with infected human beings or other primates. An interesting observation that was made but never fully explained is why the Indians native to the Amazon region do not seem to become infected by any of the malaria parasites as long as they remain in their own villages. However, when they visit settlements of Brazilians from the southern part of the country, they are subject to infection.

Without any doubt, the course of human history has been changed many times by aquatic species of insect which bore diseases to groups of people engaged in activities that eventually had to be abandoned so that the people could escape from the ravages of a disease that the insects transmitted. For example, after Count Ferdinand de Lesseps began to build a canal across the Isthmus of Panama in 1880, expended a large amount of money on the project, yet failed to show much progress due to outbreaks of yellow fever borne by the mosquito, *Aedes aegypti* (Linnaeus, 1762), among his construction workers, he was forced him to abandon his attempt in 1888.

Malaria and yellow fever are not the only fatal tropical diseases spread by insects, including aquatic ones. Many other such fatal or debilitating diseases are spread by terrestrial insect species. A few of these include sleeping sickness transmitted in Africa by tsetse flies, Chagas' disease spread in South America by a blood-sucking bug, and bubonic plague transmitted throughout Europe and Asia by fleas from rats, but they are beyond the scope of this book.

SECTION 1

Virus Vectors

Several viruses that cause serious diseases in human beings cannot be transmitted directly from an infected person to an uninfected one. They must first be consumed by suitable species of mosquito, produce infectious particles in the salivary glands of the insect, and then be injected into an uninfected person by the same mosquito, or occasionally, by one of its offspring. Although two of these diseases have produced countless fatalities among people in the tropics and warmer parts of the temperate zones over the centuries, studies have continued to provide new information about them and the mosquitoes that transmit them. Over the years, changes have been recorded in the life histories of the viruses and their vectors, indicating that both are still capable of adapting themselves to new measures undertaken by man to stop the spread of the diseases they cause.

Most of the serious virus-spread diseases are transmitted by mosquitoes in the genus *Aedes* in the family Culicidae. Most viruses can be transmitted by more than one species of vector, and some vectors can spread more than one disease. One of these mosquitoes, *Aedes aegypti*, has adapted itself to life in close proximity to human habitations. It breeds easily is small assemblages of water that collect in broken jars, old tires, hollowed out rocks, and modern plastic containers. Its close proximity to man gives it easy access to human blood infected with several dangerous viruses and uninfected people to whom it transmits the viruses. This is by no means the only mosquito species capable of transmitting the viruses, and a few species of mosquito that reproduce at locations far from human population centers function to spread the disease from populations of monkeys and apes, which appear to be the reservoirs of the viruses, to produce new outbreaks of the diseases in places where they have not occurred for many years. The versatile species, *Aedes aegypti*, has been reportedly found on bromeliads in the state of Rio de Janeiro, giving it the ability to obtain viruses from reservoirs in wild primates to start new outbreaks without the assistance of rare species from rural phytotelmata (Cunha et al., 2002).

The virus diseases discussed here include yellow fever, dengue fever, chikungunya fever, West Nile virus, Japanese encephalitis, Rift Valley fever, and the sickness caused by the zika virus, although the last of these can reportedly be transmitted by physical contact as well as by mosquito bites. Of all of these diseases spread by *Aedes* species, yellow fever is historically the deadliest. Apparently, it has afflicted human populations since prehistoric times in the tropics, but deaths from diseases with symptoms similar to those of yellow fever may actually have been caused by other diseases, some of which may have been eradicated over the centuries. Fortunately, a vaccine against yellow fever is now available, and one vaccination seems to make a person immune from the disease for life. In the past, many of the diseases were limited geographically to certain regions of the tropics, but thanks to efficient international travel, many of the diseases are greatly increasing their ranges.

A brief review of the various kinds of tropical diseases spread by aquatic insects follows. Some of the diseases are not fatal in most cases, but they can weaken those who have contracted them and cause certain life-changing handicaps, such as blindness or recurring fever. In subsequent sections of this chapter, diseases caused by protozoans and nematodes and spread by aquatic insect vectors will be discussed.

a. Yellow fever

Yellow fever, a disease caused by a virus transmitted by mosquitoes in the genus *Aedes,* caused the abandonment of the first project to build a canal across the Isthmus of Panama. After his successful project to build the Suez Canal, Count Ferdinand de

Lesseps sought to build a canal across Panama. Within a short time after the work was begun, the laborers working on the canal began to fall victim in increasing numbers to yellow fever. There was no effective remedy, and nobody knew how the disease was caused or transmitted. Hypotheses ranged from exposure to polluted air, direct transmission by contact with an infected person or his personal items, and the hot and uncomfortable climate. The fatalities caused by the disease greatly slowed the work and increased the costs. Finally, de Lesseps ended the construction work in 1888.

When the American government prepared to resume the work on a canal across Panama in 1904, Col. George Washington Goethals, who was to take charge of the construction work, was faced with the same problem that caused de Lesseps to abandon the project. He assigned Col. William C. Gorgas to find out how yellow fever could be prevented and take the necessary steps to eradicate the disease. It was correctly surmised that mosquitoes could be transmitting the disease, which had been reported by Dr. Carlos Finlay in 1881.

In 1900, Dr. Walter Reed demonstrated experimentally that Dr. Finlay's hypothesis and experimental results were correct. American soldiers volunteered for Dr. Reed's experiments and were divided into groups. Some were exposed to bedding and other fomites used by yellow fever victims, while others let themselves be bitten by mosquitoes that had previously fed on the blood of yellow fever victims. Those bitten by the mosquitoes contracted yellow fever; those who were exposed to the filthy fomites did not. Nevertheless, public opinion still supported the belief in direct contagion and objected to Col. Gorgas's mosquito eradication methods. President Theodore Roosevelt supported Col. Gorgas, and yellow fever was effectively eradicated during the construction of the canal.

The results were unequivocal and proved that yellow fever could not be directly transmitted from one person to another by contact with a victim or with his possessions. It was rather spread by mosquito bites from infected to uninfected persons. Mosquito control programs proved to be effective in greatly reducing the occurrence of yellow fever near the camps in which the laborers were temporarily quartered. The same efforts also resulted in the elimination of malaria, providing evidence that it was also a mosquito-borne disease. This made the efforts to build a canal across Panama successful without excessive loss of life.

Those other serious diseases spread by aquatic insects did not cause as many fatalities as yellow fever or the most dangerous kinds of malaria. This is partly due to the nature of the diseases and to geographic limitations on their occurrences. The deadliness of yellow fever has greatly decreased due to the successful development of a vaccine, and it appears that it will continue to decrease in the future. Malaria continues to be a major cause of fatalities in tropical countries. It will be discussed further in the section of this chapter on diseases caused by Protozoa.

In discussing the transmission of yellow fever, *Aedes aegypti* is invariably discussed as the primary insect vector transmitting the disease. This is most certainly because this species has been associated with human habitations since ancient times, and it can be encountered wherever human being live in parts of the earth where the winter temperature does not drop much below the freezing temperature of water for more than a short period of time.

The virus causing yellow fever is a flavivirus containing RNA, which undergoes an alternation of generations. When the mosquito infects itself by consuming the blood of an infected human being or animal, the virus causes the mosquito to produce infectious virus particles that are suited to infect human beings. These gather in the salivary glands of the mosquito and await the next blood meal, when they will be injected into a new host, usually an uninfected person or another primate. Before the blood of the final host is sucked in by the mosquito, the saliva of the insect must be injected into the host to thin the blood and to prevent it from coagulating. By allowing its salivary glands to be filled with virus particles, the mosquito vector ensures that the viruses are provided with the quickest and safest route into the human bloodstream.

There are reports that a variety of other species of *Aedes, Haemagogus, Mansonia, Culex,* and *Sabethes* can also be infected by the viruses and produce new particles capable of infecting human beings. Some of these mosquito species are capable of developing in forests, sometimes in tree holes or bromeliads. These species can and do infect monkeys or apes, which may then act as reservoirs for the viruses. When they are bitten by mosquitoes of the appropriate species again, they can infect these mosquitoes, which can then transmit the viruses to other human beings.

According to the World Health Organization, two different cycles can be spoken of. *Aedes aegypti* is the only vector in the urban cycle. Human beings infected by the virus pass them on to the mosquitoes that bite them, and the mosquitoes can then pass them on to other human beings they subsequently bite. The infected female mosquitoes can also pass the viruses on to their eggs, so that the next generation of mosquitoes is infected even before their first blood meal.

The second cycle is called the jungle cycle. It involves many species of mosquito, all of which are capable of transmitting the virus. In most cases, the mammalian hosts are monkeys, but in rare cases, human beings are also infected. The existence of this cycle guarantees a permanent reservoir of the viruses, preventing the final elimination of the yellow fever virus anywhere within its range.

In Africa, where most of the fatalities from yellow fever occur, a third cycle called the savanna cycle is known. Primates, including human beings, constitute the mammalian hosts of the viruses, and several species of mosquito are also capable of acting as the vectors.

Yellow fever has only mild symptoms and is quickly recovered from when contracted by some people. If it persists, it causes a yellowing of the skin, from which the disease gets its name. Some people have serious and persistent symptoms after contracting the disease, and in many cases, death results. Most deaths from the disease occur in Africa. The disease kills a smaller percentage of its victims in South America. Yellow fever is not a problem at all in Asia. Historically, it has occurred from time to time in southern Europe, and cases were reported in the late 18th century in Pennsylvania, although it would be difficult to confirm this diagnosis after more than two centuries.

Mosquito control programs can reduce the occurrence of yellow fever temporarily, but over the long term, it has proven ineffective. As already discussed in Chapter 10, the mosquito species, *Aedes aegypti,* has been associated with human habitations in tropical and subtropical countries since prehistoric times. Since the use of pesticides has become commonplace, this mosquito has apparently developed resistance to every pesticide in use and has prospects of adapting to each one introduced in the future. The use of formerly effective pesticides, such as DDT, actually increases the numbers of this mosquito found near towns and villages because it kills off local predatory insects without having any effect at all on the mosquitoes. This was personally witnessed in Cuiabá, Mato Grosso, Brazil in 1993.

An effective vaccine against yellow fever has long been in use, and where it was used in countries that were formerly parts of French West Africa, yellow fever was practically eliminated. Fighting the disease through vaccination programs is the only method of protecting people threatened by the disease in a reasonable and sensible way because it works. It is also cheap, and one injection confers immunity to a person for life. Furthermore, the vaccine causes no collateral damage to the environment.

b. Dengue fever

Aedes aegypti, the insect vector of yellow fever, transmits a second serious disease called dengue fever. *Aedes albopictus* (Skuse, 1894) is a second mosquito also capable of transmitting this disease. While *Aedes aegypti* is a species associated with cities and towns and develops in water held in discarded man-made objects, *Aedes albopictus* is a mosquito that lives in tropical forests and develops in the water held in tree holes and tropical plants.

The disease is caused by a virus, which infected mosquitoes transfer from person to person. The dengue viruses contain RNA and are in the genus *Flavivirus*, which is in the family Flaviviridae. Four distinct serological varieties have been identified, designated DENV-1 through 4. Unlike yellow fever, a dengue fever infection does not confer immunity to the victim against subsequent infections by other dengue viruses (Lanciotti et al., 1993). The four serological varieties correspond to four different strains of the virus, which have their own geographical ranges, although many of these ranges overlap. Any immunity that might be conferred by an infection of one variety of the virus apparently does not interfere with infections by any of the other three strains. Having once contracted dengue fever does seem to confer a lifetime immunity against the same serotype of the virus which caused the first infection, however.

In 2015, a vaccine was approved for dengue fever. Like flu vaccines, however, it is not likely that a vaccination would protect the person vaccinated against all present and future strains that may develop. Time will tell whether or not a vaccination program will be as successful against dengue fever as a similar program was against yellow fever, but it is likely that it will not.

At the present time, the World Health Organization estimates that there are about 390 million new infections by the dengue virus in the world each year, and all cases are not correctly diagnosed or reported. In most cases, it simply resembles a flu infection, and patients complain about a variety of symptoms. Sometimes, the infection develops into what is called severe dengue, also known as dengue haemorrhagic fever. Once the symptoms of severe dengue have appeared, the death rate from this condition becomes alarmingly high, especially among children. As the name "haemorrhagic fever" implies, excessive bleeding may occur because the production of blood platelets declines. If the condition persists, the patient may go into shock and die.

A peculiarity of the disease is that one infection usually does not cause death, but the chances of death from the disease increases if the person is infected a second time. The chances of death presumably become even greater after a third or a fourth infection. It is likely that successive infections must each be caused by a different strain, but even that is not known with certainty. Because the disease is spreading so rapidly to many different parts of the world, it can also be expected to produce new strains that do not follow the behavioral patterns typical of the old, familiar ones.

In recent years, the incidence of dengue fever has been increasing significantly, and the disease is spreading to countries where it was not known before. During the 1950s, there were outbreaks of the disease in Thailand and The Philippines. Today, the regions most affected include Latin America, Southeast Asia, and the islands in the Western Pacific. Dengue fever has recently moved into Florida, Hawaii, Laos, Yunnan in Southwest China, and warmer parts of Japan. Until recently, no occurrences of dengue fever had been reported from Japan for 70 years.

It may be that the attention dengue fever is receiving is part of the reason that its incidence seems to be increasing dramatically. Many people infected by the dengue virus may have been diagnosed with a mild case of the flu in the recent past, but now doctors have been taking a closer look and recognizing that the patient actually has dengue fever. Conversely, there are few reports of dengue fever in Africa, which may have to do with the concern for more dangerous mosquito-borne diseases that are taking more lives there than dengue fever could be expected to.

Without a promising outlook for those attempting to develop an effective vaccine to be able to control the disease, only improving mosquito control holds promise of stemming the apparent flood of new cases of dengue fever during the coming few years. The role of the mosquito in vectoring the disease is rather simple. The mosquito becomes infected by biting a person infected with the virus. The viruses produce reproductive stages in the gut of the mosquito, which exit the mid-gut and relocate to the salivary glands to produce infectious viruses, which remain in the glands until the mosquito bites another person. From then on, the infected mosquito carries the infectious virus particles in her salivary glands and infects a new person each time it takes a blood meal.

There has been much speculation about the possible spread of the dengue viruses by other species of mosquito than those reported to be the common vectors of the disease. Like many other viruses, the four known strains of the dengue virus are closely associated with a few species of mosquito in the genus *Aedes*, most frequently *Aedes aegypti*.

What should not be forgotten is the relationship between the virus and the mosquito. When any insect that feeds on blood consumes viruses with its meal, the insect can only be infected if it is susceptible to infection. That means that the nucleic acid of the virus must be able in some way to take over functions of the natural nucleic acid in the body of the host mosquito. In the case of the mosquito, the virus will assume control of certain cells and force them to produce certain new virus particles capable of attaching to the cells of its next host. These infectious virus particles congregate in the salivary glands of the mosquito and remain dormant until the mosquito bites its next victim. Saliva is immediately injected into the tissues of the victim in order to thin the blood and prevent coagulation. At this time, a great many of the virus particles enter the bloodstream and attach to the cell membranes of their next host.

Just as the mammals susceptible to infection by the dengue viruses usually belong to closely related species, primates in this case, the mosquito species capable of becoming infected and transmitting the disease will also be closely related, explaining why the mosquito vectors are all or mostly in the genus *Aedes*. Different kinds of viruses can have different spectra of suitable vectors. Some are obviously able to infect greater numbers of vectors than others. It is the close proximity in which *Aedes aegypti* lives to human beings that facilitates its transmission of so many kinds of viruses to human beings and hastens the spread of those viruses around the world. It also assures that a person travelling by aircraft from one part of the tropics to another is sure to attract populations of *Aedes aegypti* almost identical to the ones that bit him in the place he just came from. The mosquitoes at both locations will also share the susceptibility to the same viruses, the ability to produce infectious virus particles for storage in their salivary glands, their resistance to DDT and most or all of the most commonly used pesticides, and their ability to survive in any tropical, urban area of the world.

c. Chikungunya fever

A third viral disease transmitted by *Aedes aegypti*, *Aedes albopictus*, *Aedes luteocephalus* (Newstead, 1907), and possibly unspecified other mosquito species, has most frequently been encountered in Africa. *Aedes albopictus* is an important vector in Asia, where it breeds farther from human settlements than *Aedes aegypti*, while *Aedes luteocephalus* spreads the disease mainly in Africa. Other mosquitoes are also involved in the spread of the disease, but they have not yet been identified with certainty.

The virus causing the disease was first identified in Tanzania in 1952 or 1953, where it was found to cause fever and pains in the muscles and joints. It is apparently not as deadly as the two virus diseases borne by *Aedes aegypi,* that were discussed above. However, it causes much joint pain, and those infected with the virus often walk with a characteristic hunched over posture. The name for the disease is taken from the word for this posture in Makonde, an African language used by a people of the same name.

Much still has to be learned about chikungunya fever. The virus is in the genus *Alphavirus* and the family Togaviridae. It is an RNA virus, and the mosquito usually carries it from person to person, but there is also a forest cycle, which sometimes involves unspecified hosts in addition to human beings. Five strains of the virus have been identified according to their geographic distribution. Three of these are African, called the West, Central, and South and East African strains. The fourth is the Indian Ocean strain, and the fifth is the Asiatic strain.

The Indian Ocean strain has moved from island to island, producing a series of epidemics on the islands in the Indian Ocean. The worst of these outbreaks occurred in 2005 and 2006 on the island of Réunion. Smaller outbreaks were reported from Madagascar, Mayotte, the Comoro Islands, Mauritius, and the Seychelles. It then moved to many parts of India, and from there, on to the islands of the eastern Indian Ocean: Andaman and the Nicobar Islands.

In addition to the outbreaks of the disease within the main areas of distribution of each of the five strains of the virus, infected persons are being found in different parts of the world. Outbreaks have occurred on various islands in the Caribbean Sea, the State of Florida, and other isolated locations in various parts of the world where it was never before found. The World Health Organization has reported that four persons contracted the disease in Montpellier, France, in 2014, while later that same year, there were outbreaks in the Marshall Islands and the Cook Islands. Fewer cases were recorded in Samoa, American Samoa, French Polynesia, Kiribati, Senegal, and the State of Punjab, India.

Cases have been reported from various parts of the United States, Canada, and Mexico, but the virus was apparently imported with the victims from places outside of North America they had visited. However, the disease may have been spread by mosquitoes in Florida. The diseases have also swept through the Caribbean Islands, causing outbreaks on individual islands as the viruses passed through. They were first reported there in 2013 by agencies that track the movements of diseases. Outbreaks were recorded on islands visited by many vacationers from Europe and North America. The disease appeared on such vacation spots as Saint Martin, Jamaica, Barbados, Saint Lucy, and the Grenadines. It then appeared in South America for the first time, hitting Colombia hard and fast and appearing for the first time in Venezuela, Brazil, and Argentina.

No known medicine is effective against chikungunya fever, and no vaccination against it has yet been developed. The disease appears first as a sudden high fever. It then causes a variety of symptoms, usually ending with increasingly painful joints. However, the government agencies that monitor such diseases have found that chikungunya fever is fatal in fewer than 1 in 1000 cases, although the pain can persist long after the disease has run its course and is a long-term consequence of having had the disease.

In addition to transfer of the viruses by mosquitoes, it is thought that they can be transferred from an infected mother to her unborn child, even though the placental barrier is a strong one. Other means of transfer without mosquitoes include blood transfusions and organ transplants.

d. Malformation of children caused by the zika virus

The newest mosquito-borne virus to make the news is the zika virus. This is another pathogen that is less dangerous to most people than the first three discussed in this chapter. However, after the disease began to spread in Brazil, it was discovered that a seemingly mild case can affect the development of the baby of an infected, pregnant mother.

Apparently, the virus is able to cause the growth of the baby's skull to stop or significantly slow, forming a cranium too small for the brain. It is still too early to observe the long-term consequences of this condition for the baby, although these results could certainly be fatal in many cases.

The zika virus was first discovered in Uganda in 1947. It consists of a single strand of RNA, and it is classified in the genus *Flavivirus* and in the family Flaviviridae. It is thought to be spread by *Aedes aegypti* and probably closely related species of mosquito.

There is evidence that the zika virus is being spread by direct physical contact between an infected person and an uninfected one. The dangers faced by pregnant women and their babies after an infection by the zika virus are not yet fully known. If outbreaks continue, much more will be learned about the virus and the harm it can cause, information that can be expected to be tragic in most cases.

e. *Other viruses carried by mosquitoes*

Mosquitoes are capable of transmitting even more viruses, although the vector is often a species other than *Aedes aegypti*. Generally, these diseases are not rated as very great risks to public health. Many are diseases that originally affected animals and cause only light symptoms, if any, in man. Exceptions to this are few, but they constitute great public health problems. Three such exceptions are the West Nile virus, Japanese encephalitis, and Rift Valley fever. These are caused by RNA viruses in the genus *Flavivirus* and family Flaviviridae. Japanese encephalitis has been known since about 1870, and it has caused epidemics in many parts East and Southeast Asia. The other two diseases have been discovered more recently.

Perhaps the kind of encephalitis that has received the most recent publicity is the so-called West Nile virus. In 1999, it appeared for the first time in the United States and was responsible for a local epidemic in New York City and its surroundings. The death toll was almost 2000. The virus attacks organs of the nervous system, including the brain, where it causes encephalitis. The human being is considered a dead end host because, although the virus causes a disease that can be fatal, the victims of the disease do not produce enough of the viruses to infect the uninfected mosquitoes that bite them.

In this case, the disease vector was the commonest local species of mosquito, *Culex pipiens* Linnaeus, 1758. In the parts of the temperate zones with cold winter climates, this is the mosquito species most familiar in urban and suburban habitats. It lives in houses and breeds in all kinds of standing water. Its range includes most of Europe, Western Asia, and the parts of North America not too far north or south to be either in a frigid or a subtropical zone. It is usually absent from subtropical and arctic regions, including Alaska and Siberia. As a disease vector, it is displaced by other species in Eastern Asia.

Although *Culex pipiens* has not notably increased its range at the expense of other mosquito species, it can show up in almost any part of the Northern Hemisphere, having travelled in private autos, buses, trains, or aircraft. It is able to hide itself during the daytime and fly under its own power away from the vehicle it has been travelling in at night to find a home at a new location. It has introduced itself to all parts of the North Temperate Zone that are not too dry and where the climate is otherwise suitable for it. It has not been so successful in the Southern Hemisphere, where there is a native mosquito fauna that can maintain itself against invasive competitors.

Because the population of *Culex pipiens* is well adapted to life with human beings, it can at any time be utilized by a virus introduced with an animal from a distant continent, which requires a suitable vector to infect people and has already adapted itself to another species of *Culex* close enough physiologically to *Culex pipiens* to be infected. The West Nile virus could have been introduced by a bird, and the resulting large-scale outbreak in New York in 1999 was just a typical phenomenon of an intensive, short-term epidemic, like those caused periodically by bacteria before the invention of modern antibiotics. West Nile virus outbreaks on a smaller scale are familiar in Europe, which seems to be closer to the animal populations that pass their viruses on to *Culex pipiens* from time to time.

North America is not the only place where such a virus epidemic has occurred. A virus, apparently very similar to the West Nile virus, was responsible for an outbreak of other encephalitis epidemics in Japan. First officially recorded in 1870, this virus is said to cause the worst variety of viral encephalitis known in Asia. Since first being recognized in Japan, it has been found in parts of China, Korea, Taiwan, much of Southeast Asia, India, and Sri Lanka. It appears briefly, infects many people, and then ends quickly, only to reappear somewhere else at a later time.

In eastern Asia, two species of *Culex* have been responsible for several outbreaks. One of these, *Culex tritaeniorhynchus* Giles, 1901, was responsible for one bad epidemic in Japan, and later, an epidemic in Korea was brought about with *Culex bitaeniorhynchus* Giles, 1901, as the main vector. Over the years, epidemics of Japanese encephalitis have broken out in various parts of Southeast Asia, throughout India, and in many parts of China, each appearing and disappearing within a relatively short time frame, usually during the warm part of the year.

The third disease, Rift Valley fever, is a disease caused by a virus in the genus *Phlebovirus*, which is mainly a disease of livestock. It was first discovered causing a disease among sheep kept in the Rift Valley of Kenya in 1931. It was later discovered that the virus was also capable of causing a disease in human beings, which was usually mild but could also become serious, causing lesions on the retina, meningial encephalitis, hemorrhagic fever, or combinations of two or more of the conditions. It can be fatal. The virus is apparently spread among sheep by blood-sucking insects. People can become infected by direct contact with the blood of infected mammals or by bites of unspecified mosquitoes or blood-sucking flies.

f. Mosquitoes as vectors of viruses

Generally, most diseases transmitted to human beings by aquatic insects are known to be spread only by no more than a small number of insect vectors. Most mosquitoes are not known to spread any disease to human beings, and many of them bite human beings seldom, if at all. *Aedes aegypti* is an exception because it can and does transmit at least four viruses that cause serious diseases. The reason for this is the close association of *A. aegypti* populations with human settlements. The species is now found throughout the tropical and subtropical regions of the earth, and it generally reproduces in human settlements, where small amounts of waste water are almost always present, and potential predators that consume mosquito larvae are eliminated through counterproductive human activities undertaken to control mosquitoes, such as the heavy use of insecticides to which the mosquitoes have become resistant. In some places, this mosquito is even able to survive mild winters in the temperate zones. Whether new races of *Aedes aegypti* will develop in the future, which will be able to cope with cold temperate winters, remains to be seen. In the past, this mosquito has maintained populations in southern Europe.

Epidemiological studies have established that diseases almost certainly transmitted by *Aedes aegypti* have afflicted human beings throughout history. Native populations have often adapted somewhat to the dangers from these diseases, but fatalities still occurred. When European explorers and traders first began to visit and work in tropical countries, they were especially hard hit by local epidemics, as the example of the European workers who tried to build a canal through Panama clearly illustrates.

The location of so many large populations of *Aedes aegypti* within human settlements guaranteed that this species of mosquito would be the one most steadily exposed to the modern insecticides introduced to control the mosquito populations in an attempt to eliminate the diseases they were spreading. While the elimination of standing water near human habitations produced and continues to produce permanent reductions of the local populations of mosquitoes near human settlements, the introduction of DDT and its successors produced impressive but only transitory reductions.

How the resistance to a pesticide is developed can be outlined using *Aedes aegypti* as an example. Its steady exposure to a newly introduced, modern pesticide results in the development of a resistant strain, which is no longer killed by the new insecticide. Like most insects, any population of *Aedes aegypti* encompasses a small percentage of individuals with a natural resistance to any new toxic substances intended to eliminate them. When the new insecticide is first used, the majority of individuals are quickly killed immediately after it is sprayed. Those few with a form of immunity effective against the pesticide survive, mate with other survivors, and pass the immunity on to their offspring. The more the pesticide is used, the faster the sensitive mosquitoes are killed, leaving only individuals with hereditary immunity in the populations. After an immune population of mosquitoes has established itself, the pesticide is never again effective against the mosquitoes. The first generation of mosquitoes after the first spraying usually includes some individuals susceptible to the pesticides. If they mate with resistant mosquitoes, some or all of the offspring may be resistant. Subsequent sprayings will eliminate only the susceptible ones until only resistant mosquitoes are left, and the local population will eventually become fully resistant, making subsequent spraying of the pesticide futile. Because of their faster reproduction rate, the mosquitoes will reproduce many generations of offspring before the predator populations even begin to develop immunity, making the swarms of disease vectors produced in the absence of natural enemies a major problem for public health.

Although any new insecticide will turn out to be useless for the long-term control of the mosquito vectors, just as all of its predecessors did, the impression on the public that it made will persist of many years. The mass mortality of dangerous and annoying insect pests makes a strong impression on the customers who purchased and used it. Many people who witnessed hundreds of house flies fall dead on the floor after one spraying of DDT during the 1940s lived for many decades in the belief that using DDT again would solve all of the problems that insects can cause. Even now, people call for the use of DDT to control mosquitoes and other insect vectors that endanger public health.

The ban on DDT is wrongfully attributed to its toxicity to human beings rather than its environmental impact and the ever increasing amounts of it that contaminate the water and soils around the world. In fact, DDT is one of the least toxic insecticides to human beings, and no serious health problems, other than rare cases that seem to have been allergies, have been attributed to its use. Nevertheless, DDT should never be broadcast in the environment again, simply because it does not work and it never will again. There is a great deal of DDT stored by various chemical companies, and they would like to be able to sell it for profit. That is the only reasonable motive for demanding that it be reauthorized for sale and use as an insecticide. Its use would contribute nothing toward the control of disease vectors. The insects that people want to get rid of are not killed by it, and the suggestion that it will work against them again at some time in the future is a hypothesis unsupported by any sort of evidence.

DDT is very stable chemically in well oxygenated soils and water, and it will be present in detectable amounts in most parts of the world for the next few hundred years. The amounts detected in the bodies of several aquatic insect species during the early 1980s would have been sufficient to kill fully sensitive insects during the late 1940s, so it is hard to imagine that any insect could keep from being exposed to it when swimming in the same water body (Heckman, 1982b).

The case of DDT and *Aedes aegypti* is a good example of a thought process that has repeated itself countless times after a new insecticide has been put on the market. Important differences between the case of DDT and those of subsequent pesticides include the findings that the newer products are considerably more toxic than DDT to human beings and less effective against insects than DDT used to be. However, people who have witnessed the rapid death of many insects after a new chemical product is first used never forgot what they witnessed and believe that some way can be found to repeat what they witnessed before the insects became resistant to the insecticide.

This process of developing resistance to an insecticide is most rapid among insects that live in close proximity to human beings. The fact that the insecticides are still effective against insect populations that were not previously exposed to the toxic substances they contain sometimes makes spraying to control *A. aegypti* counterproductive. Aquatic environments not exposed to pesticides are usually populated by certain insect species that prey on mosquito larvae. These can greatly reduce the number of mosquito larvae that complete their development to imagoes. After a formerly effective pesticide is sprayed, the local predatory species are killed, allowing much larger numbers of resistant mosquito larvae to reach the adult stage. The frequent use of these and other toxic substances near human habitations is sufficient to delay the development of potential predators long enough for new toxic substances to be introduced and prevent the predators from obtaining a foothold in the old tires, plastic pails, and other artifacts that catch and hold rainwater for the chemical-resistant mosquitoes to develop in.

This phenomenon was witnesses personally in Cuiabá, the capital of the Brazilian state of Mato Grosso. Because dengue fever began to spread in the city after being introduced there by persons from eastern Brazil, who had migrated to Mato Grosso, where dengue fever had previously been rare, the state government ordered the spraying of DDT. This was unwise because the substance had been used much longer than most other pesticides, and insects that live in close proximity to human populations could be assumed to be completely immune to DDT.

Prior to the spraying, *Aedes aegypti* had never been abundant in the city. Personal collecting efforts to sample insect specimens during both the rainy and dry seasons in the city of Cuiabá had not yielded any specimens. Many mosquitoes of other species had been collected in the region of the Pantanal, the large, seasonal wetland to the southwest of Cuiabá, but none of them belonged to the genus *Aedes*.

Only weeks after the fogging with DDT began, species of *Aedes* began to appear around the city in large numbers, after none at all had been seen during the previous two years. The assumption that the increase was caused by the DDT eliminating the natural enemies of the mosquitoes but not the resistant *Aedes* could explain this (Heckman, 1998). Such an effect had been observed in many tropical countries. Other reports have shown that the degree of resistance of *Aedes aegypti* and *A. albopictus* to DDT have made the use of these obsolete pesticides fully counterproductive, although a small but significant difference in the resistance of these two species still exists (Vontas et al., 2012). Because the population of *A. aegypti* lives in close association with human beings, it has been exposed to larger quantities of DDT for a longer period of time, making its resistance to DDT measurably stronger than that of *A. albopictus*, a species which lives in rainforests and has no close association with people.

SECTION 2

Malaria

Malaria is the common name of a group of diseases caused by closely related protozoan species in the genus *Plasmodium*. At least five separate species of *Plasmodium* are known to cause malaria in human beings. These can usually be recognized by the nature of the diseases they cause. The most dangerous of these pathogens to human beings is *Plasmodium falciparum* (Welch, 1897). More human fatalities are thought to have been caused by this parasite than by any of the other species of *Plasmodium*. If a person contracts this form of malaria, called malaria tropica, and survives the initial outbreak of the disease, which is usually characterized by a high fever, it apparently does not recur.

Plasmodium malariae Marchiafava and Celli, 1885, and *Plasmodium vivax* (Grassi and Feletti, 1890) cause forms of malaria that are not quite as dangerous as that caused by *P. falciparum*, but the pathogenic protozoans produce a resting stage with one nucleus, known as the hypnozoite, that usually colonizes the liver. These stages remain inactive until still unknown conditions induce a metamorphosis producing new infectious stages.

Malaria quartana, caused by *Plasmodium malariae*, is a malarial disease with a remarkably long incubation period. The symptoms may first appear between two and eight weeks after the victim is bitten by the mosquito. After the patient seems to have completely recovered from the disease, relapses can occur throughout his life, and such relapses have been recorded as long as 50 years after the first attack. Anyone who contracts the disease in the tropics, recovers, and then returns to regions with colder climates may show no symptoms for decades but relapse upon returning to the tropics. The conditions triggering the processes which lead to the relapse remain unknown.

Recent research has indicated that *P. vivax* is more dangerous than it was previously thought to be. The disease it causes is called malaria tertiana, which has not been previously distinguished from a benign form of the disease caused by *P. ovale* (Table 25.1). Because it is so deadly and has a geographical distribution covering areas with large human populations, it was speculated that *P. vivax* could cause far more harm even than *P. falciparum* and should be monitored more carefully (Howe et al., 2016).

Plasmodium ovale Stephens, 1922, and *Plasmodium knowlesi* Sinton and Mulligan, 1932, cause relatively mild kinds of malaria, which are seldom, if ever, fatal. *P. ovale* causes a benign malaria tertiana, which has usually not been distinguished from the widespread and sometimes fatal malaria tertiana caused by *P. vivax*. According to Howe et al. (2016), the malaria tertiana caused by *P. vivax* can cause a swelling of the spleen, which has not been reported for the disease caused by *P. ovale*.

P. knowlesi causes malaria quotidiana, a disease of a large number of apes and monkeys with only a mild effect on man. Sometimes this kind of malaria is not classified as a disease of humans, even though it produces distinct symptoms in infected human beings. Recently, more and more researchers have come to recognize this kind of malaria as a fifth form of the disease capable of affecting human beings.

Which kind of malaria a person contracts depends on his geographical location at the time he is infected and which mosquito species are available as vectors of the disease (Table 25.1).

Table 25.1 List of the locations where each of the protozoan species that cause a specific kind of malaria in human beings occurs and other species they are known to affect.

Plasmodium species	Disease	Known Range	Other hosts
P. falciparum	Malaria tropica	Pantropical, most common in Africa	Chimpanzee
P. malariae	Malaria quartana	Amazona, sub-Sahara Aftica, S.E. Asia	None
P. vivax	Malaria tertiana	Pantropical, formerly Europe	Many primates
P. ovale	Malaria tertiana	African rainforests, Southeast Asia	Chimpanzee
P. knowlesi	Malaria quotidiana	Southeast Asia	Monkeys

a. Mosquito vectors

It is difficult to determine which mosquito species act as the main disease vectors for the individual species of *Plasmodium* inhabiting each region of the tropics where a malarial disease occurs. The problem here is that different kinds of experimental evidence are used to determine which vectors are important.

The most conclusive information is the association of the common species of *Anopheles* in a specific region with the occurrence of one of the kinds of malaria. It is then demonstrated experimentally that the bite of this mosquito can actually infect a person with the prevalent kind of malaria in the region after it has infected itself by consuming the blood of someone who has already contracted the disease. Once these two steps have been taken, the observations and experimental evidence show whether this mosquito is an important vector for the disease.

Without the association of the abundance of the mosquito species during an outbreak of the disease, it is only of academic interest whether or not the mosquito species has the ability to transfer viable *Plasmodium* species to a human host. It is rather important to know whether the mosquito species actually does so. If it does, it would then be designated as an important mosquito vector for the disease, and its control in the region would contribute to the control of any epidemic that might occur in the future.

Only five species in the genus *Plasmodium* are considered to be important protozoans for infecting human beings with malaria. Roughly 150 species of *Plasmodium* have been described, and only the 5 in Table 25.1 are known to have any kind of impact on human health by causing one of the kinds of malaria. The great majority of the rest are thought to cause diseases only in other kinds of animal.

The disease-causing protozoans can infect and are carried by a great many species of culicids in the subfamily Anophelinae and genus *Anopheles*. Some of the others show no inclination to feed on human blood, while it is suspected that the rest will feed on human blood if there is absolutely nothing else to eat. If some of the species did feed on human blood due to a lack of any other kind of food, they would probably be free of all protozoans with the ability to infect man because their sources of vertebrate blood normally belong to a completely different species of animal that can be shown to share no common diseases with human beings.

In Table 25.1, human beings and chimpanzees both appear to be susceptible to three common malarial diseases. However, the evidence that a suitable vector is present near the parts of the forests in which the chimpanzees live and actually bites champanzees to feed on their blood has not yet been obtained. Therefore, is has only been proven that the species of *Plasmodium* could live in both human beings and chimpanzees but not that they actually do under natural conditions.

A perusal of general information given out by public agencies concerning the most important insect vectors for the five species of *Plasmodium* does not usually yield complete lists of species that could act as vectors of the protozoans. At best, the names of mosquito species can be found that are thought to be the most common vectors of the disease in certain regions in which the disease is known to occur. However, there are many gaps in the knowledge of which species do and which do not transfer each of the five *Plasmodium* species to human beings.

The mosquitoes that reportedly can transmit what is arguably the most deadly malaria pathogen, *Plasmodium falciparum*, have been determined according to the geographical locations that were investigated. Along the coast of Kenya, Mbogo et al. (1995) reported that *Anopheles gambiae* Giles, 1902, *sensu lato*, and *Anopheles funestus* Giles, 1900, were collected during their research, and a small percentage of these two species tested positive for *Plasmodium falciparum*. At least one specimen of the following species was also collected, but none of them was found to be infected by *P. falciparum*: *Anopheles coustani* Laveran, 1900; *Anopheles nili* Theobald, 1904; *Anopheles squamosus* Theobald, 1901; and *Anopheles pharaoensis* (Theobald, 1901).

In the region of Cameroon where they conducted their research, *Anopheles funestus* was thought by Cohuet et al. (2004) to be the predominant vector of *Plasmodium falciparum*. They reported that while *Anopheles gambiae* was considered to act most frequently as the vector that infected people in sub-Sahara Africa with malaria tropica, they thought that *A. funestus* was the most important one in the part of Cameroon where they conducted their studies, which was located along the line separating forest from savanna. However, in the forested parts of southern Cameroon, as well as Congo, Burkino Faso, Nigeria, and Ethiopia, *Anopheles nili* has been found to be the most important vector of *Plasmodium falciparum* by Carnevale et al. (1992).

Also identified as possible vectors of *Plasmodium falciparum* by Cohuet et al. (2004) were *Anopheles moucheti* Evans, 1925, and *Anopheles hancocki* Edwards, 1929.

In South America, Laporta et al. (2015) identified the main vectors of *Plasmodium falciparum* which were infecting human beings as three species in the *Anopheles albitarsus* complex: *Anopheles darlingi* Root, 1926; *Anopheles marajoara* Galvão and Damasceno, 1942; and *Anopheles deaneorum* Rosa-Freitas, 1989. In addition, about seven other species in the same complex were found to be potential vectors.

In Southeast Asia, *Plasmodium falciparum* shows a high degree of resistance against anti-malarial drugs. Studies of the mosquito vectors of this form of malaria in Thailand have resulted in the production of keys to the species of all mosquitoes in that country and what is known about the forms of malaria they are capable of transmitting (Rattanarithikul et al., 2006). Taxonomic problems have made it difficult to interpret earlier publications. For example, Rattanarithikul et al. (2006) pointed out that the name *Anopheles balabacensis* Baisas, 1936, which was frequently cited as an important vector of malaria in the forests of Thailand, had long been misidentified, and texts reporting it are actually referring to one of several species more recently described.

A large number of *Anopheles* species are cited as vectors of *Plasmodium* species in Thailand, a few of which have not yet been given names, so it is recommended that the publication by Rattanarithikul et al. (2006) and subsequent works be consulted for up-to-date taxonomy of the vectors.

Vectors of the other malarial diseases are also known to be important in one geographical region and less important in others. The second most important species of *Plasmodium* that causes one of these malarial diseases, *Plasmodium vivax,* seems to be overtaking *P. falciparum* as the most dangerous of the species to man. This seems to be a result of the resistance of Africans to infections by *P. vivax* because they are negative for the Duffy genome, as discussed below.

A list of known vectors of *Plasmodium ovale* was provided by Collins and Jeffery (2005). These include *Anopheles gambiae* Giles, 1902, possibly the most common one in forested regions, as well as *Anopheles albimanus* Wiedemann, 1820; *Anopheles atroparvus* van Thiel, 1927; *Anopheles dirus* Peyton and Harrison, 1979; *Anopheles farauti* Lavaran, 1902; *Anopheles freeborni* Aitken, 1939; *Anopheles maculatus* Theobald, 1901; *Anopheles quadrimaculatus* Say, 1824; *Anopheles stephensi* Liston, 1901; and *Anopheles subpictus* Grassi, 1899. Although this form of malaria occurs mainly in Africa south of the Sahara, at a few sites on mainland Asia, and on islands in the western Pacific Ocean, mosquitoes native to temperate and tropical America and southern Europe are among those listed.

The fifth kind of malaria, caused by *Plasmodium knowlesi*, has previously been called simian malaria because it was believed to infect primarily monkeys and other primates and only rarely produces an infection in human beings. With the realization that it often produces infections in human beings, this form of malaria, called malaria quotidiana, has come to be recognized as a real disease of humans (Vythalingam et al., 2008), albeit one that seldom, if ever, causes fatalities. The disease itself is comparatively mild, and it is not known to cause frequent relapses. The most commonly reported vectors for this pathogen are *Anopheles hackeri* Edwards, 1921, in Thailand, and *Anopheles cracens* Sallum and Peyton, 2003, in peninsula Malaysia. *Anopheles dirus* Peyton and Harrison, 1979, was found to be the predominant vector in southern Vietnam (Marchand et al., 2011).

b. Developments in mosquito taxonomy

Of the many species of mosquito in the genus *Anopheles* that occur in tropical and temperate regions in which at least one kind of malaria occurs, only certain ones are known to be suitable for transmitting one or more of the species of *Plasmodium* capable of producing the disease in human beings. It is not always been clear how or why certain species of mosquito can transmit malaria to human beings, while others cannot. What was once considered a single species, *Anopheles albitarsis* Arribalzaga, 1878, occurs in Brazil, both in the Amazon Basin of Mato Grosso and the Pantanal, located along the upper tributaries of the Rio Paraguai. In the Amazon region, it is feared as a vector for more than one kind of malaria. In the Pantanal, however, it has not yet been found to transmit any malarial disease. The presence of tourists from other parts of Brazil in the Pantanal would suggest that the mosquitoes could easily become infected with a *Plasmodium* species which can cause malaria in human beings. However, malaria is not known from the parts of the Pantanal where *Anopheles albitarsis* sensu lato is by far the most common member of its genus (Heckman, 1998). It seems more likely that this one mosquito species has one variety capable of acting as a host for one of more species of *Plasmodium* and another which is resistant to all species. It may also be that the monkeys in the Pantanal do not act as reservoirs for any species of *Plasmodium*, while those in Amazonia do. Nevertheless, alternative systems of taxonomy are being developed, one which distinguishes recognizable morphological features as a basis for distinguishing species and another, which incorporates various biochemical characteristics to define species.

When the second of these systems is employed, the term "*Anopheles albitarsis* complex" is used. Some of the taxa in this complex have been described and given specific names according to taxonomic rules, while others are simply designated as varieties without names. More research will be required to see whether the biochemical or physiological differences relevant to the susceptibility of the insect to a disease merits status as an independent species. If it is found that two of the biochemical varieties can mate and produce reproductive offspring it could be assumed that they are conspecific. Closer studies of the members of the species complexes will be necessary to determine what the taxonomic statuses of their member-varieties should be.

If the people in a region are resistant to one of the *Plasmodium* species, the form of malaria it causes will remain absent or rare among human beings, even if suitable mosquito vectors are present and other mammals act as permanent reservoirs for the protozoan pathogens. This was found to be the case with *Plasmodium vivax* in West Africa, where the erythrocytes of a large proportion of the native population, estimated at 95%, are negative for the Duffy blood-group factors (Langhi and Bordin, 2006). Persons who are Duffy-blood-group negative are completely immune to infections by *Plasmodium vivax,* as reported based on experiments in the United States (Miller et al., 1976). This explains why *P. vivax* does not occur in West Africa, although it is likely that it originated there. Elsewhere in the world, including many places where it was unknown until fairly recently, it has become the most prevalent form of malaria.

c. *Plasmodium* life cycles

The parasites enter the human body as sporozoites together with the saliva of an infected mosquito. When the mosquito bites an uninfected human host, it injects its saliva to thin the blood and prevent coagulation. With the saliva come many sporozoites, which had gathered in the salivary glands in preparation for invading a new human host. Apparently, only some of the infectious sporozoites are injected into the new host, while the rest are reserved in case the mosquito has to bite another new host later on. Each species of *Plasmodium* goes through specific kinds of metamorphosis during their lives, some of which have probably not been described, as relatively recent discoveries indicate.

Once inside the human body, the sporozoites move actively through the dermis to reach blood vessels, through which they are quickly transported to the liver. Inside of the liver cells, each sporozoite begins to reproduce and produce a schizont, in which the sporozoite reproduces to form many small merozoites. After these have completed development, the schizonts burst, releasing the tiny immature trophozoites, which enter the blood stream and attach themselves each to one erythrocyte. Each trophozoite first forms a ring-like cell attached to the outside of the erythrocyte, and later, they mature into a round attachment to the blood cell, which provides them with food. They can then reproduce asexually by fission or sexually through the production of gametocytes. Some remain in the circulatory system for many generations, reproducing asexually to produce more trophozoites.

Many cell lines produce successive generations or trophozoites, which decrease in abundance whenever the symptoms of malaria disappear and the victim is judged to be cured. The defenses in the human body against disease are able to reduce the numbers of parasites attacking the erythrocytes but apparently not able to eliminate them completely. However, for reasons not fully understood, some of the trophozoites do not continue to produce more trophozoites but rather develop into male or female gametocytes, which are capable of infecting any suitable mosquito that feeds on the blood of the human host.

After blood containing infectious gametocytes is consumed by the mosquito, some of the gametocytes in the blood are able to find a gametocyte of the opposite sex, and the process of forming infective sporozoites which gather in the salivary glands of the mosquito begins again. The male microgamete enters a female macrogametocyte, which becomes a zygote called an ookinete. This is a motile stage that penetrates the midgut wall of the mosquito to form the oocyst. After the oocysts have grown and matured, they burst to release the sporozoites, which migrate to the salivary glands. These are typically long, narrow, and spindle-shaped so that they can pass quickly and easily through the elongated mouthparts of the mosquito.

The symptoms of malaria in its human victims are the result of the events that take place in the circulatory system. Whether or not a person who has contracted any of the forms of malaria can ever really eliminate all of the protozoans from his body is not certain. However, recurrences of the disease symptoms of infections by *Plasmodium malariae* after half a century would tend to suggest that it cannot. On the other hand, the failure of *Plasmodium falciparum* to ever recur in most cases would suggest the opposite. Obviously, although the life cycles of each of the five kinds of malaria that infect human beings are very similar, there are subtle but profound differences in their life cycles and effects on their hosts.

This outline of the life cycles of the *Plasmodium* species that cause one of the malarial diseases in humans is a general one. Each kind of malaria has its own peculiarities, complexities, and survival strategies of the pathogenic protozoan species that causes it. It is likely that more special features in the life cycles of the individual species remain to be discovered.

For example, rather than beginning the general processes of the life cycle immediately after the initial infection by the sporozoites, some of those of at least two species, *P. vivax* and *P. ovale,* remain in the liver and transform into a cell called a hypnozoite, a kind of cell which remains inactive for a long period of time until conditions are right to cause the infected person to have a relapse. At this time, the hypnozoites are reactivated and produce sporozoites which produce the schizonts and gametophytes and evoke the symptoms of malaria tertiana. The release of gametophytes into the blood stream of the infected person allows him to infect suitable mosquito vectors, which then can infect all people they subsequently bite. Thus, new epidemics of one of these two kinds of malaria after it seemed that it had been eradicated in the region can be one obvious result of the reactivation of a hypnozoite. This is a strong weapon in the arsenal of the species of *Plasmodium* that produce hypnozoites to protect themselves from extinction as a result of the efforts of physicians and public health workers to eradicate malaria tertiana. As long as the infected person lives, he remains a living reservoir of infectious protozoans, which may not reappear again as active pathogens for several decades.

In some cases, however, the infection of a human host can be what has come to be called a "dead end" infection, which does not result in the infection of any new vector. Such infections effectively result in the end of an epidemic unless non-human reservoirs of gametophytes, usually a primate, are available. If the human host is not able to produce enough gametophytes to successfully infect new mosquito hosts, the life cycle of the *Plasmodium* species is interrupted, and the spread of the disease is also interrupted.

The protozoans causing malaria must complete their life cycles in two kinds of metazoans. One is a vertebrate and the other is an insect that consumes the vertebrate's blood. Ignoring the rest of the details of the parts of the life cycle occurring in vertebrates, we will focus on the developmental processes that make the insects and the protozoans essential to the spread of a specific malarial disease that is a major cause of human fatalities in many tropical countries.

Obviously, the adaptation of the *Plasmodium* species to take advantage of the morphology and physiology of the insect vector must be comparable, even if somewhat less complex, than its adaptation to the vertebrate host. Gametes formed by stages parasitic on individual erythrocytes must fuse to produce zygotes that will modify themselves into spindle-shaped cells that fit into the salivary glands of the mosquitoes so that they can be injected into a suitable vertebrate host.

There are a great many kinds of blood sucking insects in the tropics, but only certain ones can act as vectors for each species of disease-causing protozoan. There are many more species of *Plasmodium* that cannot cause any disease in human beings than there are that can. Each of these is capable of causing diseases only in specific reptiles, birds, or mammals. This shows that malaria can be transmitted only in a suitable location where specific pathogens, vectors, and final hosts occur at the same place at the same time.

SECTION 3

Other Protozoan Parasites - *Leishmania*

There are other members of the Protozoa which are parasites of human beings and cause fatal diseases in some cases. Although they do not cause nearly as many fatalities as the worst of the malaria infections, they are widespread in the tropics and dangerous, especially if left untreated. These diseases have aquatic insect vectors, which are usually not members of the family Culicidae. Because the number of persons in which these parasites live is relatively small and because the diseases caused by the parasites are not notably severe, the knowledge of which insect vectors can spread the *Leishmania* parasites is by no means complete.

The medical treatments for these diseases are generally similar to those used to control other protozoan parasites. Some of them involve the use of drugs meant to control the reproduction and spread of the parasites in the human body. These treatments can become less effective over the years because of the increases in the resistance of the parasites to the drugs. In many cases, the complete elimination of the parasites from the body shows no promise of being successful at the present time. The most effective method of controlling some of these diseases still seems to be avoidance of the insect vectors until more promising treatments are discovered.

Most diseases discussed here are confined to the tropics and subtropics. Where freezing temperatures prevail in winter, some of the aquatic insect vectors are not present and no other insect vectors are available to take their places. Terrestrial insects with the ability to spread the tropical diseases are also absent in the temperate zones. Because persons infected in the tropics can maintain the parasites in their bodies for the rest of their lives, it is evident that the spread of the diseases in places with relatively cold climates is limited mainly by the absence of suitable insect vectors. Just as in some of the protozoans that cause malaria, these tropical pathogens seem to become dormant when the host returns to places with colder climates. It seems that the parasites causing some of the tropical diseases require a functioning ecosystem to complete their live cycles, and dormancy is a reaction when parts of this ecosystem are not functioning as they do in the tropics.

The genus *Leishmania* encompasses many tropical species of Protozoa. They are flagellates classified as trypanosomes, and each has the ability to cause a different disease. In this way, the life cycles of the individual species are reminiscent of those of the species of *Plasmodium*, each of which causes an infection with features of its own, although most produce rather similar symptoms. The diseases called by the name leishmaniosis after the generic name *Leishmania* are less dangerous than at least two of the malarial diseases, and they are less widespread. Some of the individual diseases caused by various species of these trypanosomes are kala-azar and dum-dum fever.

Unlike the insect vectors of malaria, none of those for leishmaniosis belong to the family Culicidae. Many of the known vector species belong to the genus *Plebotoma* in the Old World, while those in the New World belong to the genus *Lutzomyia*. Both of these genera belong to the family Psychodidae. Most members of this family in the temperate zones are innocuous and resemble tiny moths with hairy rather than scaly wings. They do not bite or suck blood. These are members of the suborder Nematocera, and the adults are often found in outhouses or near polluted water.

The tropical subfamily Phlebotominae, on the other hand, includes many species that bite and suck blood. They spread the diseases by infecting themselves with various species of *Leishmania* and injecting them into uninfected animals and people.

Recently, there has been an increasing number of reports that species in the family Ceratopogonidae are also vectors of *Leishmania*. Some of the suspected vectors belong to the genus *Culicoides*. These species feed on the blood of vertebrates, including human beings. Most of the diseases ceratopogonids spread affect various animals, including livestock. However, they have also been implicated in transfers of serious diseases of animals to human beings. Research concerning their ability to transfer pathogens or parasites might reveal that they can also be vectors for various tropical diseases. The larvae of many species develop in rapidly flowing water.

Reports have suggested that *Culicoides sonorensis* Wirth and Jones, 1957, is one of the species that can transmit *Leishmania*. Seblova et al. (2015) have discussed this matter in detail.

SECTION 4

Aquatic Insect Vectors of Nematodes

Nematodes, commonly called round worms, are usually spoken of as parasites rather than pathogens, although some of them are microscopic and cause diseases with symptoms not very different from those caused by protozoans or bacteria. Among the smallest of the nematode parasites are those that live and reproduce within the lymph system, which they can block after their populations reach considerable densities. Those that infect human beings can cause enormously swollen legs, blindness, and eventually death.

Some nematode parasites can reach large sizes, but these are not carried by aquatic insect vectors and will not be discussed in this book. Like viruses and protozoans, their microscopic relatives are carried by the adults of insects that develop as larvae in water. These insects occur only in the tropics, and most are confined to relatively small geographical regions of the world.

a. River blindness

One of the serious diseases caused by parasitic nematodes is Onchocerciasis, also called river blindness, caused by the species *Onchocerca volvulus* Leuckart, 1893, in the phylum Nematoda and family, Onchocercidae. It occurs mainly in sub-Saharan Africa, but it is also known from Yemen and was apparently introduced to various locations from Mexico to Brazil, where most efforts to eliminate the disease have been successful.

According to the World Health Organization, approximately 300,000 people in the world have been completely blinded by the tiny worms, while about 800,000 suffer from impairment of their eyesight due to the disease called river blindness.

The name of the disease suggests that the victims of *Onchocerca volvulus* lived at locations near a river. This suggests that contracting river blindness is usually the result of a bite by a blood-sucking aquatic insect. However, the insect is usually not a mosquito but rather a black fly, that is, a member of the order Diptera, suborder Nematocera, and family Similiidae. The larvae of almost all species in this family develop in lotic water, and many of them prefer water flowing very rapidly. Simuliid larvae build very sturdy structures, in which they are able to keep from being carried away by the current and hide from their natural enemies, as already discussed in Chapter 7.

Adult simuliids consume the blood of vertebrates, including human beings. After consuming the blood, the females produce eggs, which they deposit along suitable water courses. If a female bites a human being infected with *Onchocerca volvulus,* it ingests the microfilariae of the parasites, which infect the simuliid. These exit the digestive tract and make their way to the thoracic muscles, where they develop into first stage larvae. Growth and development continues until the larva reaches the third stage, when it migrates to the head and remains near the proboscis to wait for the vector to bite another person. It is now in the infectious stage.

The next time the simuliid bites another person, the third stage larvae enter the wound and develop into male and female worms. The male is small, while the female is long and forms a nodule beneath the skin, in which mating and production of unsheathed microfilaria take place. These proliferate and spread through the skin into the lymph ducts. Some are able to enter blood vessels, through which they can spread throughout the body. Microfilaria can sometimes be found in the blood, mucus, and urine, but most will develop dense populations in the lymph system.

The disease these tiny nematodes cause gets its name from the damage this species of worm can do to the eye. The large populations of these worms block the lymph ducts and damage the tissues of the host, especially those in the eyes. This eventually causes loss of eyesight and sometimes complete blindness. Meanwhile, the infected person can pass the parasites on to each simuliid that takes a meal of his blood.

b. Filariasis

Filariasis is a word derived from the name *Filaria*, which was long the accepted name of the genus of tiny round worms, which are known to cause the disease. Taxonomic changes have been made, and today the two species of Nematoda which are responsible for most of the cases of filariasis in human beings throughout the world are *Wucheria bancrofti* (Cobbold, 1877), and *Brugia malayi* (Brug, 1928). In addition, other species infect various birds and mammals and occasionally infect a human being, as well. Once one of these minute parasites is introduced into the human body by an insect vector, it enters the lymph and circulatory systems and begins producing large numbers of offspring. The *Filaria* parasites are introduced into the body of their human hosts by the vector, one of many species of mosquito. They will begin reproducing and continue until they have formed a massive population in the lymph system of the host.

A severe case of filariasis infests human beings causing the bizarre symptoms of elephantiasis, the name of a condition that comes about because the lymph canals are blocked by thousands of tiny worms. Severe cases of filariasis often cause one leg to swell until it reaches a diameter that makes it appear like the leg of an elephant. Not only can a host of the parasites become a cripple because of the massively swollen leg, he can die from the tissue destruction the worms cause.

The parasites are small members of the phylum Nematoda. They are less selective of their vectors than are the species of *Plasmodium* that cause malaria, and they can be transmitted to a human host by many mosquitoes belonging to the genera *Anopheles*, *Aedes*, *Culex*, *Mansonia,* and *Coquillettidia*. The American Center for Disease Control has posted a list of mosquito vectors that transmit *Wucheria bancrofti*, which can be updated on the Internet as new vectors are discovered. The list includes the following species: *Aedes aegypti* (Linnaeus, 1762); *Aedes cooki* Belkin, 1962; *Aedes kochi* (Donitz, 1901); *Aedes polynesiensis* Loos et al., 2014; *Aedes pseudoscutellaris* (Theobald, 1910); *Aedes rotumae* Belkin, 1962; *Aedes scapularis* (Rondani, 1848); *Aedes vigilax* (Skuse, 1889); *Anopheles aquasalis* (Curry, 1932); *Anopheles arabiensis* (Patton, 1905), *Anopheles bancroftii* Giles, 1902; *Anopheles bellator* Dyar and Knab, 1906; *Anopheles darlingi* Root, 1926; *Anopheles farauti* Laveran, 1902; *Anopheles funestus* Giles, 1900; *Anopheles gambiae* Giles, 1902; *Anopheles koliensis* Owen, 1945; *Anopheles melas* Theobald, 1903; *Anopheles merus* Donitz, 1902; *Anopheles punctulatus* Donitz, 1901; *Anopheles wellcomei* Theobald, 1904; *Mansonia pseudotitillans* (Theobald, 1910); *Mansonia uniformis* Theobald, 1901; *Coquillettidia juxtamansonia* Chagas, 1907; *Culex annulirostris* Skuse, 1889; *Culex bitaeniorhynchus* Giles, 1901; *Culex pipiens,* Linnaeus, 1758; *Culex quinquefasciatus* Say, 1823.

26
Insects that use People for Bait

During a four-year study of the Pantanal, probably the largest wetland in the world still in a near pristine state, an interesting phenomenon was observed and described (Heckman, 1998). During the rainy season, a particularly large mosquito, *Psorophora varipes* (Coquillette, 1904), feeds during the hottest part of the day. The adults wait in the bushes and trees on the small islands just above the water level in the flooded wetland until a large mammal passes. Human beings are apparently particularly attractive prey, especially when large areas of the skin are left exposed.

The mosquitoes suddenly emerge from the underbrush in swarms sometimes estimated to consist of considerably more than 100 individuals. Their bites cause extreme itching, forcing people not well covered by clothing to flee the area and seek shelter in vehicles or buildings.

During the same season, larvae of dragonflies of various species emerge from the water and undergo metamorphosis to adults while the water level is still high and almost all of the wetland is available for spawning. One of the species that is particularly abundant in the northern Pantanal is *Erythrodiplax castanea* (Burmeister, 1839), a dragonfly frequently seen to form large swarms of adults. The swarms of this dragonfly were often observed following researchers as they travelled though the wetland. Large swarms of the dragonflies usually hovered over their heads or over other visitors to the wetland during this season.

The dragonflies had apparently noted that human beings and other vertebrates lure the large mosquitoes out of their hiding places in the underbrush. They therefore regard the researchers, ranchers, and tourists as outstanding bait for the mosquitoes. When the human bait passes near a swarm of *Psorophora varipes*, the mosquitoes suddenly take to the air and emerge in large swarms to surround their victims, attacking from all directions. They seek exposed areas of skin to take a quick and rather painful blood meal at every opportunity. Some mosquitoes seek to approach from behind, while the rest of the swarm moves in front of the victim to divert attention. The more skin that is left exposed, the more chances they have to quickly land on the back of the neck or outer edge of the hand to take a quick blood meal. Making simple water analyses while the air temperature is above 40°C and the humidity, nearly 100%, is no simple matter and requires jackets, hats, and gloves. Apparently, the odor of perspiration attracts even more mosquitoes, and the swarm becomes larger and more aggressive as the time goes by.

While the mosquitos are attacking their human or animal victims, the dragonflies descend into the swarm of mosquitoes and pick them right out of the air. During my personal observation of this predation by the dragonflies, I noted that only about 10 minutes was required for a dragonfly swarm estimated to consist of 30 to 40 individuals to almost completely clear away a swarm of mosquitoes estimated to consist of 100 to 200 adult female *Psorophora varipes*. During the next 10 minutes, a few mosquitoes that had been overlooked came out of hiding and were also captured and quickly devoured. By standing very still while this was happening, it was possible to see mosquitoes only inches in front of the face snatched out of the air by fast flying dragonflies. This was an impressive and memorable display of how efficiently predatory insects can dispose of large numbers of pests.

The spiny fore-legs of the dragonflies seized the mosquitoes and pressed them into contact with the dragonfly's mouthparts. *Erythrodiplax castanea* is a relatively large dragonfly, and it can completely consume one of these large mosquitoes in two or three bites. One dragonfly could consume a fairly large number of mosquitoes, and it is impressive to see how fast and completely a swarm of 30 to 40 dragonflies can completely eliminate a large number of these aggressive mosquitoes once they have left their hiding places in the vegetation.

Observing this was not without discomfort until the dragonflies finished their work. As already mentioned, the reason for being in the field was to make chemical analyses of water samples. After that, photographs had to be taken of the surroundings under a bright, tropical sun. At temperatures exceeding 40°C, it was necessary to wear a thick, long-sleeved shirt, hat, and piece of cloth almost completely covering the neck and face. The only skin exposed was that on the hands, after the gloves were removed to follow the analytical procedures demanded to complete the individual tests. Continual movement was necessary to keep the aggressive mosquitos from landing on the hands and taking a quick blood meal. After the dragonflies had done their work, the analyses and photography could be continued without disturbance, and some of the clothing almost completely protecting the skin could be loosened sufficiently to permit some breezes to evaporate the perspiration and cool the skin.

After the entire swarm of mosquitoes had been consumed by the dragonflies, a small cohort of them typically continued to hover above their human bait until the work was finished. These dragonflies remained to pick off any late arrivals from among the mosquito populations still hiding in the bush.

A considerable number of dragonfly species inhabit the Pantanal, but *Erythrodiplax castanea* (Burmeister, 1839) was the only one that frequently used human beings as bait to lure the mosquitos out of the dense brush in which they remained concealed. The appearance of the swarms of *Psorophora varipes* was seasonal, and adult females of that species were only observed during the middle to late rainy season, when the floodplains were hot and humid between rainstorms. The adults of no other mosquito species flew during the middle of the day, and none were active during most of the dry season.

Two other mosquito species were observed to be abundant seasonally in the northern part of the Pantanal. One is *Anopheles albitarsis* Arribalzága, 1878, which is crepuscular and occasionally nocturnal during the dry season. In other parts of Brazil, this species is known to be an important vector of malaria, although those in the Pantanal have never been known to spread diseases caused by any of the species of *Plasmodium*. A proposed solution to the problem of distinguishing between the populations of this species that spread malaria from those that do not was the creation of the subspecies *Anopheles albitarsis albitarsis* for the members of the species that never enter houses and do not act as malaria vectors and *A. albitarsis domesticus* Galvão and Damascene, 1944, for those that enter houses and spread malaria.

The third mosquito species observed in the Pantanal was *Culex maxi* Dyar, 1928, which is mainly nocturnal and appears occasionally near permanent streams and rivers immediately after the first rainstorms of the rainy season. Because of their habits, the last two species mentioned are not subject to much, if any, predation by dragonflies. *Culex maxi* shows no inclination to attack human beings and is thought to live on the blood of animal species other than man.

The dragonflies gain several advantages by using vertebrates as bait, in addition to the obvious one of gaining access to a large food supply. Much terrestrial vegetation, including thorny plants, grow rapidly during the rainy season, which usually begins in September and continues until April. The wetland during the high water period is covered by low islands, most less than one meter above the highest water level. A few trees and some terrestrial bushes and annuals grow on the islands while the wetland fills with water. By the late rainy season, the vegetation on these islands is fairly dense. The adult female *Psorophora varipes* in the wetland seek blood in the middle of the day by hiding in the bushes on the small islands until a suitable vertebrate approaches. When a suitable animal passes by, large numbers of the mosquitoes swarm out of the bushes and swarm around the intended prey looking for a chance to land and quickly take a blood meal. Under circumstances ideal for the mosquitoes, they can feed quickly and return to the bush without encountering any flying predators. After feeding, they only have to rest in the thickets on the islands of terrestrial vegetation until their eggs have developed and then find a suitable place to deposit them. It is not possible for dragonflies or other large predators to get at them among the branches of the terrestrial plants because if they tried, they would risk physical injury from thorns or capture by larger predators in the foliage, such as spiders or tree frogs.

As a result, the mosquitoes could wait in the foliage in complete safety, while the dragonflies would have to find the most efficient way to obtain an alternate food source. The method adopted by the dragonflies to safely gain access to these large mosquitoes while avoiding the obvious dangers of trying to penetrate the thickets is to follow a large animal from which the mosquitoes could get a blood meal and wait for the mosquitoes to emerge from their hiding places to draw blood from their prey. The obvious advantages of this method, which were mentioned above, include avoiding being captured by large predators hiding in the terrestrial vegetation, avoiding physical injury to the membrane of their wings from trying to fly through stiff or thorny branches, and getting species that they could easily capture and eat to emerge from their hiding places in large numbers at the same time. In the aerial dogfights between these two insect species, the clumsy and slow mosquitoes are at a distinct disadvantage against the fast and maneuverable dragonflies. The long spines on the forelegs of the dragonflies are designed perfectly to scoop the mosquitoes out of the air and hold them against the mouth of the dragonfly as they become fast food.

Although there is no way at the present time to determine how the dragonflies came to develop these methods, some anecdotal information suggests that they have been used for at least several centuries. One of the Portuguese names for dragonflies in Brazil originates from the time they were noticed by early explorers of the inland regions of the country: *cabelleros do Diabo*. This translates as "the Devil's horsemen". It is likely that those dragonflies that use people as bait would follow the early explorers of Brazil throughout the day waiting for them to attract mosquitoes for them to eat. From their name for these insects, it is obvious that the European explorers never realized that the dragonflies were doing them a favor by keeping them safe from the local mosquitoes, some of which carried yellow fever, malaria, or other serious tropical diseases.

Such a name is not descriptive of the behavior of the great majority of Brazilian dragonfly species, which hunt alone or in groups of no more than two or three. The name is rather descriptive only for the seasonal behavior of *Erythrodiplax castanea,* which forms swarms to hunt mosquitoes in the Pantanal by day during the latter part of the rainy season, and perhaps a few other species that use people for bait.

As a footnote to this chapter, it should be mentioned that other dragonflies form small swarms to hunt using other cues for locating masses of dipterans on which to feed. Such groups of hovering dragonflies have also been observed in Southeast Asia and in North America at times. There is no sign that these groups have any interest in mating, as they seem to be simply waiting for food to come to them. Most dragonfly species in the suborder Anisoptera mate once, after which females can live for at least several weeks, feeding and producing eggs. Swarms of these species can therefore be assumed to serve purposes other than mating.

Chironomids, mayflies, and other groups of insects emerge simultaneously following cues involving the phase of the moon or the tides, local weather conditions, or seasonal changes to synchronize their emergence from the water as adults and begin mating and depositing eggs. Certain dragonfly species seem to know what these cues are and form feeding swarms in anticipation of these emergences. One such swarm of an unidentified dragonfly species was observed along the banks of the Chao Phya Estuary in Bangkok, Thailand, near the end of the rainy season. The dragonflies paid no attention to the movement of people, and seemed to be waiting over a particular spot near a park where mangrove trees had been planted in order to try to replace a mangrove swamp that had been destroyed in the past. The location was separated from the channel of the estuary by a wooden barrier, which prevented the water flow in the estuary from flowing between the trees and disturbing the sediment. It did not prevent water from entering and leaving the developing mangrove swamp according to the movement of the tides.

This location in the freshwater section of the estuary seemed to be one in which species in the family Chironomidae could be expected to develop. Because of the season and the time of day, the dragonflies in the hovering swarm could have been expecting that a mass emergence of adult chironomids near sundown would be likely.

In this case, a feeding swarm of dragonflies waiting for the midges to emerge would not have been an unexpected phenomenon.

It seems safe to say that the formation of the feeding swarms by anisopteran dragonflies, as described here, is not simply a case of mass emergences of other aquatic insects attracting all of the dragonflies in the area to feed on random assemblages of edible insects. Swarms of dragonflies were observed hovering without having any edible insects in sight. They could rather be groups of dragonflies forming in anticipation of some coordinated activity by members of a species that the dragonflies feed upon. Either they expect a group of adult insects to emerge over a section of a water body for mating, or they expect blood sucking insects to come out of their hiding places in order to attack a large animal that is passing by. In either case, their feeding indicates a degree of planning that goes beyond a simple search for edible insects that might happen to cross their paths.

27

Flies in Hot Water

Another physical barrier to the colonization of a habitat by insects is temperature. Once again, it is the family Ephydridae to which one of the species with a solution to master extreme conditions belongs. Only the species *Stratiomyia japonica* van der Wulp, 1885, in the family Stratiomyidae, has managed to equal or surpass the ephydrids in tolerance to high temperatures in Japanese hot springs. Apparently, only one species in the world belonging to the family Chironomidae can surpass these species in tolerance to high temperatures by less than 1°C in hot springs in Yellowstone National Park, Wyoming, in North America.

Life can exist only within relatively narrow temperature limits, and unusual extensions of those limits always merits attention. Other shore flies have managed to find niches in waters containing various chemicals that would make the water body uninhabitable for almost all other insects. However, temperature is another kind of barrier. Heat denatures various proteins, including vital enzymes upon which most forms of life rely. Why can one ephydrid species develop in water too hot for almost all other insects?

All over the world, there are hot springs in places where hot magma from the center of the earth comes near enough to the surface to heat water to a temperature at or near its boiling point. Sometimes, the locations of hot springs are evident from geysers or discharges of steam. In the deep sea, "black smokers" mark the locations of water being discharged into the deep sea after being heated by stones in proximity to magma. Wherever this water appears, living organisms attempt to utilize it to obtain nourishment and avoid predators. Usually, microorganisms are able to adapt to temperatures higher than any that all other forms of life can bear. Near black smokers, such microorganisms have adopted chemosynthesis to produce food, which is shared with animals that have developed in symbiosis with them. Where the water exits at terrestrial locations, cyanobacteria have adapted to surprisingly high temperatures to get the first choice of the minerals that leave the earth with the emerging hot water.

As the water leaves the hot spring, it usually flows into a stream, cooling as it flows. Animals can enter the water wherever it has cooled to a temperature at which the animal can survive. The species that can stand the highest temperature is the one that can come closest to the spring. In this way, the species can not only benefit from the first choice of the food supply that the thermophilic cyanobacteria have produced, they are protected from all predators by the high temperature of the water.

The thermophilic insects have been studied well at three hot springs in different parts of the world: those on the Japanese island of Hokkaido, where the hot water is impounded to create warm baths; Yellowstone National Park in the State of Wyoming, U.S.A., and near the volcanoes on Iceland. The findings obtained from all three of these studies indicate that the family Ephydridae in the order Diptera has been able to adjust to higher water temperatures more successfully than any other kind of insect.

SECTION 1

Hotsprings of Hokkaido

The island of Hokkaido, Japan, is famous for hot springs, which contain water in which most animals cannot live. People visit the springs in order to bathe at temperatures at the limit of human endurance. In winter, a species of monkey native to Japan, *Macaca fuscata* (Blyth, 1875), has also adapted to the waters of the springs so that they can remain comfortable at the very low temperatures prevailing on that island. The local spas are able to adjust the water temperature to a tolerable level for people and monkeys. Most insects avoid the springs and first approach the water after it has flowed far enough away to have cooled to a tolerable temperature. However, Matsumura (1915) discovered that a small fly also inhabited the water from the hot springs while it was still too hot for other insects to live in.

An insect species that has displayed a much greater tolerance to high temperature than others is *Scatella calida* Matsumura, 1915, the adults of which were found in a hot spring, in which its larvae feed on species of algae also adapted to high water temperatures (Sakagami, 1960). Miyagi (1977) suspected that *S. calida* may be conspecific with the nearly cosmopolitan species, *Scatella stagnalis* (Fallén, 1813), which is very abundant on Hokkaido. However, *S. stagnalis* is a widespread species that is not known to be adapted to hot water anywhere else. The wing of *S. calida* is also one of the darkest colored wings in the genus and much darker than *S. stagnalis* from any other habitat (Fig. 27.1).

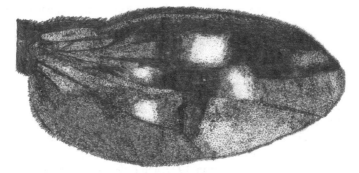

Figure 27.1 The wing of *Scatella calida,* which is almost entirely dark with only a few paler spots on the membranes. The larvae develop in the water from hot springs. Redrawn and modified from Miyagi (1977).

The adaptation to high temperatures by species inhabiting water from hot springs is not easy to investigate because most hot springs emerge from the ground at high temperatures but cool rather quickly after they emerge from the ground. The rate of cooling is usually determined by recording the temperature of the water at regular distances from the source. However, the impact of the water on any insects that may be present depends only on the microclimate in which the insect remains. The farther the water moves from the source, the more heterogeneous is the temperature within the moving body of water. The water cools quickly at the surface due to evaporation. The sediment of the streams through which the water from the spring passes is also cooler than the spring water, and the water is also cooled by contact with the sediment at a rate that can vary considerably with the season. At the same time, the central part of the moving water mass could be expected to retain its elevated temperature longer, and routine recording of the temperature as the water mass moves downstream would probably result in making a record only of these somewhat higher values.

In fact, recording the presence of insect species along a temperature gradient in a stream draining water from a hot spring should reveal the maximum water temperature at which each insect species present can survive. The species that can survive at the highest water temperature will be the species that appears closest to the water source.

The maximum temperature at which the ephydrid larvae can survive in the water will have to be determined experimentally because of the differences in the microclimates along the course of the stream. This is also difficult because artificially flowing water habitats must be created that truly correspond to the natural habitats in which the larvae develop.

SECTION 2

Geothermal Springs at Yellowstone

Recent studies of hot springs have shown that species of microorganisms can survive at much higher temperatures than previously considered possible. The ephydrid species have been reported only with certainty at temperatures considerably lower than those at which certain Cyanobacteria have been found to be active. They are much higher, however, than almost any other insect would be able to tolerate. In streams flowing from the hot springs, there is a measurable thermocline beginning at a higher temperature than any known insect could tolerate. Any insect can enter at any point voluntarily, so the uppermost point at which a particular insect species is first found is at a temperature chosen for development of the next generation of larvae by the female that lays her eggs there.

Brues (1924) reported that the hot water from the springs at Yellowstone National Park were populated by various animals, which could survive and even thrive at higher temperatures than most other animals. Brock et al. (1969) identified two species in the family Ephydridae, which could survive at higher temperatures than other insects and feed on the thermophilic algae and cyanobacteria, including an unidentified species of *Synechococcus*, which formed mats on the surface of the water. These were identified as *Paracoenia turbida* Curran, 1927, and *Ephydra bruesi* Cresson, 1934. The authors also provided evidence that the larvae of *Paracoenia turbida* could nourish themselves satisfactorily on the thermophilic primary producers, which floated on water at temperatures between 30°C and 40°C. The imagoes deposited their eggs on these mats at temperatures within that range, and the larvae undergo metamorphosis to the pupal stage in the same habitat.

Ephydra bruesi develops at somewhat lower temperatures, usually up to about 35°C. The imagoes lay their eggs on parts of the mats of algae and cyanobacteria that extend above the water, allowing some cooling of the eggs themselves. The egg masses were found to maintain their temperature at a maximum of about 31.5°C (Brock et al., 1969).

There is one other fly that sometimes shares the warm water at Yellowstone with the two ephydrids. It is a member of the family Dolichopodidae, which does not feed on the algae but is rather a predator that feeds on the ephydrid larvae (Brock et al., 1969). Unfortunately, the species of Dolichopodidae was not identified.

Yellowstone National Park is not the only location of hot springs in North America, but it is a place at which a great many such springs are found in a relatively small area. The fauna of hot springs at other locations in North America were studied by Brues (1928, 1932).

To put these temperatures in perspective, studies should determine the water temperatures at which the adult fly lays its eggs in the water, at which the larvae develop over the long term, and the maximum temperature which an insect can tolerate voluntarily for short periods of time.

SECTION 3

Flies of Iceland

Although the islands of Iceland are located far to the north of other parts of the world considered to be reasonably comfortable for people to live, it has natural geothermal heating of the ground. It is therefore not uncomfortably cold to live there throughout the year, and the natural springs supply both hot and cold running water. However, the natural heating under the feet has its price. The citizens there must live with constant rearrangement of the countryside through the activity of volcanoes, which produce wonderful but changing landscapes.

The geothermal heating supplies the population with ample amounts of hot water for bathing, and the natural hot springs provide a habitat for a very interesting flora and fauna of thermophilic species. Tuxen (1944) studied the fauna of the hot springs and discovered that still another member of the family Ephydridae has developed the highest degree of tolerance to the warm water of almost all other known insects, which was reported to be 49°C by Wesenberg-Lund (1943). He identified the species as *Scatella thermarum* Collin, 1930.

The larvae of this species feed on the thermophilic algae and cyanobacteria, which seem to be common primary producers in hot springs all over the world. Besides *Scatella thermarum* and the Japanese species, *Stratiomyia japonica*, which has been found to live in water as hot as 48.8°C, only larvae of the North American chironomid, *Chironomus tentans* Fabricius, 1805, can live in water as warm as c. 49°C (Wesenberg-Lund, 1943). This chironomid species has been cultured for many years as the standard test species for the toxicity of sediments (Lee et al., 2006).

Like the brachyceran family, Ephydridae, the nematoceran family Chironomidae has managed better than any other family in its suborder to adapt to extreme environments (Thienemann, 1954).

28
Ants and Water

Ants are generally regarded as terrestrial, social insects, which have as little to do with water as possible. However, ants have some amazing habits and abilities, which are seldom reported and poorly understood. Among these are two species of ant that behave differently from what would be considered typical behavior of ants. One of these is understandable but surprising; the second must be considered amazing and motivated by instincts that are not understandable. In either case, the second of these could be concluded to be a faulty observation or, most likely, a behavior more complex than the reports reflect and part of a much more amazing pattern of behavior. Both of the ant species involved live in the tropics of South America.

SECTION 1

Ants that Live on a Floodplain

In the tropics, most species of ant build their nests either in the ground or in trees. Almost all of them choose places that will assure protection against flooding from heavy rains. One unidentified species that lives in the Pantanal of Mato Grosso, moves the location of its colony from underground to a location in emergent plants and back again once each year in a conspicuous way.

As described earlier in the book, the northern part of the Pantanal floods rapidly during the first weeks of the rainy season until large areas that were completely dry before the freshet are covered by a layer of water more than 1 m deep in places. Almost immediately after the water begins to fill the wetland, emergent aquatic plants begin to grow. These plants continue to grow, flower, and produce seeds while the water covers the floodplain and then die off and are replaced by short-lived terrestrial plants after the water flows slowly away during the dry season.

One species of small, reddish brown ant lives in underground colonies during the dry season. When the floodplain becomes covered by water, which will remain there continuously for six months or more, the ants leave their underground nests together and form a small floating island on top of the water, usually near some plants which emerge above the surface of the water and remain as emergent plants throughout the rainy season.

Such floating colonies of ants, which are buoyed up by the bodies of the ants themselves, are in constant motion allowing the ants pressed below the surface of the water to take their turns on the top of the mass of ants in order to breath for a while before descending again below the water level. These floating colonies usually drift to a group of emergent plants, to which some of the ants attach themselves in order to anchor the whole colony. If disturbed, however, they let go again so that the colony can float to a less disturbed location. They do not immediately climb into the plants because the water is rising, and they do not find a new nesting site until the water level has almost stabilized and the emergent aquatic plants have grown to remain above the water level. After a few days, the plants have grown quickly to form a rich vegetation above the water level.

At this time, the ant colonies disappear into the emergent vegetation, in which they build new nests above the water level in concealed locations. At the end of the rainy season, the water recedes from the flood plain, and the ants disappear below the ground, where they presumably remain until the next rainy season.

This behavior of relocating the nest to cope with seasonal flooding is not seen in the temperate zones, where seasonal floods do not occur. The rare floods that occur, caused by unusually heavy rains, are typically prepared for by the ants, which build branches of their underground tunnels almost vertically upwards to chambers that can hold an air supply to meet the needs of the ants for several days until the flood waters have receded. If this is not sufficient, the ants might drown.

In regions of the tropics, where rainfall can be far more intense than in the temperate zones, several species of ant, primarily terrestrial in their habits, have to cope with water from time to time. The so-called army ants, which are continually migrating, are known to build bridges from the bodies of the worker ants or take advantage of floating objects to cross small bodies of water. They also can form balls of individuals that enter the water and propel themselves to the other side of a stream or a canal. Care has to be taken to transport the queen ant across.

SECTION 2

Ants that Plant Bromeliads

It was observed long ago that certain species of ant formed symbiotic relationships with certain species of plant, including both green plants and fungi, somewhat reminiscent of the relation between human beings and various crops raised by farmers or plants cultivated by gardeners. In rainforests and various kinds of tropical habitat, it was noted that certain species of ant associated themselves with species of "antplants" (Wheeler, 1942). Bromeliads themselves and other plants capable of catching rainwater and storing it are involved in large and highly complex associations of species ranging from prokaryotic organisms through vertebrates and encompassing many insects belonging to diverse orders (Céréghino et al., 2010). It was revealed that certain bromeliads, including *Aechmea martensii* (G. Mey.) Schultes and Schultes f., are cultivated by arboreal ants in "ant gardens," where plants that the ants grow provide reservoirs of rainwater inhabited by a great variety of aquatic organisms (Dejean et al., 1995; Darby and Chaboo, 2015). The species studied by the authors cited in this paragraph were located in tropical rainforests in French Guiana, Peru, and Mexico. However, there are undoubtedly many other such symbiotic associations contributing to the great diversity of species present in tropical rainforest communities waiting to be discovered through intensified research.

The two arboreal ant species found to cultivate the bromeliads are *Pachycondyla goeldii* (Forel, 1912) and *Camponotus femoratus* (Fabricius, 1804). Apparently, either of the ant species can plant and tend *Aechmea martensii,* which does not grow unless it is planted and tended by one of the two ant gardeners. Once the bromeliad has grown, it has a sufficient number of leaves modified sufficiently so that they can hold the rainwater that falls on them long enough for insects to complete their larval stages before it evaporates. In northern South America, it is likely that subsequent rain storms would replenish the water in the bromeliads before they would ever dry up.

This is clearly a three stage symbiosis. The aquatic insects that develop in the *Aechmea martensi* are directly dependent upon the plant for the existence of their habitat. The plant, in turn, is completely dependent on one of the two ant species for the planting of its seeds and probably for its cultivation. Presumably, the plants supply nutrients to the ants or attract prey on which the ants feed. This would make it a true case of symbiosis, in which the ants receive real benefits from the plants, while the plants depend on the "ant-gardeners" for their cultivation. What is still missing is a demonstrated advantage that the ants receive from the plants or the biotic community of aquatic species supported by the plants. Clearly, a great many hypothetical advantages can be proposed as such advantages, but could these advantages not also be provided by any bromeliad species that did not have to be tended by the "ant-gardeners".

The aquatic larvae depend on the plant for a suitable habitat, while the plant presumably receives inorganic nutrients from the insect larvae that release them when they capture and digest their prey. The question is whether the aquatic larvae developing in the plant could develop just as well in other species of bromeliad. Another question is whether the ants could live just as well if they did not cultivate *Aechmea martensii.*

The ants contribute much work to the planting and tending of the bromeliads, but what do they receive in return? Human beings can recognize aesthetic improvements that they make without gaining any real material advantage. There has been evidence of such activities since the stone age, when hunters drew pictures on the walls of their caves showing the animals that they hunted.

A very popular activity today is placing water in a glass aquarium to keep attractive fishes and other animals that live in freshwater bodies or in the oceans. Such projects are often thought of as activities unique to human beings. Somewhat akin to such activities are the construction of colorful displays or dances by male birds involving displays of feathers, but these follow an instinctive choreography used to attract females of the same species for mating. Such activities are essential for the survival of the species of birds that perform them.

Raising plants that produce aquatic habitats for many insects by storing rainwater would be understandable if it were done by an animal motivated to improve the community as a habitat for many different forms of life. But ants?

If there is no understandable reason for the ants to cultivate a large plant, could this be classified as a "hobby" of the ants in the colony? Human beings spend a large amount of time and expense on various kinds of hobby. Little has been written to speculate about the possibility of animals, including insects, spending time and effort on hobbies. Could it be that raising bromeliads to hold water in which various aquatic insects develop once served an important function for survival of the ant species, which is not any longer necessary for the survival of the ant? More complex relationships between the ant species that grow bromeliads and the plants themselves will probably come to light after more studies have been undertaken. Meanwhile, the question remains open concerning the possibility of ants pursuing hobbies or doing useful things to preserve the ecosystems in which they live.

The formation of symbiotic relationships with many different forms of life without gaining any clear benefits from the work performed to accomplish it is now being studied as parts of research projects focused on phytotelm communities in epiphytic plants deliberately grown by ants (Céréghino et al., 2011). As yet, the necessary evidence has not been provided to understand the activity of the ants, although unproven hypotheses are not at all difficult to formulate.

As a source of drinking water for the ants would be one advantage if the locations of the plants were not in rainforests, where rainfall is frequent and plentiful. The aquatic invertebrates living in the water would provide the ants with a source of protein, which would be helpful for improving the ants' diet. However, nobody has shown yet how the ants catch the insects living below the surface of the water. The plants also attract herbivorous terrestrial insects, but the rainforests in which the nests of the ants are located are full of terrestrial insects of all kinds. Hence, the problem of finding an advantage the ants gain by using the bromeliad as a garden plant remains unsolved.

It could be speculated that the ants gained a permanent aquarium inhabited by a biotic community that was important for the survival of the gardiner ants at some time in the distant past, when climatic conditions, the species in the water of the plant, or the amount of annual rainfall was substantially different from now. The ants that survived under the unfavorable conditions that existed at the time because they were able to cultivate the bromeliad species may have inherited the instincts to plant seeds of *Aechmea martensii* near their nests and continue to do it to this day. The continued work in their gardens would then bring the ants no advantages, but there is no mechanism for them to abandon their inherited pattern of behavior. Formulating hypotheses such as this one is easy, but providing enough evidence to prove them without having a machine for time travel would be nearly impossible.

This again leaves the open questions concerning the apparent possibility that the ants' activity of cultivating plants that hold water is motivated by altruistic sentiments to provide conditions for the survival of whole communities of aquatic invertebrates which benefit directly from the plants.

Regardless of what the motivation of the ants might be, Céréghino et al. (2011) provided an impressive list of invertebrates that inhabit the water trapped in the plants and are able to contribute to the ecosystem only because the ants make the efforts to grow the bromeliads. Unfortuntely, they were not able to identify the invertebrates to species, but they were able to find members of the insect families developing in the phytotelm (Table 28.1).

Table 28.1. The families represented in the aquatic community living in the water trapped in *Aechmea martensii*, an epiphytic plant in the family Bromeliaceae planted and cared for by ants either in the species *Pachycondyla goeldii* or *Camponotus femoratus*. The study was conducted at a site in French Guiana by Céréghino et al. (2011). The number of species listed is the minimum number represented because in some cases, all species were not distinguished.

Order	Family	Number of species present
Odonata	Coenagrionidae	1
Hemiptera	Veliidae	1
Coleoptera	Scirtidae	3
	Dytiscidae	1
	Hydrophilidae	1
Diptera	Limoniidae	1
	Cecidomyiidae	1
	Psychodidae	1
	Chironomidae	2
	Ceratopogonidae	4
	Culicidae	more than 4
	Tabanidae	1
	Syrphidae	1

In addition to the insects listed in Table 28.1, at least one species of Acari, four species of Naididae, and one species of Aelosomatidae were present. In addition, a great many species of microorganisms, including species of Bacteria, Fungi, Protozoa, and Rotifera would be expected in the water. Because the insects leave the bromeliad after reaching the adult stage, they leave and fly to new bromeliads to spawn or are fed upon by predators in other communities. In that way, they enter the vast and diverse food web of the tropical rainforest after having utilized the phytotelm prepared through the work of the ant gardeners that produced the habitats in which they developed.

In summary, the greater biotic community that encompasses the species from the phytotelm forms a broad and diverse network extending throughout Neotropical forests. Whether similar situations can be encountered in other parts of the world remains to be discovered. Plant species in the family Bromeliaceae are planted and tended by ants, and these trap rainwater in which a whole community of aquatic species develops. Trapped detritus is decomposed by bacteria and other members of saprobic communities of microorganisms, which are available as food for insect larvae. These larvae are consumed by a variety of predatory insects, a few of which remain in the community as adults. The adults of other species in families, such as Culicidae and Tabanidae, fly away from the community and seek blood meals from vertebrates. They then return and deposit their eggs containing much rich nutrient material for the developing larvae in the phytotelm. A second interaction with species that are not aquatic occurs when terrestrial insects come to bromeliads to eat the leaves. Two beetles in the family Chrysomelidae have been identified as such insects. In Peru, *Calliaspis rubra* (Olivier, 1808), one such terrestrial species, feeds on the leaves of *Aechmea nallyi* L.B. Smith, as reported by Burgess et al. (2003). The second species, *Acentroptera pulchella* Guérin-Méneville, 1830, feeds on an unspecified species of bromeliad in Neotropical forests (Mantovani et al., 2005).

Among the terrestrial insects that feed on bromeliad leaves, species in the order Lepidoptera seem to be even more numerous than beetles, but, of course, only their caterpillars eat leaves.

Frank and Lounibos (2005) summarized the reports of aquatic insects inhabiting water trapped in plants but left many of the species unidentified. The results help to confirm, however, just how large and varied the insect fauna involved in some way with a phytotelm actually is. It is also remarkable how the activity of ants is involved in the maintenance of this community.

29

What We Still Need to Discover

A perusal of the literature on the ecology and behavior of aquatic insects reveals the dearth of detailed information on most species. Furthermore, most of what we know about the biology and natural history of aquatic insects was learned by studies of species that inhabit only a small part of the world. Whether what we have learned from studies of individual species hold true for species in the same family on other continents remains to be learned in most cases. Even some of the aquatic insects that transmit diseases or cause damage to rice and other aquatic crops in many parts of the world are still poorly known. Published results of investigations often reveal additional gaps in the knowledge of these species. It is safe to say that information about most aquatic insects is available only phrased as generalities. Specific details about the functioning of individual organs, chemical reactions, and the physiological adaptations of the insects are only rarely reported. Scientific studies of these species are needed, and they would most certainly provide not only interesting but also very important findings for agriculture, public health, and technology.

To date, the greatest amount of information about aquatic insects with especially unusual adaptations to their environments and other organisms with which they interact has been obtained about insects of obvious economic importance, such as species which transmit diseases to human beings or domestic animals and species which either benefit or damage plants important for agriculture. Geographical factors also determine how much information is probably available about individual aquatic insect species. Species from northern and western Europe are very likely better known than those from other parts of the world. Species that inhabit urban and suburban habitats are better known, in most cases, than those found only in remote mountain regions, standing water bodies in deserts, and small assemblages of water in tropical rainforests.

We know from those studies which have been completed that there are surprisingly many complex processes involved in the lives of insect disease vectors, agricultural pests, and insects useful for biological control of such pests. We also can see that our knowledge is far from complete, and much is still to be learned that would permit us to better control problems of public health, cultivation of plants, and animal husbandry.

Much less is known about the aquatic insects that do not directly impact the human population enough to gain the attention of scientists. The fact that we do not have important information about the adaptations which permit survival of insect species not directly affecting the economy or spreading diseases does not mean that the knowledge we would gain by studying them would not greatly assist us in many ways. For example, knowing how insect larvae are able to cope with heavily polluted water and how they are able to subsist on the microorganisms which contribute to the breakdown of organic pollutants and consumption of the available oxygen might help in the selection of species to employ in wastewater treatment.

Biological control of the larvae of blood-sucking insects and pests that feed on aquatic crops, such as rice, taro, sweet potato, yam, or lotus, is another field of investigation that has been neglected until very recently. Even when much information is available about the pests themselves, the natural enemies that would control them are still poorly known. Aquatic insects that could be employed to greatly reduce the numbers of destructive insects by feeding on their aquatic larvae are increasing in esteem as more and more people come to realize that the value of pesticides is decreasing rapidly as the destructive insect species develop resistance against their toxicity.

While many people still think of insects only as economic factors, for good or for ill, more perceptive scientists realize that the secrets of their success in specific habitats are still often unknown. Learning these secrets will open great fields to science and technology, which are inconceivable to most people at this time.

What permits a population of small water striders to thrive on the high seas, surviving without shelter through storms, rough seas, and long, unbroken periods of tropical sunshine? Species inhabiting freshwater bodies, which appear similar to the species on the high seas, may drown due to heavy rainfall or minor agitation of the water. How do pelagic marine water striders meet their needs for fresh water? Must they wait for rain, or do they have physiological adaptations permitting them to separate the salt from seawater?

Not only are the details of their individual life cycles, adaptations to their habitats, ecology, and behavior of most aquatic insects in the world poorly known, many such insects remain completely unknown to science and still lack names. Aquatic entomology is a field of science that is still rich in opportunity for research in many fields. This chapter will make just a few suggestions for research topics that have excellent prospects for yielding valuable results.

SECTION 1

The Basics

To add to the useful information available about the ecology of aquatic insects does not require a research ship, intercontinental air ticket, or extremely expensive equipment. The study of the Hexapoda is particularly rich in opportunities because this group of animals encompasses more species than all others combined. In some parts of the world, they have been studied using the methods of the natural sciences for half a millennium. In other places, they have hardly been studied at all.

Even in those parts of the world where insects and collembolans have been studied the longest, it requires only a relatively short period of observation to begin noticing things that cannot immediately be explained. When observing aquatic insects, the things that are not yet fully understood about their biology are notably more numerous than the things we already know, except in the cases of insects that are harmful to public health or agriculture.

If an entomologist is fortunate enough to have an opportunity to travel to a part of the world where little entomological research has been done, he will have the chance to observe insects with behavior patterns that have never been described before. If the place he visits happens to be in a tropical wetland, the chances are excellent to encounter species that have never been named or described before. During the first half of the 20th century, many long monographs, some several volumes long, were written to describe species collected during expeditions or extended studies at locations far from any universities or research institutes. The results of some of these studies were written by highly competent taxonomists, who had little or nothing to do with collecting most of the specimens. Lundblad (1933), for example, provided excellent descriptions of new aquatic species in the order Hemiptera, but no information at all was provided about the habitats in which the specimens were found and no mention of anything concerning the role of the species in the biotic community was made. This is not surprising because the author probably had little to do with observing or collecting them in the field and had little or no communication with those who did.

The specimens had been collected in the Dutch East Indies, now the Republic of Indonesia. The taxonomic studies were undertaken to give names to and describe the endemic species inhabiting two of the largest islands in the country: Java and Sumatra, and a smaller one, Bali.

At the time the volumes were being prepared for publication, most species of aquatic insect in the European countries had already been described more than 50 years earlier, and in the countries where major museums and university research institutes were located, most of the endemic species had been described more than a century earlier. Today, it is still possible to find parts of the world in which most of the invertebrate species have not been described and catalogued as well as they were in the Dutch East Indies in 1933. As a result, it is still possible to perform pioneering work on the flora and fauna and find undescribed species that must be discovered and given names. At the same time, discoveries about the ecology of a great many species can still be made almost anywhere in the world.

Most research opportunities for entomologists specializing in aquatic insects will continue to focus on the species inhabiting water bodies without unusual physical or chemical particularities. The vast majority of aquatic insects inhabit such water bodies. Habitats that are not extremely small and trapped in terrestrial plants or broken artifacts, not saline or hypersaline, not extremely hot or cold, and not periodically desiccated or anaerobic simply have a considerably larger variety of insects inhabiting them than those that are. Such habitats are also plentiful in most parts of the world where the human population is concentrated. It is obviously much cheaper and requires much less travel time to visit a local water body than to make an intercontinental excursion to a place far off the beaten track.

Nevertheless, it is the exceptional aquatic insect species that has acquired traits that few, if any other aquatic insect has. To live in any typical aquatic habitat, thousands of species are able to breathe atmospheric air carried beneath the water on a plastron. Thousands of others are able to breathe through a snorkel or breathing tube. Thousands more use some sort of gills to change gases directly with the water while developing as larvae. It is the one that uses a still poorly understood method of respiration in a chemically exceptional habitat in a remote region of the world that will yield the information on a previously unknown respiratory pathway.

When the principles of the natural sciences were established in Europe during the Renaissance, initially to accurately establish the date to celebrate Easter, it represented a revolutionary change from deductive to inductive reasoning. During the Middle Ages, when information was sought, it had to be found in books written during ancient times. The theory was that the ancient Greek and Roman civilizations were more advanced than the contemporary society was, so their experts knew better about almost all things. Therefore, the writings of the ancient experts had provided all of the answers, and by using deductive logic, the answers to current questions could be provided.

By establishing new rules of logic, the natural sciences were substituted for reliance on ancient texts. The Church in Europe reasoned that ignorance is a physical evil and that the pursuit of the natural sciences was a virtuous occupation because it facilitated an advancement of knowledge without relying on the expertise of ancient writers. In this case, the natural scientists were required to display the highest level of honesty and integrity.

The first step in the activities of natural scientists involved observation, sometimes by the use of instruments to magnify or detect invisible radiation. This was not as easy as it sounds, as was discovered when applying the second step, that is, the interpretation of the observations. The products of the second step were the hypotheses, which are plausible explanations for the observations. Once one or more explanations were available, each had to be tested, either by more observations under controlled conditions or by experiments, under which all variables except one were carefully controlled. After one of the hypotheses passed as many tests as possible while others had failed one or more, the successful one was raised to the level of a theory, which could be accepted or abandoned after more relevant evidence was obtained.

The natural sciences have been plagued by one mistake in logic almost since their methods were first accepted. From time to time, scientists made observations and then looked for the reasons for what they observed, and accepted a hypothesis that had been well formulated but never proven with actual evidence. Either they heard the hypotheses of others or convinced themselves that their own must be correct. Once scientists convinced themselves that one hypothesis was correct, they would argue in its favor even though physical evidence to prove it was still lacking. This continues to the present day, but now economic factors rather than evidence are often used to justify its acceptance. Just because a hypothesis sounds very logical and sensible and conforms well with a scientist's personal view of the world does not mean that it is true. Yet a scientist's personal view of the world does not influence the validity of his hypotheses. Raising an idea from a hypothesis to a theory requires convincing proof. When government or industry wishes the public to accept a hypothesis as true, they can pay a scientist to tell them what they want to hear. This takes us back to the

Middle Ages, when kings could pay soothsayers for the advice to do whatever they know the king already wants to do.

There is a fantastic amount of basic research work yet to be done just to get past the phase of observation before we can begin to consider hypotheses. For studies of insects, the first step is to describe and name the species we observe. With a considerable amount of descriptive information, we can just begin to ask intelligent questions and begin to find hypothetical answers to them. This is most sensible to do in countries in which insects have been investigated by many scientists for well over a century. Elsewhere, it is still not possible to identify all of the species encountered with certainty, assign observations to the correct species, or identify the other species of organisms with which they interact.

The taxon Hexapoda accounts for more species than any comparable higher taxon of animals. Insecta alone accounts for more species than all other classes of animal, combined. The class Collembola is estimated to account for more individual organisms than any other class. At the present time, there is enough basic information about the common species of aquatic hexapod in many countries to formulate and perform experimental studies and other advanced work in fields such as biochemistry, physiology, ecology, behavior, and a large number of other specialized fields. Practical studies to evaluate species to use in biological control of pests in agriculture and disease control are also required throughout much of the world.

In geographical regions that have not been well studied in the past, research is needed to catalog the aquatic collembolan and insect fauna and to compile information on their ecology with a view toward establishing data banks for correct identification of all stages of each species present and to prepare for more advanced studies as soon as enough basic information is available. In order to record changes in the water bodies of such regions, monitoring programs should also be initiated. Because the organisms inhabiting such water bodies interact with each other, entomologists should cooperate with specialists studying other groups of animals to discover more general changes occurring. For example, a considerable decline in the number of mayflies in a stream could be expected to cause declines in the number and size of certain fish species, unless the decline is compensated for by great increases in insects in other orders.

This chapter of the book has outlined some of the more esoteric research studies concerning unexplained observations of survival by aquatic collembolans and insects in extremely stressful aquatic habitats where most such species would not be expected to survive. For various reasons, these studies would require special costs to overcome various difficulties, such as travel to hard-to-reach locations, long periods of time to complete, or special laboratory equipment to perform. In comparison to most studies of aquatic insects, which require little more than travel to nearby water bodies, many of these special studies of unusual species have been attempted at various times, but lack to time or the complexity of the problems to be elucidated have created fields for continued research. Finding the necessary funding for research is certainly a key problem that has to be solved before any scientific program can be initiated. It is therefore necessary to make sure that the institute or agency supporting the research knows why the study will add to the basic knowledge in the scientific field and might lead to some fundamental discoveries that will open new fields of research.

SECTION 2

Changing Ecology of Lotic Habitats

Most lentic habitats are notably easy for aquatic hexapods to adapt to. Only those that are notably saline, polluted, deficient in oxygen, or contaminated by various toxic substances may have an impoverished fauna. Lotic habitats, on the other hand, are more likely to show declines in the numbers of aquatic animals and in species diversity.

There are various reasons for this. Streams are sometimes used for disposal of toxic wastes, organic pollutants, or various fertilizers. The water flow distributes these substances along the whole length of the stream downstream from the point of entry of the contaminants. At other locations, the water from natural streams is sometimes diverted for irrigation or dammed to create ponds or lakes, which causes the stream to dry up or become deficient in oxygen. Construction along the course of the stream can also cause fine suspended matter to contaminate the water far downstream from the site.

During the 20th century, many dams were built along the streams and rivers in the Pacific Northwest of the United States. These produced more electric power than the region needed, and the interference with the water flow in the streams and rivers were associated with a gradual but significant reduction in the salmon populations returning to spawn each year. As a result, programs were developed to improve the survival chances of the spawning salmon, such as completely eliminating several of the dams.

The lack of success in restoring the salmon populations highlighted the general ignorance concerning the complex relationships between the fishes and the aquatic insects in the streams. After spawning, the Pacific salmon all die, leaving a great deal of protein, bone, and muscle to decompose at the spawning sites, some of which are long distances from the sea. This material moves downstream, where it nourishes a great many larvae of insects which feed on it. These insects provide prey for the hatchling salmon as they move downstream to the sea. After spending several years in the sea, the salmon have greatly increased in size and weight. They then begin what can be a very long upstream journey to the headwaters of streams in which they had hatched. While the dams were in place and other hindrances decreased the number of salmon reaching the spawning sites, the amount of nutrient material left far upstream in the carcasses of the salmon also decreased from year to year. It is often postulated that this deficiency in nutrients moving downstream in the water reduces the number of insect larvae that are able to survive, which, in turn, reduces the number of hatchling salmon able to successfully complete the downstream trip to the sea. Obviously, the size of the salmon populations could not immediately begin to increase after the obstructions were removed from the streams and rivers, and they continue to decrease.

Other explanations for the decline in the salmon populations include decimation of the small salmon shortly after their arrival in the sea by the oversized ships in the modern fishing fleets, which have repeatedly shown that they can fish the most abundant species almost completely from the sea after only a few years. Because the small salmon wind up being sold as "sardines" in cans, they have no chance to reach spawning size. Obviously, after a great many years of modifying the spawning streams with dams, intensifying the fishing efforts on the high seas, and going through the motions of performing research, there is still a lack of enough basic information to prove which of the hypotheses is the real explanation for the decline of the Pacific salmon species and what role the decline of the insects with larvae that develop in the streams plays in the gradual elimination of an extremely valuable food fish.

The design of artificial streams in which insect species that develop in lotic water can be raised in the laboratory would be a first step to facilitate studies of the larvae. Such artificial streams have long been used in The Netherlands to test the effects of toxic substances and other pollutants on insect larvae selected as standard test organisms. The information gained would contribute much to an understanding of the relationships between the insect species and the value of the insects as foods for other animals living in the streams, especially fishes. The results of such research would certainly provide more reliable information concerning such problems as the decline of the salmon populations. Guesswork to decide among two or more alternative hypotheses does not facilitate forming effective solutions to environmental problems.

SECTION 3

Freshwater Epineuston

In small, shallow water bodies, the community living on the surface tension layer is rich and varied in species. Especially in the tropics, this community merits special studies. Almost all reports concerning their ecological requirements are written in general terms. Even the diets of the individual species mentioned in publications are not reported, although one or two casual observations are occasionally made. For example, the diet of a collembolan species that lives on top of the surface tension layer is sometimes reported to be "pollen", based on a casual observation. However, during seasons when pollen is not being produced, what foods do they switch to?

In some cases, authors do not even provide enough information about a hexapod species they describe to determine whether it should be classified as aquatic or terrestrial. This is especially a problem with species that live on the surface tension layer and rest on small, floating plants. The fact that these species feed on a variety of microorganisms as well as pollen and probably also detritus make them important links in the food web starting with both the detritus from large tracheophytes and microscopic primary producers, such as algae.

Clearly, there is still much to learn about the biology, behavior, and natural history of the species of Collembola and minute insects that rest on the surface tension layer of various kinds of water bodies during at least one of their developmental stages. More research on these organisms in Europe and North America would be welcome, but a considerably greater amount of information on the epineuston living on the surface of the water in large tropical wetlands is urgently needed to determine how and where they spend extended dry seasons, the seasons during which they reproduce, the places where each such species leaves its eggs during dry seasons, and what triggers the development and hatching of the eggs. Unfortunately, the fact that the tiny hexapods among the epineuston cause no notable damage to crops and spread no diseases to human beings or domestic animals gives research on this very interesting group a low priority for funding.

SECTION 4

Terrestrial Habits but Aquatic Food

In contrast to the tiny hexapods living on the surface tension layer of open water, the insects that feed on the emergent parts of aquatic plants suffer from no lack of attention by scientists because of their perceived economic impact on the human population in rural areas, both for good and for ill. For research meant to provide results with great relevance for the local economy and public health, this is still one of the most promising fields. In addition, the research does not require more than a minimal knowledge of the modifications of insects for life in or on water.

Rice is one of the most important staple crops in the parts of the world that have a warm or temperate climate and appreciable annual rainfall. Taro is one of the oldest crops in the tropics. Sweet potatoes, yams, water chestnuts, and lotus are aquatic plants with emergent parts, which were local food items that were transported to other parts of the world to be grown as aquatic crops. These and other plants that grow in water and are used for food are all eaten by insects which appear to be typical terrestrial species, except in their choices of plants to feed on.

Not only do entomologists and other natural scientists perform research on such insects, people working in agriculture to improve methods and increase yields are continually confronted by promoting the development of certain insects and eliminating others that associate themselves with such crops. Some of the insects eat the leaves or spoil the commercially exploited parts of the aquatic plants, while others prey upon the species that cause this damage. The methods are constantly changing as pests develop resistance to chemicals and adjust to changing biological conditions. Research will be funded for a long time to come in order to facilitate the development of new methods of controlling the pests damaging aquatic crops.

SECTION 5

Insects of the Pseudoterrestrium

Many standing water bodies in the tropics are covered with densely packed masses of large floating plants, which almost completely obscure the surface of the water. A casual observer viewing one such mass of plants might well be convinced that he is looking at a moist meadow covered by plants rooted in wet soils. Occasionally, people have attempted to walk on this false terrestrial habitat and found themselves standing in one meter or more of water. Winds blowing across a lake for a while can press floating plants close enough together to fortify this false appearance of a terrestrial community bordering a lake and make it even more convincing.

A characteristic of a pseudoterrestrial community is that it has both an upper and a lower side. The upper side is formed by a mass of densely packed leaves, stems, and flowers, while the lower side is a submerged mass of roots, rhizoids, and underwater stems, which can have deep water below them. When not blown close to the shore by the wind, the pseudoterrestrium can form plant masses that appear to be floating islands (Heckman, 1994). While the weight of a person can only rarely be supported by such a false island, wading birds are frequently seen walking on top of the pseudoterrestrium hunting for the many small animals that live in such communities.

Pseudoterrestrial communities appear in all of the warm regions of the world, especially on lakes and large rivers in tropical regions. A special kind of pseudoterrestrium has been called the sudd in the Nile watershed and neighboring river systems in central Africa. The relatively large plants that grow on them, such as *Cyperus papyrus* Linnaeus, makes them appear to be real islands. They are so stable that small trees and woody bushes can grow on them. They will even support the weight of a person if he walks carefully enough.

The underside of such aquatic habitats is a source of a great many aquatic insects for the descriptive entomologist, especially in large water bodies of the tropics, in which there are many relatively large predatory fishes. The submerged parts of many of the aquatic plants provide shelters and hiding places for the adults of aquatic beetles, water bugs, and even grasshoppers, as well as for the larvae of species belonging to most aquatic insect orders. Without habitats that can conceal relatively large insects, many of the species could not survive under the extreme predation by the many voracious fish species that inhabit large tropical water bodies.

There are many reasons for undertaking extensive research on the pseudoterrestrium, at all levels of scientific complexity. To start with, the richness of the flora and fauna living among the floating islands of plants provides considerable material for taxonomic and systematic studies. Depending on the geographical regions in which the pseudoterrestrial community is located, naming and describing previously unknown aquatic insect species can still produce great pioneering publications. There are simply so many places in the tropics that have not yet been fully explored that studies in such places by competent specialists cannot help but produce spectacular results.

In addition, discoveries in the fields of insect ecology, physiology, biochemistry, behavior, zoogeography, and related fields are likely in places that have rarely been visited before. However, to set up research stations at remote sites is not always easy and inexpensive. Finally, there are opportunities for applied sciences in such fields as control of floating water plants, finding ways to limit the intercontinental spread of plant and insect species that threaten to become serious pests at the locations to which they are introduced. The water hyacinth alone has already cost countries around the world vast sums of money to control or eliminate it after it was introduced to new places. All such efforts have been to no avail. Research to find ways to monitor other introduced species will continue to be pursued, and many of the controls tested involve the use of herbivorous insects.

SECTION 6

Predatory Insects in Lentic Water

How useful studies of predatory insects will be, whether they feed on microscopic animals or larger prey, depends mainly of the geographical location of the studies. The less the location has been sampled in the past, the more there remains to be found out about the aquatic insect fauna. In Europe, for example, the most useful results could be expected from food web studies, through which the good or the harm caused by an insect species to local agriculture or other economic activities could be analyzed. Before making meaningful prognoses of changes to be expected due to the elimination of a species or the introduction of a new one from another part of the world, it is necessary to know what each of the species in the habitat eats. Studies of this nature are still required in almost every part of the world.

Where aquatic crops are grown, it is useful to know with certainty which insect species feed on the crop, which feed on the insects that damage the crops, which develop in the water together with the crop and as adults become disease victors, and which feed on the disease vectors as larvae or as adults. As it becomes more and more evident that long-term insect control cannot be achieved by chemical insecticides, studies that provide well defined food webs will yield the information vital to formulating biological control methods that can work well over the long term without harming human beings or the environment as a whole.

At locations that are remote and hard to reach, there is still much taxonomic and systematic work to be done, even on the most conspicuous and well studied groups of insects. However, there is not much more that can be learned about the respiration methods used by the predatory species in the water column because a great many insects use these same basic methods, and those that do demonstrate that these methods are extremely successful in lentic fresh water bodies. Thousands of species all over the world use air carried underwater in a plastron on the external body surface for respiration as adults, and some also carry a plastron of air as larvae. Even more species breathe using gills as larvae and leave the water as adults. Still others breathe atmospheric air through a breathing tube while resting just below the surface. These methods are simple and follow the basic rules of physics, especially the physics of forming a surface tension layer and simple rules of gas exchange between water and air.

The only basic information that still needs to be supplemented concerns the morphology of the tracheal systems of the individual species, the chemicals that assist the respiratory pigments, if any, extract and store oxygen, and any supplemental methods to improve gas exchange in water containing various kinds of pollutants.

SECTION 7

Shelters Constructed by Aquatic Insects

Many kinds of shelters are constructed by aquatic insects, and these are of value for the identification of the individual species, even when the insect itself is absent. Some exhaustive studies have been made of the cases constructed by caddisflies in limited areas of the world, and other works have been completed on the shelters constructed by members of the dipteran family Simuliidae. The publications on such case or tube-building insects are extremely useful for determining local species, and their use within a limited area makes the identifications reliable.

Species of Trichoptera are harder to identify using only the cases they construct. Many do not construct portable cases, and some do not seem to construct cases at all. Some cases by species of different families appear very similar, and other species belonging to the same family appear surprisingly different. Relying on identification of the shelters constructed by species of Trichoptera are not yet useful except in those places where there are reasonably complete sets of descriptions of the larvae and their cases. An example of an exemplary work on the larvae of Trichoptera is the study of caddisflies in Great Britain, which, of course, cannot encompass the shelters of the caseless caddisflies because there are none (Hickin, 1967).

There is usually no requirement to distinguish insect species on the cases they build, but being able to do so can be very helpful when surveying local insects. For describing and identifying species, the anatomical features of the larva and adult are still the definitive features. At the present time, there are two problems when trying to identify a species using only the cases. The first is that some of the cases appear superficially similar. The second is that all of the cases have not yet been described. Therefore, if a case that has never been described appears similar to one that is well-known, there is a good probability that the undescribed one will be mistaken for the one that is known. Furthermore, caseless caddisflies will then be left out of fauna lists for the location.

Unlike Trichoptera, the family Simuliidae in the order Diptera is characterized by larvae that build very sturdy cases firmly attached to stable substrates. They do not disintegrate soon after the metamorphosis of the larva, and they have a somewhat similar shape but with features that are characteristic of the species. That means that in most parts of the world, all local species could be correctly identified using only keys to the cases of the species as soon as suitable keys have been prepared. The cases can be identified long after the larvae have become pupae and then emerged as adults. This makes the correct identification of the local species in large genera, especially *Simulium*, for example, possible as long as illustrated keys have been prepared. Preparing such keys is therefore useful and should be completed in all parts of the world in which some *Simulium* species act as vectors of diseases.

Other studies of insects that construct cases or shelters certainly should include the methods of construction used by the insects and the material out of which the structures are made. The insects studied should also include those few members of the order Lepidoptera with caterpillars that live beneath the surface of the water. Many books were prepared in different parts of the world to provide keys to identify aquatic insects. Some of these were prepared many years ago, such as the one on the aquatic insects of California, including the aquatic ones (Usinger, 1956), and they are still in use. These books require updating because new insects have since been described or reported in California, while a few nominal species have been found to be junior synonyms of other species.

SECTION 8

Overcoming the Problems of Living in Fast-Flowing Streams

The conditions encountered in streams, especially those sections of streams flowing down steep mountains, provide certain important conditions that make the habitat particularly attractive to aquatic insects that have modifications for overcoming the disadvantages of living in rapidly flowing water. However, most species of aquatic insect cannot overcome all of the adverse conditions and must select a habitat with either a slow water flow or no flow at all.

Streams flowing steeply downhill through the mountains usually contain relatively cold water for the region in which they are located, much dissolved oxygen, and far fewer aquatic insect predators than nearby standing water bodies. Most aquatic insects do not swim fast enough to keep from being carried continuously downstream, except those that are hydrodynamically assisted by the shape of their bodies or can utilize the sediment as shelter or points of attachment. These methods are already known for most species that have been carefully examined. Those species that are skilled at moving on the surface tension layer of the water have also developed the skill of hanging onto sections of the surface tension to keep from being carried away with the current.

Research on the insect larvae in lotic water does not usually go much farther than the general biology of the species. However, some of the species are limited to individual watersheds in their areas of distribution. Therefore, many watersheds support an aquatic insect fauna unique in the structure of their biotic communities. In more arid region of the world, the pattern of distribution of the aquatic insect species reveals something of the history of the fauna during climatic changes from wet to dry or vice versa. In colder regions, alternations between ice ages and warmer periods can sometimes also be traced by observing the structures of the different biotic communities in the drainage areas of the streams and small rivers. Once a hypothesis is developed for the effects of climatic changes on the aquatic insect fauna, the challenge of finding evidence for the validity of the hypothesis begins.

In planning studies of the aquatic insect fauna encountered in mountain streams, it should be expected that the more remote and off the beaten path a habitat is, the more undescribed and poorly known species will be encountered. In preparing for research in such places, as much literature as possible concerning the streams to be visited should be carried along as part of the equipment. Look for species of larvae with hydrodynamically distinctive shapes developing in the streams. Such shapes help force the body of the larva ventrad. Look for special modification of the tarsi and tarsal claws for holding strongly to the substrate, and look for modifications of the fore-legs for digging into coarse sediments.

SECTION 9

Looking for Aquatic Orthoptera in Tropical and Subtropical Water Bodies

Do not be surprised if confronted by a species of Orthoptera swimming in the water column or hiding at the bottom of a stream the next time you visit tropical South America. Amphibious species of grasshopper are fully adapted to life on the floating islands of vegetation covering large water bodies and seasonally flooded areas. One species is seen swimming underwater, resting on an emergent leaf of a water plant, or flying from one group of plants to another. Another species is concealed underwater among the thick mass of rhizoids of a water fern, on which it also feeds. Beneath the submerged sediment at the bottom of the water body, a mole cricket is digging a tunnel through the coarse gravel.

Only general information is available about these species. Observations are made, and some of them are reported in the scientific literature. How is the life cycle of each aquatic grasshopper or cricket regulated by the local climate? If their favorite food plant is not available, what else will they accept as food? What are their natural enemies? How do they keep from drowning when swimming underwater?

There are many questions about their behavior, nutrition, and relationships with other organisms in their habitats. To answer these, a great deal of research will be necessary. If the orthopterans can be kept in aquaria, these questions can be answered sooner and with a greater degree of certainty. They can be given various foods, and each food item can be judged to be preferred, acceptable but not preferred, rejected except to prevent starvation, or rejected under all circumstances. The seasonal activity pattern can be recorded, and the seasons of mating, egg laying, and larval development can be determined in each climatic zone in which the insect lives.

An even greater challenge would be a detailed study of the weta species, one of which is known to dive into streams when threatened by a predator. They hide among debris at the bottom and can remain underwater for at least 12 minutes until the predator leaves. The wetas are apparently the heaviest insects, and there are many species inhabiting New Zealand. Only one is known to enter streams and hide at the bottom, but few of the others have been well observed in their natural habitats. Studies of these species should include everything from fundamental studies of the differences between them to produce up-to-date keys to distinguish them along with thorough treatises on their biology, ecology, and behavior.

In South America, there are also species of cricket that live among the roots and rhizoids of the plants of the pseudoterrestrium. They are not uncommon, but they are seldom seen. As best as could be determined, there are undescribed species among these crickets. In addition, many South American species have been described based on specimens collected over a century ago and deposited in European or North American collections without any notes by the collectors concerning the locations at which the specimens were collected or about the habitats in which the insects were found. It may well be that many of the crickets known only from collections of preserved specimens may actually be aquatic or semi-aquatic without any specialist knowing about it.

Risking adding to the lore of mythical animals, a closing word could be added concerning stories of insects that were seen but not collected deep in Neotropical rainforests. One of these legends concerns an insect resembling an aquatic walking stick that was sighted in streams, where it blended in perfectly with the riparian vegetation of dangling branches and adventitious roots in the water. An insect resembling this description would almost certainly belong to the order Orthoptera, and it might be the largest aquatic insect in the world. If this insect is real rather than legendary, it might even be the largest, or at least the longest insect species ever found. According to the tales of the giant walking stick, the species typically lives along the edges of rainforest streams and rivers with part of its body underwater and part above the water level, where it presumably feeds on the foliage of the littoral plants. Because of its excellent camouflage making it look like part of a woody plant, this species was presumed to have been seen by many people without most of them knowing that they were actually looking at a giant insect.

SECTION 10

Aquatic Insects that Develop in Odd Places

In recent years, studies of insects that develop in rainwater that collects in unusual places have become commonplace. There are several reasons for this. The first is that a large number of insects which develop in places other than ponds, streams, lakes, or rivers have been found to be important vectors of diseases threatening human beings and domestic animals. Another reason is that habitats in which rainwater collects are far more plentiful in tropical rainforests than they are in most parts of the temperate or frigid zones, and identifying the insects that develop as larvae in such places has become a very rich field of study. Pools of rainwater trapped in places other than bodies of surface water support biotic communities called phytotelmata, and interest in research on the biota of such communities has been increasing rapidly in recent years.

Rainwater is notably poor in nutrients needed by insect larvae to grow and develop. Perhaps because of this, many of the species that develop as larvae in broken containers, old tires, hollowed out rock, tree holes, bamboo internodes, and similar places belong to species that live as adults on the blood of animals, mainly vertebrates, including human beings. The nutrients and trace elements that the larvae might fail to procure from the phytotelm are provided instead by the blood of the animals the adult insects live on. Some of these are then stored in the eggs and in this way passed on to the next generation of insects by the adult females.

A major advantage insects that develop in trapped rainwater enjoy is the exclusion of many large predators from the habitat in which they live. Some of the predators are excluded by their large size, which makes it difficult or impossible for the adult females to enter the small containers of rainwater, which has not yet been able to mix with nutrient rich surface water. Some of these predators are large enough to completely eliminate an entire generation of a small, blood-sucking dipteran species in a matter of a few days or even hours.

SECTION 11

Rainwater in Mosses and Lichens

The tropics is not the only part of the world in which rainwater provides an important habitat for aquatic insects. In cold rainforests, both in North America and in New Zealand, the branches of the evergreen conifers or their southern equivalents are covered with epiphytic mosses and lichens, which are inhabited by a great variety of invertebrates. The species present are predominantly aquatic or semi-aquatic members of the classes Collembola and Insecta.

Many mosses have the ability to form dense clumps that soak up rainwater like a sponge. Tiny insects, mostly in the orders Diptera and Coleoptera, live in these clumps of moss. Many have been overlooked, and in any case, their activities inside of the clumps of mosses have seldom, if ever been described. The larvae of several unidentified insect species were personally observed briefly moving in the spaces between the shoots and tiny, scale-like leaves of the mosses. These insects were typically submerged in rainwater held semi-permanently in the mosses, except where the evergreen trees on which the mosses were growing were located too far inland from the cold rainforest to receive almost daily rainfall. During July and August, there is almost no rainfall in western Washington State, so only the forests close to the coast can be regarded as true cold rainforests. In such coastal rainforests, water reaches the trees in summer as mist or fog, even when there are few periods of heavy rain storms. Substantial amounts of rain otherwise fall almost daily, both along the coast and inland west of the Cascade Mountains during the rest of the year.

Most of the insect larvae observed appeared to belong to the order Diptera, while most beetles present were the adults of very small species. In addition to insects, Collembola were common in and on the moss plants. By remaining inside the moss branches and on the wet lichens, especially those aquatic lichens that could maintain microclimates at or near 100% saturation most of the time, the springtails were able to keep themselves almost indetectable from predators. The thalli of many lichen species were large and flat. Many epiphytic lichens completely covered the branches of the trees in the rainforests, and some of them also appeared on trees farther inland from the sea.

An assessment of the research opportunities for studying aquatic insects in cold rainforests indicated that the opportunities are excellent for both faunal surveys and the likely discovery of new aquatic and semi-aquatic species, especially those in families of predominantly terrestrial insects. Some of the mosses in the rainforests that grow on the ground but not in contact with water bodies have also escaped the attention of specialists for aquatic hexapods. The same opportunities are open to researchers specializing in studies of the insects living in water trapped by mosses on the ground as those just described for scientists choosing to specialize in the species living in the arboreal mosses and lichens.

The biology and life cycles of the hexapods in the cold phytotelmata are certain to include some modes of life new to science. How these species manage to cope with the cold temperatures during the winter is one of the problems that will need to be elucidated. Locations at which these studies could be undertaken extend from Oregon to Alaska, including forests along the coasts and on the many islands just offshore from the mainland in southeastern Alaska. Alternative locations include forests along the coasts of southern New Zealand. Any forests farther from the sea but displaying a thick vegetation of epiphytic mosses and lichens on their branches should also be investigated. In addition, extensive growths of mosses and lichens on the forest floors of the cold rainforests should also be surveyed. It should then be determined to what extent the hexapod fauna differs according to their locations in the forest.

For more than a century, intensive logging has destroyed many forest habitats in the coastal regions in which the cold rainforests occur. Old growth forests are now hard to find, except in a few parts of southern Alaska. The trees in many secondary forests often suffer from depletion of the soil and do not grow with as much vigor as the original forest showed. The best locations for studies are usually found in parks, where logging has been prohibited. One of these that appears to be most suitable at the time of this writing is located in the Hoh National Forest on the Olympic Peninsula in northwestern Washington State. A greater selection of research sites can be found in southern Alaska, but they would be hard to reach because of a lack of roads leading to most of the isolated islands.

Generally, the true rainforests in the Pacific Northwest are located close to the coast. They exist because there are both warm and cold ocean currents, with the warm currents running close to the coast. This assures that relatively warm and moist air is present throughout the year. Only during the summer does the sky remain clear most of the time, and although the weather may be warm and dry during the day, radiation fog develops quickly over the land near the coast soon after sunset. Frosts may occur occasionally during most of the summer due to a rapid cooling of the air at night. This unusual combination of rapid cooling on the ground after sunset under a warm, moist atmosphere creates the conditions for the very moist forests encountered.

As an extra footnote concerning a special characteristic of the cold rainforests on the Olympic Peninsula in Washington is a surprisingly great weight of epiphytic mosses and lichens that grow on the branches of the evergreen conifers in the forest. The hexapod fauna inhabiting these epiphytic communities is worth a special study of its own.

SECTION 12

Tropical Phytotelmata

Until recently, it would have been considered odd for an entomologist to have to climb a large tree in a rainforest in order to collect the larvae of aquatic insects. However, considering the impact on public health that the vectors of especially dangerous tropical diseases have on people living in or visiting the tropics, this activity is proving its value to science. The great importance that insects of the phytotelm presently play in tropical medical research is accounted for by the fact that arboreal primates have been found to serve as reservoirs for certain viruses that can also infect human beings. Mosquitoes that develop in epiphytic plants have only a short flight to the nearest monkey or ape. In addition, the paucity of predators feeding on insects in the phytotelm is one of the most important features making it an attractive biotic community for small larvae of insect disease vectors to develop in.

Much basic research on the insect larvae from phytotelmata throughout the world is being done in order to compile lists of species likely to act as vectors of serious diseases and to determine which are actually transmitting the pathogenic organisms causing the diseases. In addition, other species of vertebrates serving as reservoirs for diseases affecting human beings, the geographical ranges of the insect vectors, and their potential for spreading a serious disease to other parts of the world should also be determined. Although studies involving the insect vectors of tropical disease organisms are extremely important considering the death rate worldwide resulting from the sicknesses they cause, scientific research to help eliminate these diseases has a history of being underfunded in comparison to the money granted to investigate less serious diseases in the temperate zones. Therefore, funding is available, but it may not be the easiest research support to obtain.

Judging by the number of scientific publications that have appeared on insect species from phytotelmata in many parts of the tropics, once a research project is launched, it is very likely to be successful and provide important information about insect species that develop in the water trapped in epiphytic plants. While it has been known for more than a century that mosquitoes spreading viruses often develop in broken artifacts, it has only more recently been learned that the failure of efforts to completely wipe out diseases by cleaning up their breeding places close to human habitations has been due to the presence of other mosquito species that develop in epiphytic plants in forests far from human population centers. These mosquitoes become infected by a virus when they bite a monkey and then bite a human victim, who is given the disease. From then on, the disease victim can spread the disease to the common species that breed in broken artifacts near human habitations, and an epidemic breaks out, often in a place where nobody had been infected by the disease in many years.

The first goals of this research should be to make lists of all mosquito species that breed in the bromeliads in the region of the study. These lists can then be used not only to determine the vectors of human diseases but also those known to spread a disease to a domestic animal or any other animal native to the region. From there, the information on the mosquito species can be used to find out whether the pathogens that cause diseases in animals can also infect human beings and learn what kind of symptoms they cause.

With increased travel opportunities, tropical diseases can spread around the world faster than ever before, making basic information on the distribution of disease vectors and the nature of the illnesses they spread extremely important for public health. Also important are notes on the breeding places of the mosquitoes and their known distribution to predict the probability of an epidemic if an infected person transports the pathogen from another part of the world in his blood. There are both national and international organizations with the responsibility for doing these jobs, so cooperation with them will be a good start to performing research in this field.

SECTION 13

How Insects Help Plants Digest Their Meals

Research on the insects with larvae that live in pitcher plants is important for the understanding of the biochemical adaptations of insects to potentially toxic substances. First, however, it is important to know whether or not the fluid in the pitcher plant contains toxic or digestive substances and what these chemical substances are. This might sound easy, but there are many species of pitcher plant in two plant families, one present in the Old World and the other in the New World. It is not likely that the chemical nature of the fluid in the pitcher is the same in each of the species.

The second problem is that there are apparently many insect larvae, as well as ants, that are able to take their food from the carcasses of insects trapped in the pitchers. Each of these must have a method of surviving in a plant that is constructed as a death trap for insects. How do these species avoid being killed while living comfortably in a fluid that is deadly for a large number of other insect species?

Among entomologists, it is still debatable whether or not the insects that live in the pitchers are living in symbiosis with the plants, and if so, what the nature of the symbiosis is. Pitcher plants live in places that most other species of plants cannot. The soil or water in which the plants live are simply too poor to supply enough inorganic nutrients to permit the plants to live by photosynthesis. The missing nutrients must be supplied by the insects that the plants trap in their pitchers. Before the nutrients can be utilized by the plants, the trapped insects must first be digested. If the insects living in the pitchers are symbionts, then they must accelerate the digestion process to release the inorganic nutrients in time for the plants to utilize them. If they are parasites, then they would simply feed on some of the dead insects and contribute nothing to the host plant.

Research into this problem would raise more questions. There are several species of insect living in many pitcher plants. Are they all symbionts? Are they all parasites? Are some of them parasites and others, symbionts? Is there a more complex kind of symbiosis involved?

Next comes the question regarding how the insect larvae can survive. Does it have an immunity against poisons produced by the plant? Does the plant provide the insects living in its pitcher with oxygen-rich water sufficient to meet their respiratory needs?

In the scientific literature, there are contradictory answers to some of these questions. Research to clarify the processes actually going on in the pitcher plants is needed, and it is likely that surprising answers will be found.

SECTION 14

Hexapods Active at Subfreezing Temperatures

Vast regions of Alaska, Canada, Siberia, and other countries in the North Frigid Zone, as well as islands near Antarctica, have winter temperatures that fall so far below the freezing point of water that lakes and rivers are covered with several meters of ice and even adjacent areas of the ocean form pack ice thick enough to stop most large ships. Some aquatic insects living on the tundra must cope with such conditions for six months or more and then proceed with a life cycle that involves a method of obtaining nutrients quickly, often by obtaining a blood meal from a vertebrate, and then laying eggs that must remain dormant under ice for half a year or more. There are a great many insect species, mainly members of the order Diptera, that have to survive under the same conditions. Only a few of them have been studied, and it is unlikely that all species in the tundra use exactly the same strategy for survival. Detailed studies will be required to determine the variations in the life cycles of these species.

The second group of insects must survive in taiga and the northern parts of the temperate zones. These are the winter craneflies in the family Trichoceridae. They are active during the cold season rather than during the summer. A few entomologists have observed these insects for many years, mainly in Scandinavia, and still have only a sketchy idea of where they live and how they survive. They might be called aquatic, although their contact with water seems to be in its solid form. They are active throughout the winter, but apparently not when the temperature is too cold. An enterprising researcher might do well by building refrigerated aquaria for these insects containing various possible foods and maintaining them at temperatures slightly below the freezing temperature of water. It would be interesting to learn the complete life cycle of at least one species and to know how the adults and the larvae remain active at freezing temperatures.

Several species of the class Collembola are also active on ice and snow. Some of these species are called "snow fleas" because of their ability to spring relatively long distances on packed snow on the ground. How they maintain activity on or under ice and snow remains to be learned.

In summary, there is a considerable number of questions to be answered to assemble sufficient information about aquatic and semi-aquatic hexapods in the coldest parts of the world. Such studies would probably receive low priority for funding, although one argument for performing extensive studies on the activity of insects on ice and snow might be the fact that many of the adult insects in the Arctic feed on human blood as adults if it is available, and some of them attack people on the tundra in swarms large enough to significantly reduce the number of tourists willing to visit Alaska, northern Canada, or Siberia during the summer. Knowing their exact seasons of activity might provide clues for scheduling visits to the Arctic at times in the warmer seasons when the attacks by the blood sucking mosquitoes, black flies, and gnats are at a minimum.

SECTION 15

Short-Lived Adults

Much of the information concerning the insects that do not eat after metamorphosis to the imago stage is already available. However, there are still a great many tropical species about which little is known concerning reproduction. There have been some contradictory reports concerning whether the males of several species feed on the same items as females, feed only on the nectar from flowers, or do not eat at all, and these need some clarification.

In any case, the ability of the imagoes to reproduce without feeding and other information concerning the mating behavior of adults should be reported whenever pertinent observations have been made.

SECTION 16

Respiration in Anaerobic Water

The most fundamental modification that every aquatic insect has in order to live in water is the ability to breathe while in the aquatic environment. When the water contains a sufficient amount of dissolved oxygen at all times, gills are sufficient. Otherwise, insects have developed a variety of ways of obtaining air directly from the atmosphere and often carrying it with them underwater.

During temporary shortages of oxygen, a few insects have respiratory pigments for storing the oxygen. Otherwise, the insect must have a snorkel or breathing tube for exchanging gases with the atmosphere while remaining underwater. When the water is overburdened with organic detritus, a permanent oxygen shortage is created by the microorganisms breaking it down. In some cases, the water can become anaerobic and remain so for weeks at a time. In such cases, sulfur replaces oxygen for photosynthesis and respiration, and sulfur bacteria take over as the predominant microorganisms in the water. Substances toxic to many insects may be present, and the insect may have to possess a snorkel or breathing tube for exchanging gases with the atmosphere. The larvae of some insects possess short breathing tubes that are inserted into the aerenchyme tissues of aquatic plants.

What is still needed is information about how the larvae that develop in totally anaerobic habitats are able to avoid the toxic substances that are formed in such waters, especially those produced by the sulfur bacteria usually present in such waters.

Any research on how insects can survive the effects of the toxic substances present in anaerobic waters would have to be primarily studies on biochemistry. One apparent necessity is for the larva to have an integument completely impervious to all of the toxic substances produced by the microorganisms in the water. The chemical composition of this integument should be determined, and its physical characteristics should be investigated, especially its permeability to water and substances dissolved in it.

SECTION 17

Pelagic Insects

The two pelagic insects mentioned in this book have each been described and named by taxonomists more than once, and revisions have reduced some of the names to junior synonyms. The two species, one a member of the family Chironomidae and the other a member of the Gerridae, seem to be widespread in the pelagic regions of three oceans: the Atlantic, Indian, and Pacific.

The problems of insects trying to survive on the surfaces of the high seas would be daunting, and the fact that only two species of insect have succeeded in doing this shows how difficult it is. There are more species of insect known from the neritic zones of the oceans, but they have many advantages over the two that live on the high seas. For example, they can enter a mangrove swamp or take refuge on a beach during a severe storm. They can find shelter in a heavy rain, and they can keep the shore in sight to resist drifting aimlessly due to winds and ocean currents.

A main problem in planning a study of their behavior and survival strategy is finding them on the open ocean and taking the time to observe and examine them. Research ships typically carry groups of researchers, each with his own projects. Observing one water strider or one midge for several days while groups wait to make studies of dozens of rare marine taxa might be regarded as a waste of resources. The alternative would be searching for the insects in a small boat, several hundred nautical miles from the nearest land.

Information on the "sea skaters" in the family Gerridae have become better known in recent years, but basically everything known about them is from casual observation. It is hard to imagine how these insects deal with a heavy storm. This insect has been reported to feed on various marine organisms, including jelly fish, but what are its main foods and its preferred foods? Other reports allege that the insects are benefitting from all of the plastic debris on the oceans because it provides them with resting places.

The benthic chironomid is assumed to live as a larva slightly below the surface of the open water. How deep can the larvae go? There is little known concerning its feeding or behavior other than accounts of foods it seems to eat. However, there are descriptions of its mating and production of eggs. It is assumed that mating would be synchronized, perhaps with the phase of the moon. However, it has never been demonstrated what triggers metamorphosis to the imago stage. A good research program would require enough time to observe the actions of the insects for a long enough time to determine when the adults spawn, what the larvae eat, and how long it takes for a generation to complete its life cycle. It is also important to know what organisms, if any, consume the larvae. It has already been reported that the maximum life span for an adult is less than two days.

Any project to investigate one of the insects on the high seas should allow enough time to overcome any problems that arise and plan to have sufficient time to complete the necessary observations. Finding a sponsor for the research might be difficult.

SECTION 18

Research on the Seashore

Unlike the study of one or two species on the high seas, any research project along a beach or seashore of any kind would yield a large number of specimens as well as many observations of their activities in return for much less money for funding. The intertidal habitat is typically one of the richest in insects that are usually considered rare. Although most insect species known from seashores are abundant somewhere in the sub-tidal or intertidal zones, they are seldom found in any other kinds of habitat, except perhaps along the shores of large rivers, occasionally long distances inland from their mouths. A research study in these zones would be economical and productive. However, it would provide only basic results on the insects found, and it would encompass studies only of marine insects that inhabit the neritic zone.

The insects of the intertidal zone have some economic importance on beaches used for swimming or resting. When a row of drying seaweeds lines the upper margin of the intertidal zone, there are sometimes swarms of small flies resting on it until disturbed. Sometimes, these flies are accompanied by many collembolans, which form small swarms of springing hexapods. Visitors to the beach are sometimes frightened by the sight of so many insects. They also fear that the insects will bite them, although in most cases, they do not bother people at all.

Entomological surveys of the species on the beach usually dispel the fears of being bitten, but some people want the insects removed. Only occasionally are species of Diptera present that actually bite in order to obtain blood. A few of them belong to the family Tabanidae, that is, a horsefly.

The smallest flies in the intertidal zone usually belong to such families as Ephydridae, Sphaeroceridae, Dolichopodidae, and others common is locations that are inhospitable to most other insect species. All of these are harmless to people, and it will take some of the fear away from the visitors at the beach to know this.

SECTION 19

The Tolerance of Certain Flies for Salt

The small flies in the family Ephydridae are particularly tolerant of salt at concentrations above that of seawater. In various parts of the world, there are water bodies with particularly high concentrations of salt, such as the Dead Sea and the Great Salt Lake. In addition, there are artificial water bodies with excessively high concentrations of salt, especially coastal salt pans used to producing sea salt. These water bodies normally contain few insects.

The insects present are typically small flies that walk on the surface of the water. Most frequently, these flies belong to the family Ephydridae. They are often found in water bodies in which flies of most other families cannot survive. Finding the reasons for the tolerance to high salt concentrations by ephydrids living in hypersaline water would be an excellent research project, especially in cooperation with biochemists and physiologists. The first step in a project of this nature would be taking the selected ephydrids from a hypersaline lake and raising then in a laboratory culture.

After starting with suitable ephydrid species, similar observations would be made on species of other families of flies that seem to have similar tolerances for high salt concentrations. Knowledge of physiological modifications which can make insect species tolerant of hypersaline water could provide a possible explanation of why the family Ephydridae is particularly adept at developing tolerances to extreme conditions in the water bodies in which it lives. In addition, it should be determined whether the physiological methods employed by species in the family Ephydridae to avoid harmful effects from hypersaline water are the same as those employed by other families of Diptera.

SECTION 20

Tolerance to Petroleum

Another species in the family Ephydridae was discovered at the La Brea tar pits in Los Angeles in the early 20th Century. This small fly was appropriately named the petroleum fly because it can live in a mixture of tar, petroleum seeping out of an oil well, and water. The sticky petroleum traps many insects, which die quickly in the black oil. It has frequently been reported that the bodies of the dead insects are eaten by the petroleum fly larvae together with the oils, but the larvae seem to digest the dead insects but not the petroleum.

While the petroleum flies can feed on the same crude oil that kills almost every other insect which falls into it is a mystery which should be investigated. Crude oil is not only sticky, it contains a whole variety of substances that are toxic. Therefore, the tolerance of the petroleum fly larvae to this toxic material as it passes through its digestive system of hard to explain.

What the research should seek to elucidate is how the petroleum fly larvae can come into contact with petroleum and not be killed by it and how they can eat the petroleum without showing any ill effects. It should also be determined whether the bodies of the insects are actually digested and assimilated or whether the larvae are nourished on other items that they eat.

SECTION 21

Looking Inside Sponges

The spongillafly is not really a fly but rather an aquatic member of the order Neuroptera. The adults are very small and inconspicuous. They lay their eggs on or inside one of several species of freshwater sponge in the genus *Spongilla*. After the larva of the spongillafly enters the sponge, it reportedly enters the flagellated chambers and either eats the cells of the sponge or protects the sponges by feeding on microorganisms that enter the sponge as parasites. There are also other possible relationships that have been suggested for the real relationship between the sponge and the spongillafly.

In order to perform the research, it would be necessary place a probe inside the sponge to observe what the spongillafly larva does inside of the coanocyte chamber of the sponge. The coanocyte, also called the collar cell, is flagellated and keeps a current going through the chambers of the sponge by constant movement of the flagellum synchronized with the movements of the other cells. The food particles in the water are filtered out by the cells in the coanocyte chambers.

One hypothesis is that the larva feeds on the cells of the sponge with its piercing and sucking mouthparts. The alternative hypothesis is that it uses its mouthparts to pipette microorganisms that are sucked into the sponge, thereby protecting the sponge from damage or blockage of the chamber by the harmful microorganisms that the cells of the sponge cannot overpower. If the first hypothesis is correct, then the spongillafly larva is a parasite. If the second is correct, it is a useful symbiont.

Today, there are probes available which can send a magnified image from inside of a small tube or organ. There are also small instruments that can send highly magnified images from the open water of a pond or lake. With such an instrument, it would be easy to monitor the activities of the larva inside of the sponge. The only remaining problem would be to construct an aquarium and fill it with clean, oxygen-rich water, which could be circulated by a small pump. If done correctly, a sponge could be raised in the aquarium, and a spongillafly larva could be placed inside of it. With some patience, the answer to weather a spongillafly is a parasite or a sybiont of the sponge could finally be answered. However, each species of spongillafly would have to be tested separately to make sure that there are not some species that are parasites and others that live as symbionts. In addition, other information could be gained about the activities of both the spongillafly larvae and the sponges. To confirm the results, a harmless dye could be placed in the sponge to color the cells, and the presence of the dye in the larva later on would indicate that it was feeding of the cells of the sponge.

SECTION 22

Parasitoid Hymenoptera

A study of the parasitoids in the superfamilies Ichneumonoidea and Chalcidoidea has become more of a career field than a project. There are believed to be tens of thousands of species worldwide, although the great majority of them develop in terrestrial insects.

One of the major problems in studying the species in the families of parasitoids is that there is no known way to identify the living host of a parasitoid larva before the parasitoid starts to kill it and leave. That means that the only time an entomologist can collect a specimen of a parasitoid is during the brief period of time when the female is hunting for a host on which to lay one or more of its eggs. If an infected host could be identified, it would be possible to collect only those hosts and keep them until the adult parasitoid emerges. That would give the researcher a complete specimen of the adult parasitoid, the identity of an actual host of the parasitoid species, and the time of year during which development of the larvae of the parasitoids takes place. It would also give the researcher a chance to dissect a larva out of a host in order to get its description.

This would greatly accelerate the studies of the parasitoids and provide the taxonomist writing the descriptions of new species with actual specimens of its larvae to describe and illustrate, as well as possible new specimens of the imago.

Without a clue to which living host insect is actually infected by a parasitoid, it is not possible to do anything but maintain as many potential hosts as possible in an aquarium and wait for an adult parasitoid to emerge. Only if the dead host is found or insects of only one species are kept in the aquarium can the host species be matched with its parasitoid. In no case will a specimen of a parasitoid larva be available for examination. With so many thousands of unknown species of parasitoid still waiting to be described and named worldwide, more efficient methods of collecting specimens and identifying their hosts would greatly shorten the time necessary to compile a reasonably complete catalogue of the hymenopteran parasitoid species of the world.

SECTION 23

Flies that Eat Snails

Research in this field would best be classified as studies to permit biological control of snails in agriculture and to promote public health. This kind of investigation would probably have a good chance of finding a sponsor. Aquatic snails are regarded as major pests for certain crops grown in flooded fields, especially rice.

In the tropics, snails can be dangerous vectors of parasites that cause serious illnesses in human beings. Some of the most dangerous of these are blood flukes, also called schistosomes. These are parasites in the genus *Schistosoma* that live in the blood stream of the vertebrate host and constantly produce eggs, which are excreted from the infected human or animal host and hatch to release infectious miracidia larvae that can enter and infect snails, in which the development continues through the sporocyte stage until free swimming carcaria are produced that can infect human beings by boring through the skin or being accidently swallowed. Cercaria attack farmers wading in rice fields or other freshwater bodies or swimmers who disregard the signs often placed around ponds warning against contact with the water. The infectious cercariae swim through the water seeking a host. Those of the blood flukes are recognized by their forked tails, which are shed at the time the cercaria enters a host.

Flies that are large enough to feed on large snails are good choices for predators to control them. They can fly from one flooded field to another and seek out the most abundant snail species. For some of the larger snails, such as those in the family Ampulariidae, a larger insect is sometimes required, even though small flies are known to open large snails. Large members of the family Belostomatidae in the order Hemiptera are often capable of forcing open the opercula of large snails to feed on the soft parts. These are among the insects that are being tried as agents of biological control over freshwater snails. The goal of the biological control programs for schistosomes and other parasites that employ snails as vectors is to control the parasites by eliminating the intermediate hosts. Convincing people to avoid swimming in dirty water would work better, but the fact that an estimated 200,000,000 people are infested by schistosomes worldwide indicates that this method has not been working very well.

SECTION 24

Courtship Behavior of Empidid Flies

Even in countries that have long had universities and institutions in which biological research has been conducted for more than 2½ centuries, there are still questions concerning animal behavior that have never been conclusively answered. The information about certain behavior patterns is still disputed in various ways.

The methods of male flies in the superfamily Empidoidea of courting females during the breeding season include giving her an insect wrapped in a silken package secreted by the male. This has been observed many times by persons observing mating swarms of the individual species. Unfortunately, the facts reported after various observations conflict with each other in certain ways. In spite of the multitude of reports on the courtship of these species, there is no list showing which of them are known to court females by giving them small captured insects wrapped in silk and which apparently do not use this method. Some reports would tend to indicate that this has been observed for all species in the family Empididae and even by members of other families in the superfamily Empidoidea. Others indicate that only a few species have been observed practicing this custom, and it has not been reported whether other species in the family give courtship presents at all.

Remarkable behavior by any group of insects should be reported and added to a file so that the species known for this practice could be kept on a list and distinguished from other closely related species that do not, if there are any.

SECTION 25

Medical Entomology

Most research into matters concerning public health and diseases that infect human beings is much better funded than most studies concerning entomology alone. In this kind of research, the main goal is to find ways of preventing, curing, and eventually eliminating specific diseases. The people performing the research are typically medical doctors together with specialists in virology, bacteriology, protozoology, and parasitology. In many of the publications, reports about the insect vectors of the diseases are less thorough and precise than the information concerning the other participating organisms. The general impression is that the roles of the disease vectors in the epidemiology of the diseases are seriously underestimated. This is surprising in view of the fact that so many projects to limit the spread of insect borne diseases are based on control of mosquitoes and other members of the Diptera. As research on the diseases progresses, knowledge of their vectors must keep pace in order to focus on attempts at the elimination of each disease only by focusing on control of the vector species that are actually capable of spreading it.

The danger of being infected by a dangerous disease spread by a particular species of insect increases with the proximity of the breeding places of that mosquito to human habitations. On the other hand, epidemics are more likely to be initiated by insect vectors that usually live far from human habitations but closer to populations of primates and other mammalian species, which act as reservoirs for the pathogens and parasites causing the diseases.

The role of populations of vertebrates permanently infected with pathogens for chronic diseases is to facilitate new infections of the disease at any time in the future, preventing the complete elimination of the pathogen. Today, one of the effects of such a permanent population of a pathogen maintained in the bodies of infected monkeys located in a forest not too far from a human population center is to infect mosquitoes in the forest with infections of the pathogens. Occasionally, a human being goes into the forest and is bitten by an infected mosquito. He, in turn, passes the pathogen on to a mosquito species residing in his town, causing a local epidemic of the disease by biting more people who spread the pathogen to more local mosquitoes. Because the pathogen is capable of infecting two or more species of mosquito, at least one residing in a forest and another in towns and cities, the permanent elimination of a dangerous disease remains almost impossible.

In the case of the malarial diseases, the increasing resistance of the pathogens to the quinines that were previously effective in suppressing them has been causing a resurgence of some of these diseases. This growing resistance by the pathogens to the traditional medicines for suppressing the symptoms of the disease are being paralleled by the increasing resistance of mosquitoes against the pesticides being used to keep their populations in check. These problems have been exacerbated by the ease and reliability of air traffic, which is now delivering people in just a few hours to all parts of the world, including many tropical countries where endemic tropical diseases are making a come-back.

The only way to finally stop new epidemics of many tropical diseases is to develop and introduce vaccinations. The problem is that those diseases for which it is possible to develop a vaccine have already been suppressed by immunization programs. The diseases that have not been suppressed are those for which no effective vaccine can be developed. There are simply too many varieties of the viruses for a single vaccine to be effective against all of them, and these viruses seem to be capable of mutating to still more varieties if their survival calls for this.

Without question, the continued struggle against tropical diseases will remain the province of large bureaucracies and giant pharmaceutical firms. Entomologists will continue to play only a secondary role. Nevertheless, some research opportunities will continue to exist as long as tropical diseases remain the serious problems that they were found to be during the construction of the Panama Canal.

SECTION 26

Cooperative Strategies of Insects

Many strategies are utilized by groups of insects to take advantage of food supplies which one insect alone would not be capable of doing. The behavior of birds is frequently recorded by ornithologists in the minutest detail, but remarkable behavior by groups of insects is seldom reported. Personal observations of dragonflies forming swarms and hovering over people walking in wetlands were frequent before there was any way to make sense of it. Even early Portuguese explorers in Brazil had noticed being followed by regular formations of large dragonflies, accounting for the name, "*caveleiros do Diabo*", given to the insects.

The first tactics by groups of an insect species to improve their chances of survival were those used by diurnal mosquitoes, which hid in dense bushes, making them unavailable to large flying predators, such as dragonflies. They waited until a large, vulnerable mammal walked close to the bushes, and then they would all emerge within a period of a few seconds and attack their prey at the same time, from all directions. For lightly clothed victims walking at temperatures above 40°C in the shade, flight to a protective vehicle was the only escape.

At this point, my personal escort of dragonflies swooped in and made short work of the more than 100 attacking mosquitoes. The only way to get at the large prey they could not extract from the dense shrubbery was to follow a person they knew the mosquitoes would leave their protective bushes in order to ambush.

For some reason, entomologists have failed to report seemingly intelligent behavior by insects, with the exception of certain practices of social species of Hymenoptera. Researchers should take more time to report such behavior, the purposes for which are not always obvious to human beings.

SECTION 27

Thermal Limits for Life

Until recent investigations of particularly hot springs, it was thought that the limiting temperature for life was not much more than 50°C. Later, prokaryotic organisms were found which can live in water at temperatures approaching 80°C, at which insects that could feed on them at cooler temperatures were excluded. What are the temperature limits of life? In the deep ocean, steam flows out of "black smokers" at temperatures well above 100°C into water below 0°C, which is kept from freezing by the extremely high pressure of the ambient water. In the water in which temperatures well over 100° apart come into direct contact, an extremely rich and diverse fauna exists. Chemosynthesis by bacteria replaces photosynthesis as the process of primary production, and the arthropods that live around the "black smokers" seem to be nourished directly by symbiotic, chemosynthetic bacteria.

Although no insects have yet been found in such extremely hot habitats, it is well to question the perceived temperature limits at which suitably adapted animals can survive. A fundamental question about how insects can survive at high temperatures might be answered by studies of the members of the order Diptera which have been the most successful at developing larvae with adaptations making it possible for them to develop at temperatures above those that any other known insects can survive at.

Studies of thermophilic and kryophilic hexapods could well reveal the still unknown methods by which such insects can survive at the assumed temperature limits for life. For the study of survival at particularly high temperatures, the larvae of species in the dipteran family Ephydridae would be good subjects of study. This family encompasses species living at many physiological limits for eukaryotic life, including temperature, salinity, oxygen deficiency, presence of petroleum on and in the water, and concentrations of various pesticides. The biochemistry and physiology of these species would certainly yield some clues to how such species can survive beyond what would be the limits for survival of almost any other taxon of hexapods.

SECTION 28

Ants and Water

Ants are seldom mentioned in works about aquatic insects, yet they are among those few animals which design structures to modify their living spaces in ways that allow them to control the aquatic milieu around them. Ants that live in nests in the ground can almost comfortably survive during floods, heavy rainstorms, tornados, forest and brush fires, and earthquakes. Only human beings can compete with ants in the art of building shelters to keep themselves safe during natural catastrophes, although ants sometimes have equal or better records of building structures that protect them better during floods, tornados, and forest or brush fires.

Ants have an even better record of building secure structures for themselves in trees. Of course, they have an advantage of weighing much less, allowing them to construct much lighter structures that the trees are able to support. There has certainly been enough reported about how ants can build tunnel systems in ways that keep water out during floods, store air for use during forest fires, and be heated during the colder seasons by the bodies of the ants themselves, which are warmed by the sun during winter days and brought into the nests with the ants in the evening.

Certainly, the designs of the nests built by each species of ant are different in various ways, yet detailed studies of these designs are few. The remarkable observations that ants are able to bring aquatic habitats to their own doorsteps by planting and tending bromeliads add new dimensions to the activities of these insects. This is a rich field for research, although it is one which is somewhat out of fashion at the present time. If is certainly easier keeping track of what one insect, working alone, is doing than to observe and keep track of the simultaneous actions of at least several hundred worker and soldier ants in nests where one large female ant can be producing roughly one egg each second.

Literature

Adis, J.E. 1983: Eco-entomological observations from the Amazon. IV. Occurrence and habits of the aquatic caterpillar *Palustra laboulbeni* Bar, 1873 (Arctiidae: Lepidoptera) in the vicinity of Manaus, Brazil. *Acta Amazonica* 13(1): 31–36.

Adis, J., E. Bustorf, M.G. Lhano, C. Amedegnato and A.L. Nunes. 2007: Distribution of *Cornops* grasshoppers (Leptysminae: Acrididae: Orthoptera) in Latin America and the Caribbean Islands. *Journal of Neotropical Fauna and Environment* 42(1): 11–24.

Aditya, G. and S.K. Raut. 2001: Predation of water bug *Sphaerodema rusticum* Fabricius on the snail *Pomacea bridgesi* (Reeve), introduced in India. *Current Science* 81(11): 1413–1414.

Aditya, G. and S.K. Raut. 2002: Predation potential of water bugs *Sphaerodema rusticum* on the sewage snails *Physa acuta* (Reeve), introduced in India. *Memórias Instituto Oswaldo Cruz*, Rio de Janeiro 97(3): 531–534.

Adlassnig, W., M. Peroutka and T. Lendl. 2011: Traps of carnivorous pitcher plants as a habitat: composition of the fluid, biodiversity and mutualistic activities. *Annals of Botany* 107(2): 181–194.

Adler, P.H., D.C. Currie and D.M. Wood. 2004: *The Black Flies (Simuliidae) of North America.* Cornell University Press, Ithaca, NY.

Ahee, J.E., B.J. Sinclair and M.E. Dorken. 2013: A new species of *Stenodiplosis* (Diptera: Cecidomyiidae) on florets of the invasive common reed (*Phragmites australis*) and its effects on seed production. *The Canadian Entomologist* 145: 235–246.

Aladin, N.V. 1995: Ecological state of the fauna of the Aral Sea during the last 30 years. *Geojournal* 35: 29–32.

Alexander, C.P. 1912: A bromeliad-inhabiting crane-fly (Tipulidae: Dipt.). *Entomological News* 23: 415–417.

Alexander, C.P. 1926: The Trichoceridae of Australia. *Proceedings of the Linnean Society of New South Wales* 51: 299–304.

Alexander, C.P. 1929: Trichoceridae. In: *Diptera of Patagonia and South Chile*. Part 1: i-xvi, 1-240+12 pl.

Alexander, C.P. 1967: Family Trichoceridae. *A Catalogue of the Diptera of the Americas South of the United States*. Dept. of Zoology, Series Agric., São Paulo.

Andersen, N.M. and T.A. Weir. 2004: Australian Water Bugs: Their Biology and Identification (Hemiptera-Heteroptera, Gerromorpha & Nepomorpha). *Enterograph* 14, Apollo Books, CSIRO Publishing. 344 pp.

Anon, 2010: Heuschrecken, Fangschrecken, Schaben und Ohrwürmer, Rote Liste und Artenliste Sachsens. Freistaat Sachsen, Landesamt für Umwelt, Landwirtschaft und Geologie, Dresden.

Arnett, R.H. 2000: *American Insects: A Handbook of the Insects North of Mexico*. CRC Press.

Bachmann, A.O. 1995: Insecta Lepidoptera. pp. 1232–1235. *In*: E.C. Lopretto and G. Tell (eds.). *Ecosistemas de Aguas Continentales*. Vol 3. Ed. Sur, La Plata.

Baquero, E. and R. Jordana. 1999: Species of *Anagrus* Haliday, 1833 (Hymenoptera, Chalcidoidea, Mymaridae) in Navarra (Spain). *Miscellània Zoològica* 22(2): 39–50.

Bar, M. 1873. Sur un genre nouveau de Lépidopteres de la tribu de Bombycides et dont la chenille est aquatique. *Annales de la Société Entomologique de France* (5)3: 297–306.

Barthlott, W., K. Riede and M. Wolter. 1994: Mimicry and ultrastructural analogy between the semi-aquatic grasshopper *Paulinia acuminata* (Orthoptera, Pauliniidae) and its food plant, the water-fern *Salvinia auriculata* (Filicatae, Salviniaceae). *Amazoniana* 13: 47–58.

Bastos, J.A.M. 1977: Caracterização de especies de cachorr-d'agua (Orth.: Gryllotalpidae) do Estado do Ceará, Brasil. *Fitassanidade* 2(2): 48–49.

Belizario, V.Y., G.G. Geronilla, M.B.M. Anastacio, W.U. de Leon, A.P. Suba-an, A.C. Sebastian and M.J. Bangs. 2006: *Echinostoma malayanum* infection, The Philippines. *Emerging Infectious Diseases* 13(7): 1130–1131.

Belmonte, G., S. Moscatello, E.A. Batogova, T. Pavlovskaya, N.V. Shadrin and L.F. Litvinchuk. 2012: Fauna of hypersaline lakes of The Crimea (Ukraine). *Thalassia Salentina* 34: 11–24.

Bischof, M.R. and D.L. Deonier. 1985: Life history and immature stages of *Parydra breviceps* (Diptera: Ephydridae). *Proceedings of the Entomological Society of Washington* 85: 805–820.

Blatch, S.A., F.C. Thompson and M.A. Zumbado. 2003: A review of the Mesoamerican flower flies of the genus *Meromacrus* (Diptera: Syrphidae) including the description of a new species. *Studia Dipterologica* 10: 13–36.

Blickle, R.L. 1958: Notes on *Aegialomyia psammophila* (O. S.) (Tabanidae: Diptera). *The Florida Entomologist* 41(3): 129–131.

Bohn, H.F. and W. Federle. 2004: Insect aquaplaning: *Nepenthes* pitcher plants capture prey with the peristome, a fully wettable water-lubricated anisotropic surface. *Proceedings of the National Academy of Sciences of the United States of America* 101(39): 14138–14143.

Boix, D., S. Gascon, J. Sala, A. Bavosa, S. Brucet, A. Lopez-Flores, M. Martinov, J. Gifre and X.D. Quintana. 2008: Patterns of composition and species richness of crustaceans and aquatic insects along environmental gradients in Mediterranean water bodies. *Hydrobiologia* 597: 53–69.

Bonnet, D.D. and H. Chapman. 1956: The importance of mosquito breeding in tree holes with special reference to the problem in Tahiti. *Mosquito News* 16(4): 301–305.

Bowles, D.E. 2006: Spongillaflies (Neuroptera: Sisyridae) of North America with a key to the larvae and adults. *Zootaxa* 1357: 1–19.

Brock, M.L., R.G. Wiegert and T.D. Brock. 1969: Feeding of *Paracoenia* and *Ephydra* (Diptera: Ephydridae) on the microorganisms of hot springs. *Ecology* 50(2): 192–200.

Brues, C.T. 1924: Observations on animal life in the thermal waters of Yellowstone National Park, with a consideration of the thermal environment. *Proceedings of the American Academy of Arts and Sciences* 59: 371–437.

Brues, C.T. 1928: Studies on the fauna of hot springs in the Western United States and the biology of thermophilous animals. *Proceedings of the American Academy of Arts and Sciences* 63: 139–228.

Brues, C.T. 1932: Further studies on the fauna of North American hot springs. *Proceedings of the American Academy of Arts and Sciences* 67: 185–303.

Brug, S.L. and J.W. Tesch. 1937: Parasitaire wormen aan het Lindoer Meer (Oe. Paloe. Celebes). *Geneeskundig Tijdschrift voor Nederlandsche-Indie* 77(36): 2151–2158.

Burgess, J., E. Burgess and M. Lowman. 2003: Observation of a beetle herbivore on a bromeliad in Peru. *Journal of the Bromeliad Society* 53: 221–224.

Campos, R.E. and L.A. Fernández. 2011: Coleopterans associated with plants that form phytotelmata in subtropical and temperate Argentina, South America. *Journal of Insect Science* 11(147): 1–18.

Carbonell, C.S. 1957: Observaciones bio-ecológicas sobre *Marellia remipes* Uvarov (Orthoptera, Acridoidea) en el Uruguay. *Investigaciones y Estudios de la Faculdad de Humanides y Ciencias* (Montevideo). 21 pp + 6 pl.

Carnevale, P., G. le Goff, J.–C. Toto and V. Robert. 1992: *Anopheles nili* as the main vector of human malaria in villages of southern Cameroon. *Medical and Veterinarian Entomology* 6: 135–138.

Carrera, M., H. Lopes and J. Lane. 1947: Contribuição ao conhecimento dos "Microdontinae" neotrópicos e descrição de duas novas espécies de "*Nausigaster*" Williston (Diptera, Syrphidae). *Revista Brasileira de Biologia* 7: 471–486.

Caspers, H. 1951: Rhythmische Erscheinungen in der Fortpflanzung von *Clunio marinus* (Dipt. Chiron.) und das Problem der lunaren Periodizität bei Organismen. *Archiv für Hydrobiologie*, Supplement 18: 415–594.

Caspers, H. and C.W. Heckman. 1981: Ecology of orchard drainage ditches along the freshwater section of the Elbe Estuary. Biotic succession and the influence of changing agricultural methods. *Archiv für Hydrobiologie/Supplement 43* (Untersuchungen Elbe-Aestuar 4): 347–486.

Caspers, H. and C.W. Heckman. 1982: The biota of a small standing water ecosystem in the Elbe flood plain. *Archiv für Hydrobiologie/Supplement 61* (Untersuchungen Elbe-Aestuar 5): 227–316.

Céréghino, R., C. Leroy, A. Dejean and B. Corbara. 2010: Ants mediate the structure of phytotelm communities in an ant-garden bromeliad. *Ecology* 91(5): 1549–1556.

Céréghino, R., C. Leroy, J.F. Carrias, L. Pelozuelo, C. Ségura, C. Bosc, A. Dejean and B. Corbara. 2011: Ant-plant mutualisms promote functional diversity in phytotelm communities. *Functional Ecology* 25: 954–963.

Chang, S.L. 1966: Some physiological observations on two aquatic Collembola. *Transactions of the American Microscopical Society* 85(3): 359–371.

Chaplin, M. 2009: Theory versus experiment: What is the surface charge of water? *Water* 1: 1–28.

Cheng, L. 1981: *Halobates* (Heteroptera: Gerridae) from Micronesia with notes on a laboratory population of *H. mariannarum*. *Micronesica* 171: 97–106.

Cheng, L., M.A. Baars and S.S. Oesterhuis. 1990: *Halobates* in the Banda Sea (Indonesia): monsoonal differences in abundance and species composition. *Bulletin of Marine Science* 47: 421–430.

Christiansen, K. 1964: Bionomics of Collembola. *Annual Review of Entomology* 9: 147–178.

Chvála, M. 1994: Empidoidea (Diptera) of Fennoscandia and Denmark. III. Genus *Empis*. *Fauna Entomologica Scandinavica* 29: 1–192.

Coe, R.L., P. Freeman and P.F. Mattingly. 1950: Diptera. Nematocera and Brachycera. *Handbooks for the Identification of British Insects*. Royal Entomological Society, London.

Cohuet, A., F. Simard, C.S. Wondji, C. Antonio-Nkondjio, P. Awono-Ambene and D. Fontenille. 2004: High malaria transmission intensity due to *Anopheles funestus* (Diptera: Culicidae) in a village of savannah-forest transition area in Cameroon. *Journal of Medical Entomology* 41(5): 901–905.

Collins, W.E. and G.M. Jeffery. 2005: *Plasmodium ovale*: Parasite and disease. *Clinical Microbiology Reviews* 18(3): 570–581.

Corbet, P.S. 1983: Odonata in phytotelmata. pp. 29–54. *In*: J.H. Frank and L.P. Lounibos (eds.). *Phytotelmata: Terrestrial Plants as Hosts for Aquatic Insect Communities*. Plexus, Medford, New Jersey.

Cordo, H.A. 1996: Recommendations for finding and prioritizing new agents for biological control of the water hyacinth control. pp. 181–185. *In*: R. Charudattan, R. Labrada, D.T. Center and C. Kelly-Begazo (eds.). *Strategies for Water Hyacinth Control*. FAO, Fort Lauderdale.

Coulson, S.J., I.D. Hodkinson, N.R. Webb and J.A. Harrison. 2002: Survival of terrestrial soil-dwelling arthropods on and in seawater: implications for trans-oceanic dispersal. *Functional Ecology* 16: 353–356.

Cowley, D.R. 1978: Studies on the larvae of New Zealand Trichoptera. *New Zealand Journal of Zoology* 5: 639–750.

Cunha, S.P., J.R. Carreira Alves, M.M. Lima, J.R. Duarte, L.C.V. Barros, J.L. Silva, A.T. Gammaro, O.S. Monteiro Filho and A.R. Wanzeler. 2002: Presença de *Aedes aegypti* em Bromeliaceae e depósitos com plantas no Município do Rio de Janeiro, RJ. *Revista de Saúde Pública* 36: 244–245.

Curran, C.H. 1941: New American Syrphidae. *Bulletin of the American Museum of Natural History* 78: 243–304.

Dahelmi, E.M. and S.S. Syamsuardi. 2015: Detection of transovarial dengue virus with *RT-PCR* in *Aedes albopictus* (Skuse) larvae inhabiting phytotelmata in endemic DHF areas in West Sumatra, Indonesia. *American Journal of Infectious Diseases and Microbiology* 3(1): 14–17.

Dahl, C. 1970a: Distribution, phenology and adaptation to arctic environment in Trichoceridae (Diptera). *Oikos* 21: 185–202.

Dahl, C. 1970b: Diptera: Trichoceridae of South Georgia. *Pacific Insects Monograph* 23: 271–273.

Darby, M. and C.S. Chaboo. 2015: *Phytotelmatriches*, a new genus of Atrotrichinae (Coleoptera: Ptiliidae) associated with the phytotelmata of Zingiberales plants in Peru. *Zootaxa* 4052: 1–4.

Dean, D.A. (ed.). 1973: *Lange's Handbook of Chemistry*. 11th ed. McGraw-Hill, New York. 1600 pp.

Dejean, A.I., C. Olmsted and R.R. Snelling. 1995: Tree-epiphyte-ant relationships of the low inundated forest in Sian Ka'an Biosphere Reserve, Quintana Roo, Mexico. *Biotropica* 27: 57–70.

DeLoach, C.J. and H.A. Cordo. 1978: Life history of the moth, *Sameodes albiguttalis*, as a candidate for biological control of waterhyacinth. *Environmental Entomology* 7(2): 309–321.

DeLoach, C.J., D.J. DeLoach and H.A. Cordo. 1978: Observations on the biology of the moth, *Samea multiplicalis*, on waterlettuce in Argentina. *Journal of Aquatic Plant Management* 17: 42–44.

DeMarmels, J. 1985: La náyade de *Leptagrion fernandezianum* Rácinis, especie bromelicola (Odonata: Coenagrionidae), y consideraciones sobre la posible relación filogenética del género *Leptagrion* Selys. *Boletín de Entomología Venezolana* N.S. 4(1): 1–8.

DeMarmels, J. and R.W. Garrison. 2005: Review of the genus *Leptagrion* in Venezuela with new synonymies and descriptions of a new genus, *Bromeliagrion*, and a new species, *B. rehni* (Zygoptera: Coenagrionidae). *Canadian Entomologist* 137: 257–273.

DeOliveira, C.S., F.L. Da Silva and S. Trivinho-Strixino. 2013: *Thalassomya gutae* sp. n., a new marine chironomid (Diptera: Chironomidae: Telmatogetoninae) from the Brazilian coast. *Zootaxa* 3701: 589–595.

Derraik, J.K.B. 2005: Mosquitoes breeding in container habitats in urban and peri–urban areas in the Auckland region, New Zealand. *Entomotropica* 20: 93–97.

Doležil, Z. 1972: Developmental stages of the tribe Eristalini (Diptera: Syr-phidae). *Acta Entomologica Bohemoslovaca* 69: 339–350.

Downs, W.G. and C.S. Pittendrigh. 1946: Bromeliad malaria in Trinidad, British West Indies. *American Journal of Tropical Medicine* 26: 47–66.

Drechsel, U. 2014: Aquatic habit of larval instars of *Paracles palustris* (Joergensen, 1935) (Lepidoptera: Erebidae: Arctiinae). *Paraguay Biodiversidad* 1(18): 85–94.

Drechsel, L.J. and S. Drechsel Garcia. 2017: The early stages of *Paracles aurantiaca* (Rothschild, 1910) (Lepidoptera: Erebidae: Arctiini). *Paraguay Biodiversidad* 4(1): 1–7.

Dumbleton, L.J. 1963: The classification and distribution of the Simuliidae (Diptera) with particular reference to the genus *Austrosimulium*. *New Zealand Journal of Science* 6: 320–357.

Edwards, F.W. 1926: On marine Chironomidae (Diptera), with a description of a new genus and four new species from Samoa. *Proceedings of the Zoological Society of London* 96: 779–806.

Edwards, F.W. 1928: Diptera. Trichoceridae. *In*: P. Wytsman (ed.). *Genera Insectorum,* Brussels 190: 30–37.

Eisener, T., M.A. Goetz, S.R. Hill, D.E. Smedley and J. Meinwald. 1997: Firefly "femme fatales" acquire defensive steroids (lucibufagans) from their firefly prey. *Proceedings of the National Academy of Sciences* 94(18): 9723–9728.

Ellison, A.M., N.J. Gotelli, J.S. Brewer, A.L.D.L. Cochran-Stafira, J.M. Kneitel, T.E. Miller, A.C. Worley and R. Zamora. 2003: The evolutionary ecology of carnivorous plants. *Advances in Ecological Research* 33: 1–74.

Feitosa, M.C.B., I.B. Querino and N. Hamada. 2016: Association of *Anagrus amazonensis* Triapitsyn, Querino & Feitosa (Hymenoptera, Mymaridae) with aquatic insects in upland streams and floodplain lakes in central Amazonia, Brazil. *Revista Brasileira de Entomologia* 60: 267–269.

Fincke, O.M. 1992: Interspecific competition for tree holes: consequences for mating systems and coexistence in Neotropical damselflies. *The Anerican Naturalist* 139(1): 80–101.

Fincke, O.M. 1999: Organizations of predator assemblages in Neotropical treeholes: effects of abiotic factors and priorities. *Ecological Entomology* 24(1): 13–23.

Fischer, M. 1971: Hym. Braconidae. Opiinae. *Index of Entomophagous Insects.* Paris. 189 pp.

Fish, D. and S.J. de Soria. 1978: Water-holding plants (phytotelmata) as breeding sites for ceratopogonid pollinators of cacao. *Revista Theobroma* 8: 133–146.

Flint, O.S. 2006: New species and records of Neotropical Sisyridae with special reference to *Sisyra* (Insecta: Neuroptera). *Proceedings of the Biological Society of Washington* 119(2): 279–286.

Flint, O.S. Jr., R.L. Hoffman and C.R. Parker. 2008: An annotated list of caddisflies (Trichoptera) of Virginia. Part II. Families of Interpalpia. *Banisteria* 31: 3–23.

Foote, B.A. 1995: The biology of shore flies. *Annual Review of Entomology* 40: 417–442.

Forattini, O.P., G.R.A.M. Marques, J. Kakitani, M. Brito and M.A.M. Sallum. 1998: Significado epidemiológico dos criadouros de *Aedes albopictus* em bromélias. *Revista de Saúde Pública* 32: 186–188.

Fowler, H.G. and H.L. Fasconcelos. 1989: Preliminary data on life cycles of some mole crickets (Orthoptera, Gryllotalpidae) of the Amazon Basin. *Revista Brasileira de Entomología* 33(1): 139–141.

Frank, J.H. 1983: Bromeliad phytotelmata and their biota, especially mosquitoes. pp. 101–128. *In*: J.H. Frank and L.P. Lounibos (eds.). *Phytotelmata: Terrestrial Plants as Hosts for Aquatic Insect Communities.* Plexus, Medford, New Jersey.

Frank, J.H. 2008: Insectivorous plants. pp. 1995–2008. *In*: J.L. Capinera (ed.). *Encyclopedia of Insects.* Springer.

Frank, J.H. and G.A. Curtis. 1981: On the bionomics of bromeliad-inhabiting mosquitoes. VI. A review of the bromeliad-inhabiting species. *Journal of the Florida Anti-Mosquito Association* 52: 4–23.

Frank, J.H. and G.A. Curtis. 1982: Bionomics of the bromeliad-inhabiting mosquito *Wyeomyia vanduzeei* and its nursery plant *Tillandsia utriculata*. *Florida Entomologist* 91: 1–8.

Frank, J.H. and L.P. Lounibos. 2008. 2009: Insects and allies associated with bromeliads: a review. *Terrestrial Arthropod Reviews* 1: 125–153.

Frank, J.H. and G.F. O'Meara. 1984: The bromeliad *Catopsis berteroniana* traps terrestrial arthropods but harbors *Wyeomyia* larvae (Diptera: Culicidae). *Florida Entomologist* 67: 418–424.

Frank, J.H., S. Sreenivasan, P.J. Benshoff, M.A. Deyrup, G.B. Edwards, S.E. Halbert, A.B. Hamon, M.D. Lowman, E.L. Mockford, R.H. Scheffrahn, G.J. Steck, M.C. Thomas, T.J. Walker and W.C. Welbourn. 2004: Invertebrate animals extracted from native *Tillandsia* (Bromeliales: Bromeliaceae) in Sarasota County, Florida. *Florida Entomologist* 87: 176–185.

Frank, J.H., J.P. Stewart and D.A. Watson. 1988: Mosquito larvae in axils of the imported bromeliad *Billbergia pyramidalis* in southern Florida. *Florida Entomologist* 71: 33–43.

Frith, D.W. 1979: A twelve month study of insect abundance and composition at various localities on Aldabra Atoll. *Philosophical Transactions of the Royal Society of London, Series B, Biological Sciences* 286(1011): 119–126.

Frohne, W.C. 1954: Mosquito distribution in Alaska with especial reference to a new type of life cycle. *Mosquito News* 14(1): 10–13.

Fu, X.N. and L. Ballantyne. 2006: *Luciola leii* sp. nov., a new species of aquatic firefly (Coleoptera: Lampyridae: Luciolinae) from mainland China. *The Canadian Entomologist* 138(3): 339–347.

Fu, X., N. Ohba, F.V. Venci and C. Lei. 2006: Life cycle and behaviour of the aquatic firefly *Luciola leii* (Coleoptera: Lampyridae) from Mainland China. *The Canadian Entomologist* 138(6): 860–870.

Gaume, L. and Y. Forterre. 2007: A viscoelastic deadly fluid in carnivorous pitcher plants. *Plos One* 2(11) e1185: 1–7.

Giribet, G., G.D. Edgecombe and W.C. Wheeler. 2001: Arthropod phylogeny based on eight molecular loci and morphology. *Nature* 413: 157–161.

Giribet, G., G.D. Edgecombe, J.M. Carpenter, C.A. d'Haese and W.C. Wheeler. 2004: Is Ellipura monophyletic? A combined analysis of basal hexapod relationships with emphasis on the origin of insects. *Organisms, Diversity & Evolution* 4: 319–340.

Gisin, H. 1960: Collembolenfauna Europas. *Museum d'Histoire Naturelle*, Geneva. 312 pp.

Gopal, B. 1987: *Water Hyacinth*. Elsevier, Amsterdam. 471 pp.

Goulet, H. and J.T. Huber (eds.). 1993: *Hymenoptera of the World: An identification guide to families*. Canada Communications Group, Ottawa, Canada.

Greeney, H.F. 2001: The insects of plant-held waters: a brief review and bibliography. *Journal of Tropical Ecology* 17(2): 241–260.

Guenther, K. 1913: Die lebenden Bewohner der Kannen der insektenfressenden Pflanze *Nepenthes distillatoria* auf Ceylon. *Zeitschrift der wissenschaftlichen Insektenbiologie* 9: 122–125.

Haber, W.A., D.L. Wagner and C. de la Rosa. 2015: A new species of *Erythrodiplax* breeding in bromeliads in Costa Rica (Odonata: Libellulidae). *Zootaxa* 3947(3): 386–396.

Haddow, A.J. 1948: The mosquitoes of Bwamba County, Uganda. VI. Mosquito breeding in plant axils. *Bulletin of Entomological Research* 39: 185–212.

Hågvar, S. 2000: Navigation and behaviour of four Collembola species migrating on the snow surface. *Pedobiologia* 44(3-4): 221–233.

Hågvar, S. and E. Krzeminska. 2007: Contribution to the winter phenology of Trichoceridae (Diptera) in snow-covered southern Norway. *Studia Dipterologica* 14(2): 271–283.

Halgoš, J. 2005: Types of Areas and the Origin of Black Flies Fauna in Mongolia (Diptera: Simuliidae). *Erforschung Biologischer Ressourcen der Mongolei* 2005(9): 461–464.

Hamada, N. and P.H. Adler. 1998: A new species of *Simulium* (Diptera: Simuliidae) from open areas in central Amazonia, Brazil. *Memórias Instituto Oswaldo Cruz*, Rio de Janeiro 93(3): 317–325.

Hamilton, S.W. and R.W. Holzenthal. 2011: Twenty-four new species of *Polycentropus* (Trichoptera, Polycentropodidae) from Brazil. *Zookeys* 76: 1–53.

Hardy, D.E. 1956: Diptera: Coelopidae (Phycodromidae). *Insects of Micronesia* 14(1): 41–46.

Heckman, C.W. 1979: *Rice Field Ecology in Northeastern Thailand*. W. Junk, The Hague. 228 pp.

Heckman, C.W. 1982a: Ecophysiological and phylogenetic characterization of a wintertime biotic community in shallow water habitats near Hamburg. *Internationale Revue der Gesamten Hydrobiologie* 67: 361–386.

Heckman, C.W. 1982b: Pesticide effects on aquatic habitats. *Environmental Science and Technology* 16(1): 48A/57A.

Heckman, C.W. 1983: Comparative morphology of arthropod exterior surfaces with the capability of binding a film of air underwater. *Internationale Revue der Gesamten Hydrobiologie* 68: 715–736.

Heckman, C.W. 1984a: Erstfund von *Lemna turionifera* Landolt 1975, in Europa: Haseldorfer Marsch. *Kieler Notizen zur Pflanzenkunde in Schleswig-Holstein und Hamburg* 16: 1–3.

Heckman, C.W. 1984b: Effects of dike construction on the wetland ecosystem along the freshwater section of the Elbe Estuary. *Archiv für Hydrobiologie, Supplement* 61 (Untersuching Elbe Aestuar 5): 397–508.

Heckman, C.W. 1990: The fate of aquatic and wetland habitats in an industrially contaminated section of the Elbe Floodplain in Hamburg. *Archiv für Hydrobiologie/Supplement* 75 (Untersuch. Elbe Aestuar 6): 133–245.

Heckman, C.W. 1994: New limnological nomenclature to describe ecosystem structure in the tropical wet-and-dry climatic zone. *Archiv für Hydrobiologie* 130: 385–407.

Heckman, C.W. 1998: *The Pantanal of Poconé*. Kluwer, Dordrecht. 622 pp.

Heckman, C.W. 2001: *Encyclopedia of South American Aquatic Insects: Collembola*. Kluwer, Dordrecht. 408 pp.

Heckman, C.W. 2002: *Encyclopedia of South American Aquatic Insects: Ephemeroptera*. Kluwer, Dordrecht. 419 pp.

Heckman, C.W. 2003: *Encyclopedia of South American Aquatic Insects: Plecoptera*. Kluwer, Dordrecht. 329 pp.

Heckman, C.W. 2011: *Encyclopedia of South American Aquatic Insects: Heteroptera - Hemiptera*. Springer, Dordrecht. 679 pp.

Heckman, C.W. 2017: *Encyclopedia of South American Aquatic Insects: Neuroptera, Including Megaloptera*. Springer, Dordrecht. 621 pp.

Herring, J.L. 1961: The genus *Halobates* (Hemiptera: Gerridae). *Pacific Insects* 3(2-3): 223–305.

Hickin, N.E. 1967: *Caddis larvae*. Hutchinson, London.

Hirayama, T., T. Yoshida and O. Nagasaki. 2014: The life history and host-searching behavior of the aquatic parasitoid wasp *Apsilops japonicus* (Hymenoptera: Ichneumonidae), a parasitoid of the aquatic moth *Neoshoenobia testacealis* (Lepidoptera: Crambidae). *Journal of Natural History* 48(15-16): 959–967.

Holzenthal, R.W., R.J. Blahnik, A.L. Prather and K.L. Kjer. 2007: Order Trichoptera Kirby, 1813 (Insecta), Caddisflies. *Zootaxa* 1668: 639–698.

Holzenthal, R.W. and A.R. Calor. 2017: Catalog of the Neotropical Trichoptera (caddisflies). *Zookeys* 654: 1–566.

Holzenthal, R.W. and A.M. Oliveira Pes. 2004: A new genus of long-horned caddisfly from the Amazon basin (Trichoptera: Leptoceridae, Grumichellini). *Zootaxa* 621: 1–16.

Howes, R.E., K.E. Battle, K.N. Mendis, D.L. Smith, R.E. Cibulskis, J.K. Baird and S.I. Hay. 2016: Global epidemiology of *Plasmodium vivax*. *The American Journal of Tropical Medicine and Hygiene* 95(6): 15–34.

Huang, D. and L. Cheng. 2011: The flightless marine midge *Pontomyia* (Diptera: Chironomidae): ecology, distribution, and molecular phylogeny. *Zoological Journal of the Linnean Society* 162: 443–456.

Huang, D., P.S. Cranston and L. Cheng. 2014: A complete species phylogeny of the marine midge *Pontomyia* (Diptera: Chironomidae) reveals a cosmopolitan species and a new synonym. *Invertebrate Systematics* 28: 277–286.

Hull, F.M. 1942: The flies of the genus *Meromacrus*. *American Museum Novitates*, 1200: 1–12.

Ikawa, T., A. Okubo, H. Okabe and L. Cheng. 1998: Oceanic diffusion and the pelagic insects *Halobates* spp. (Heteroptera: Gerridae). *Marine Biology* 131: 195–201.

Ivanov, V.D. 2011: Caddisflies of Russia. Fauna and biodiversity. *Zoosymposia* 5: 171–209.

Kadavy, D.R., B. Plantz, C.A. Shaw, J. Myatt, T.A. Kokjohn and K.W. Nickerson. 1999: Microbiology of the oil fly, *Helaeomyia petrolei*. *Applied and Environmental Microbiology* 85(4): 1477–1482.

Kehl, S. 2005: *Thremma gallicum* McLachlan, 1880: Zur Biologie, Ökologie und Verbreitung einer faunistischen Besonderheit Deutschlands. *Entomologie Heute* 17: 13–26.

Kitching, R.L. 2004: *Food Webs and Container Habitats*. Cambridge University Press, Cambridge.

Knab, F. and J.R. Malloch. 1912: A borborid from an epiphytic bromeliad. *Entomological News* 23: 413–415.

Knutson, L.V. and J.-C. Vala. 2011: *Biology of snail-killing Sciomyzidae flies*. Cambridge University Press.

Krzeminska, E. and C.W. Young. 1992: Description of *Nothotrichocera chiloe* new species, first representative of the genus in South America (Diptera: Trichoceridae). *Journal of the Kansas Entomological Society* 65(2): 185–189.

Kula, R.R., J.J. Martinez and G.C. Walsh. 2009: Supplement to revision of New World *Chaenusa* Haliday sensu lato (Hymenoptera: Braconidae: Alyciinae). *Proceedings of the Entomological Society of Washington* 111(3): 641–655.

Kula, R.R. and G. Zolnerowich. 2008: Revision of the New World *Chaenusa* Haliday *sensu lato* (Hymenoptera: Braconidae: Alysiinae), with new species, synonymies, hosts, and distribution records. *Proceedings of the Entomological Society of Washington* 110(1): 1–60.

Lagrange, E.B. 1992: Revision de las especies argentinas del grupo *vinetorum* del genero *Palpada* Macquart 1834 (Diptera, Syrphidae). *Revista de la sociedad Entomologica Argentina* 50: 145–166.

Lanciotti, R.S., J.G. Lewis, D.J. Gubler and D.W. Trent. 1994: Molecular evolution and epidemiology of dengue-3 viruses. *Journal of General Virology* 75(1): 65–75.

Lane, J. 1953: *Neotropical Culicidae*. Universidade de São Paulo, São Paulo.

Lane, J. and C. d'Andretta. 1956: Brazilian Blephoceridae (Diptera, Nematocera). *Annals and Magazine of Natural History* (12)9: 177–204.

Langhi, D.M., Jr. and J.O. Bordin. 2006: Duffy blood group and malaria. *Hematology* 11(5): 389–398.

Laporta, G.Z., Y.–M. Linton, R.C. Wilkerson, E.S. Bergo, S.S. Nagaki, D.C. Sant'ana and M.A.M. Sallum. 2015: Malaria vectors in South America: present and future scenarios. *Parasites & Vectors* 8(426): 13 pp.

Leader, J.P. 1972: Osmoregulation in the larva of the marine caddisfly, *Philanisus plebeius* (Walk.) (Trichoptera). *Journal of Experimental Biology* 57: 821–838.

Lee, C.Y., D.G. Kim, M.J. Baek, L.J. Choe and Y.J. Bae. 2013: Life history and emergence pattern of *Cloeon dipterum* (Ephemeroptera: Baetidae) in Korea. *Environmental Entomology* 42(6): 1149–1156.

Lee, S.-M., S.-B. Lee, C.-H. Park and J. Choi. 2006: Expression of heat shock protein and hemoglobin genes in *Chironomus tentans* (Diptera, chironomidae [sic]) larvae exposed to various environmental pollutants: A potential biomarker of freshwater monitoring. *Chemosphere* 65: 1074–1081.

Leinaas, H.P. and L. Sømme. 1984: Adaptations in *Xenylla maritima* and *Anurophorus laricis* (Collembola) to lichen habitats on alpine rocks. *Oikos* 43: 197–206.

Lloyd, J.E. and L.A. Ballantine. 2003: Taxonomy and behavior of *Photuris trivittata* sp. n. (Coleoptera: Lampyridae: Photurinae); redescription of *Apisoma trilineata* (Say) comb. n. (Coleoptera: Lampyridae: Lampyrinae: Cratomorphini). *The Florida Entomologist* 86(4): 464–473.

Lock, K. and P.L.M. Goethals. 2011: Distribution and ecology of the mayflies (Ephemeroptera) of Flanders (Belgium). *Annales de Limnologie* 47(2): 159–165.

Lounibos, L.P., J.H. Frank, C.E. Machado-Allison, J.C. Navarro and P. Ocanto. 1987: Seasonality, abundance and invertebrate associates of *Leptagrion siqueirai* Santos in *Aechmea* bromeliads in Venezuelan rain forest (Zygoptera: Coenagrionidae). *Odonatologica* 16: 193–199.

Lounibos, L.P., G.F. O'Meara, N. Nishimura and R.L. Escher. 2003: Interactions with native mosquito larvae regulate the production of *Aedes albopictus* from bromeliads in Florida. *Ecological Entomology* 28: 551–558.

Lüderwaldt, H. 1915: Insekten und sonstigen Tierleben an brasilianischen Bromeliacean. *Zeitschrift für wissenschaftliche Insektenbiologie* 11: 78–84.

Lundblad, O. 1933: Zur Kenntnis der aquatilen und semi-aquatilen Hemipteren von Sumatra, Java, and Bali. *Archiv für Hydrobiologie* 74, Supplement 12: 1–194, 263–489.

Lutz, A. 1920: Dipteros da familia Blephoceridae, observados no Brasil. *Memórias Instituto Oswaldo Cruz*, Rio de Janeiro 12: 21–40: 16–40. pl. 1–7.

Madl, M. and C. van Achterberg. 2014: A catalogue of the Braconidae (Hymenoptera: Ichneumonoidea) of the Malagasy subregion. *Linzer Biological Beiträge* 46(1): 5–220.

Mantovani, A., N. Magalhães, M.L. Teixeira, G. Leitão, C.L. Staines and B. Resende. 2005: Contributions to systematics and biology of beetles. *Papers celebrating the 80th birthday of Igor Konstantinovich Lopatin.* Pensoft Series Faunistica No. 43. Pensoft; Sofia, Bulgaria, pp. 153–157.

Marchand, R.P., R. Culleton, Y. Meino, N.T. Quang and S. Nakazawa. 2011: Co-infections of *Plasmodium knowlesi, P. falciparum,* and *P. vivax* among Humans and *Anopheles dirus* Mosquitoes, Southern Vietnam. *PubMed.* 1232–9. doi 10.3201 eid 1707.101551.

Maros, A., A. Louveaux, M.H. Godfrey and M. Girondot. 2003: *Scapteriscus didactilus* (Orthoptera: Gryllotalpidae), predator of leatherback turtle eggs in French Guiana. *Marine Ecology Progress Series* 249: 289–296.

Marques, G.R.A.M., R.L.C. dos Santos and O.P. Forattini. 2001: *Aedes albopictus* em bromélias de ambiente antrópico no Estado do São Paulo, Brasil. *Revista de Saúde Pública* 35: 243–248.

Mathis, W.M. and L. Marinoni. 2011: A review of *Parydra* (Diptera: Ephydridae) from Brazil. *Zoologia* 28(4): 505–512.

Mathis, W.M. and K.W. Simpson. 1981: Studies of Ephydrinae (Diptera: Ephydridae), V: The genera *Cirrula* Cresson and *Dimecoenia* Cresson in North America. *Smithsonian Contributions to Zoology* 329: i-iii, 1–51.

Mathis, W.M., T. Zatwarnicki and M.G. Krivosheina. 1993: Studies of Gymnomyzinae (Diptera: Ephydridae), V: A revision of the shore-fly genus *Mosillus* Latreille. *Smithsonian Contributions to Zoology* 548. 38 pp.

Matsumura, S. 1915: On *Scatella calida* Matsumura in hot spring. *Conchu Sekai* 19: 223–224. (in Japanese).

Mbogo, C.N.M., R.W. Snow, C.P.M. Khamala, E.W. Kabiru, J.H. Ouma, J.I. Githure, K. Marsh and J.C. Beier. 1995: Relationships between *Plasmodium falciparum* transmission by vector populations and the incidence of severe disease at nine sites on the Kenyan coast. *American Journal of Tropical Medicine and Hygiene* 52(3): 201–206.

McAlpine, D.K. 1990: A new apterous micropezid fly (Diptera: Schizophora) from Western Australia. *Systematic Entomology* 15(1): 81–86.

McAlpine, D.K. 1991: Review of the Australian kelp flies Diptera: Coelopidae. *Systematic Entomology* 16(1): 29–84.

McAlpine, D.K. 1998: Review of the Australian stilt flies (Diptera: Micropezidae) with a phylogenetic review of the family. *Invertebrate Taxonomy* 12(1): 55–134.

McCreadie, J.W. and P.H. Adler. 2006: Ecoregions as predictors of lotic assemblages of blackflies (diptera [sic]: Simuliidae). *Ecography* 29: 603–613.

McLachlan, R. 1868: Occurrence in England of a terrestrial trichopterous insect, probably *Enoicyla pusilla* Burmeister. *Entomological Monthly Magazine* 5: 43–44.

McPherson, S. and D. Schnell. 2011: *Sarraceniaceae of North America.* Redfern Natural History Productions, Poole. 808 pp.

McPherson, S., A. Wistuba, A. Fleischmann and J. Nerz. 2011: *Sarraceniaceae of South America.* Redfern Natural History Productions, Poole. xii + 561 pp.

Mello, R.P. de and S.J. de Oliveira. 1992: Contribution to the knowledge of the genus *Dimecoenia* Cresson, 1916 – V. On new species from Rio de Janeiro, RJ (Diptera: Ephydridae). *Memórias Instituto Oswaldo Cruz,* Rio de Janeiro 87, Suppl. 1: 137–144.

Melnychuk, M.C. and D.S. Srivastava. 2002: Abundance and vertical distribution of a bromeliad-dwelling zygopteran larva, *Mecistogaster modesta,* in a Costa Rican rainforest (Odonata: Pseudostigmatidae). *International Journal of Odonatology* 5: 81–97.

Meneses, A.R., M.Y.O. Bevilaqua, N. Hamada and R.B. Querino. 2013: The aquatic habit and host plants of *Paracles klagesi* (Rothschild) (Lepidoptera, Erebidae, Arctiinae) in Brazil. *Revista Brasileira de Entomologia* 57(3): 350–352.

Mengual, X. and F.C. Thompson. 2008: A Taxonomic review of the *Palpada ruficeps* species group, with the description of a new flower fly from Colombia (Diptera: Syrphidae). *Zootaxa* 1741: 31–36.

Meyer, H. and S. Speth. 1995: Empidoidea (Diptera: Dolichopodidae, Empididae, Hybotidae) aus Emergenzfängen eines norddeutschen Tieflandbachsystems im Bereich von Osterau und Rodenbek (Holsteinische Vorgeest). *Mitteilungen der deutschen Gesellschaft für allgemeine und angewandte Entomologie* 9: 649–655.

Mezenev, N.P. 1990: [Blood-sucking biting midges (Ceratopogonidae) in north central Siberia]. *Parazitologiia* 24(1): 24–36. (in Russian).

Miller, A.C. 1971: Observations on the Chironomidae (Diptera) inhabiting the leaf axils of two species of Bromeliaceae on St. John, U. S. Virgin Islands. *Canadian Entomologist* 193: 391–308.

Miller, L.H., S.J. Mason, D.F. Clyde and M.H. McGinniss. 1976: The resistance factor to *Plasmodium vivax* in blacks – The Duffy-blood-group genotype FyFy. *The New England Journal of Medicine* 295: 302–304.

Minhas, M.S., C.L. Brockhouse and P.H. Adler. 2005: The black fly (Diptera: Simuliidae) fauna of Prince Edward Island, Canada. *Northeastern Naturalist* 12(1): 67–76.

Miranda, G.F.G. 2017: Identification key for the genera of Syrphidae (Diptera) from the Brazilian Amazon and new taxon records. *Acta Amazonica* 47(1): 32–62.

Miranda, G.F.G., A.D. Young, M.M. Locke, S.A. Marshall, J.H. Skevington and F.C. Thompson. 2013: Key to the genera of Nearctic Syrphidae. *Canadian Journal of Arthropod Identification* 23: 1–351.

Miyagi, I. 1977: *Fauna Japonica, Ephydridae (Insecta: Diptera).* Keigaku Publishing, Tokyo. 113 pp. + 49 plates.

Morales, M.N. and L. Marinoni. 2008: Immature stages and redescription of *Lejops barbiellinii* (Cersa) (Diptera: Syrphidae) found in bromeliads in Brazil. *Zootaxa* 1830: 37–46.

Morales, M.N. and L. Marinoni. 2009: Cladistic analysis and taxonomic revision of the *scutellaris* group of *Palpada* Macquart (Diptera: Syrphidae). *Invertebrate Systematics* 23: 301–347.

Moreira, F.F.F., J.R.I. Ribeiro and J.L. Nessimian. 2009: A synopsis of the *Hydrometra* (Hemiptera, Heteroptera, Hydrometradae) from the Amazon River floodplain, Brazil, with redescription of *Hydrometra argentina* Berg. *Revista Brasileira de Entomologia* 53(1): 69–73.

Mundahl, M.D. and A.D. Mundahl. 2015: Habitats and Foraging Movements of Larvae of *Molanna uniophila* Vorhies (Trichoptera: Molannidae) in Pratt Lake, Michigan, USA. *Journal of Insects* 2015(475467): 1–8.

Navarro, J.C., J. Liria, H. Piñango and R. Barrera. 2007: Biogeographic area relationships in Venezuela: a parsimony analysis of Culicidae-phytotelmata distribution in national parks. *Zootaxa* 1547: 1–19.

Neboiss, A. 1978: Atriplectididae, a new caddisfly family (Trichoptera: Atriplectididae). pp. 67–73. *In:* M.I. Crichton (ed.). *Proceedings of the 2nd Annual Symposium on Trichoptera.* Junk, The Hague.

Ngai, J.T. and D.S. Srivastava. 2006: Predators accelerate nutrient cycling in a bromeliad ecosystem. *Science* 314: 963.

Niesiołowski, S. 2006: Morphology, biology, phenology and occurrence of the genus *Empis* Linnaeus (Empididae, Diptera) in Poland. *Fragmenta Faunistica* 49(1): 1–39.

Ohba, N. 2004: Flash communication systems of Japanese fireflies. *Integrative and Comparative Biology* 44: 225–233.

Oláh, J. and K.A. Johanson. 2010: Contributions to the systematics of the genera *Dipseudopsis, Hyalopsyche* and *Pseudoneureclipsis* (Trichoptera: Dipseudopsidae), with descriptions of 19 new species from the Oriental Region. *Zootaxa* 2658: 1–37.

Parfin, S.I. and A.B. Gurney. 1956: The spongilla-flies, with special reference to those of the Western Hemisphere (Sisyridae, Neuroptera). *Proceedings of the United States National Museum* 105: 421–529 + 3 pl.

Paterson, C.G. 1971: Overwintering ecology of the aquatic fauna associated with the pitcher plant *Sarracenia purpurea* L. *Canadian Journal of Zoology* 49(11): 1455–1459.

Penny, N.D. 1981: Neuroptera of the Amazon Basin. Part 1. Sisyridae. *Acta Amazonica* 11(1): 157–169.

Perez Goodwyn, P.J. 2006: Taxonomic revision of the Lethocerinae Lauck & Menke (Heteroptera – Belostomatidae). *Stuttgarter Beiträge zur Naturkunde, Serie A (Biologie)* 695: 1–71.

Pes, A.M., N. Hamada, J.L. Nessimian and C.C. Soarez. 2013: Two new species of Xiphocentronidae (Trichoptera) and their bionomics in central Amazonia, Brazil. *Zootaxa* 3636: 561–574.

Pescador, M.L., A.K. Rasmussen and S.C. Harris. 1995: *Identification Manual for the Caddisfly (Trichoptera) Larvae of Florida.* Florida Department of Environmental Protection, Tallahassee. 132 pp + unnumbered Appendices.

Pichard, S. 1973: Contribution à l'étude de la biologie de *Podura aquatica* (Linné) Collembole. *Bulletin Biologique de la France et de la Belgique* 107(4): 291–299.

Polhemus, D.A. 2014: Two new Peruvian species of *Paravelia* (Heteroptera: Veliidae) from water-filled bamboo internodes, and distributional notes for other *Paravelia* species. *Tijdschrift voor Entomologie* 157(2-3): 151–162.

Polhemus, J.T. and D.A. Polhemus. 1991: A review of the veliid fauna of bromeliads with a key and description of a new species (Heteroptera Veliidae). *Journal of the New York Entomological Society* 99: 204–216.

Preston-Mafham, K.G. 1999: Courtship and mating in *Empis (Xanthemphis) trigramma* Meig., *E. tessellata* F. and *E. (Polyblepharis) opaca* F. (Diptera: Empididae) and the possible implications of "cheating" behavior. *Journal of Zoology* 247: 239–246.

Proches, S. and S. Ramdhani. 2012: The world's zoogeographical regions confirmed by cross-taxon analyses. *BioScience* 62: 260–270.

Rapoport, E.H. and L. Sanchez. 1963: On the epineuston or the superaquatic fauna. *Oikos* 14(1): 96–109.

Rattanarithikul, R., B.A. Harrison, R.E. Harbach, P. Panthusiri and R.E. Coleman. 2006: Illustrated keys to the mosquitoes of Thailand. IV. *Anopheles*. *Southeast Asian Journal of Medicine and Public Health* 37(Supplement 2): 1–128.

Rattanarithkul, R. and C.A. Green. 1986: Formal recognition of the *Anopheles maculatus* group (Diptera: Culicidae) occurring in Thailand, including the descriptions of two new species and a preliminary key to females. *Mosquito Systematics* 18(3-4): 246–278.

Redekop, P., E.M. Gross, A. Nuttens, D.E. Hofstra, J.S. Clayton and A. Hussner. 2016: *Hygraula nitens*, the only aquatic native caterpillar in New Zealand, prefers feeding on an alien submerged plant. *Hydrobiologia* https://doi.org/10.1007/s10750-016-2709-7.

Ree, H.-I. and J.H. Yum. 2006: Redescription of *Chironomus salinarius* (Diptera: Chironomidae), nuisance midges that emerged from brackish water Jin-Hae Man (Bay), Kyongsangnam-do, Korea. *The Korean Journal of Parasitology* 44(1): 63–66.

Rentz, D.C.F. 1995: The Changa mole cricket, *Scapteriscus didactylus* (Latreille), a New World pest established in Australia (Orthoptera: Gryllotalpidae). *Austral Entomology* 34(4): 303–306.

Resch, V.H. and R.T. Carde. 2003: *Encyclopedia of Insects*. Academic Press/Elsevier, San Diego.

Ricarte, A., M.A. Marcos-García, G. Hancock and G.E. Rotheray. 2012: Revision of the New World genus *Quichuana* Knab, 1913 (Diptera: Syrphidae), including descriptions of 24 new species. *Zoological Journal of the Linnean Society* 166: 72–131.

Richardson, B.A., C. Rogers and M.J. Richardson. 2000: Nutrients, diversity, and community structure of two phytotelm systems in a lower montane forest, Puerto Rico. *Ecological Entomology* 25: 348–356.

Riek, E.F. 1968: A new family of caddis-flies from Australia (Trichoptera: Tasimiidae). *Journal of the Australian Entomological Society* 7: 109–114.

Rotheray, G.E. 2003: The predatory larvae of two *Nepenthosyrphus* species living in pitcher plants (Diptera, Syrphidae). *Studia Dipterologica* 10(1): 219–226.

Rubtsov, I.A. 1990: Diptera. Blackflies (Simuliidae). Second Edition. *Fauna of the USSR* 6(6): i-xii, 1–1047. (English translation from Russian). Brill, New York.

Sakagami, F.S. 1960: Über *Scatella calida* Matsumura, eine heissen Quellen bewohnenden Fliegenart (Dipt. Ephydridae). *Journal of the Faculty of Science Hokkaido University Zoology* 14: 293–298.

Sandground, J.H. and C. Bonne. 1940: *Echinostoma lindoensis* n. sp., A new parasite of man in The Celebes with an account of its life history and epidemiology. *The American Journal of Tropical Medicine and Hygiene* 20(4): 511–535.

Santos, N.D. dos. 1966: Notas sôbre coenagriideos que se criam em bromélias. *Atas da Sociedade de Biologia do Rio de Janeiro* 10(3): 83–85.

Santos, N.D. dos. 1978: Descrição de *Leptagrion vriesianum* sp. n. cenagrionideo bromelicola (Odonata: Coenagrionidae). *Boletim do Museu Nacional, Nova Serie, Rio de Janeiro, Zoologia* 292: 1–6.

Schärer, M.T. and J.H. Epler. 2007: Long-range dispersal possibilities via sea turtle—a case for *Clunio* and *Pontomyia* (Diptera: Chironomidae) in Puerto Rico. *Entomological News* 118(3): 273–277.

Schelkle, B., R.S. Mohammed, M.P. Coogan, M. McMullan, E.L. Gillingham, C. van Oosterhout and J. Cable. 2012: Parasites pitched against nature: Pitch Lake water protects guppies (*Poecilia reticulata*) from microbial and gyrodactylid infections. *Parasitology* doi:10.1017/S0031182012001059.

Seblova, V., J. Sadlova, B. Vojkova, J. Votypka, S. Carpenter, P.A. Bates and P. Volk. 2015: The biting midge *Culicoides sonorensis* (Diptera: Ceratopogonidae) is capable of developing late stage infections of *Leishmania enriettii*. *Neglected Tropical Diseases* 9(9): e0004060 doi:10.1371/journal.pntd.0004060.

Shaw, M.R. and T. Huddleston. 1991: Classification and biology of braconid wasps (Hymenoptera: Braconidae). *Handbooks for the Identification of British Insects* 7(11): 1–126.

Sinka, M.E., Y. Rubio-Palis, S. Manguin, A.P. Patil, W.H. Temperley, P.W. Gething, T. van Boeckel, C.W. Kabariafi, R.E. Harbach and S.I. Hay. 2010: The dominant *Anopheles* vectors of human malaria in the Americas: occurrence data, distribution maps, and bionomic précis. *Parasites and Vectors* 3(72): 1–26.

Snider, R.J. and F.J. Calandrino. 1987: An annotated list and new species descriptions of Collembola found in the Project ELF study area of Michigan. *The Great Lakes Entomologist* 20(1): 1–20.

Soong, K., G.F. Chen and J.R. Cao. 1999: Life history studies of the flightless marine midges *Pontomyia* spp. (Diptera: Chironomidae). *Zoological Studies* 38: 466–473.

Srivastava, D.S., M.C. Melnychuk and J.T. Ngai. 2005: Landscape variation in the larval density of a bromeliad-dwelling zygopteran, *Mecistogaster modesta* (Odonata: Pseudostigmatidae). *International Journal of Odonatology* 8: 67–79.

Stoltze, M. 1989: The Afrotropical caddis family Pisuliidae, systematic, zoogeography, and biology (Trichoptera: Pisuliidae). *Steenstrupia* 15: 1–49.

Stone, A. 1952: The Simuliidae of Alaska (Diptera). *Proceedings of the Entomological Society of Washington* 54(2): 69–96.

Svensson, B.G., E. Petersson and M. Frisk. 1990: Nuptial gift size prolongs copulation duration in the dance fly *Empis borealis*. *Ecological Entomology* 15: 225–229.

Szerlip, S.L. 1975: Insect associates (Diptera: Chironomidae, Sphaeroceridae) of *Darlingtonia californica* (Sarraceniaceae) in California. *Pan-Pacific Entomologist* 51: 169–170.

Talveres, P.C., D. Alves and M. Shapouri. 2009: Structural changes in macroinvertebrate communities associated with reduction in the management of coastal saltpans. *Journal of Marine Biology* 2009: 1–13.

Thienemann, A. 1932: Die Tierwelt der *Nepenthes*-Kannen. *Archiv für Hydrobiologie,* Supplement 11: *Tropische Binnengewässer* 3: 1.54.

Thienemann, A. 1934: Die Tierwelt der tropischen Pflanzengewässer. *Archiv für Hydrobiologie* (Supplement 13): 1–91+5 pl.

Thienemann, A. 1954: *Chironomus:* Leben, Verbreitung und wirtschaftlichen Bedeutung von Chironomiden. *Die Binnengewässer* 20: 1–834.

Thompson, F.C. 1981: The flower flies of the West Indies (Diptera: Syrphidae). *Memoirs of the Entomological Society of Washington* 9: 1–200.

Thompson, F.C. 2007: A new Costa Rican flower fly (Diptera: Syrphidae) and a replacement name for a neotropical flower fly genus. *Studia Dipterologica* 14: 167–170.

Thorpe, W.H. 1934: The biology of the petroleum fly (*Psilopa petrolei* Coq.). *Transactions of the Entomology Society of London* 78: 331–344.

Tokunaga, M. 1932: Morphological and biological studies on a new marine chironomid fly, *Pontomyia pacifica,* from Japan. Part I. Morphology and taxonomy. *Memoirs of the College of Agriculture. Kyoto Imperial University* 19: 1–56.

Tokunaga, M. 1964: Diptera: Chironomidae. *Insects of Micronesia* 12: 485–628.

Triapitsyn, S.V., R.B. Querino and M.C.B. Feitosa. 2008: A new species of *Anagrus* (Hymenoptera, Mymaridae) from Amazonas, Brazil. *Neotropical Entomology* 37: 681–684.

Tsukamoto, M. 1989: Two new mosquito species from a pitcher plant of Mt. Kinabalu, Sabah, Malaysia, *Culex rajah* and *Toxorhynchites rajah* (Diptera: Culicidae). *Japanese Journal of Tropical Medicine and Hygiene* 17(3): 215–228.

Tuxen, S.L. 1944: The hot springs of Iceland, their animal communities and their zoogeographical significance. *The Zoology of Iceland,* Vol. 1, Part 11. Einar Munksgaard, Copenhagen.

Untawale, A.G., S. Wafar and T.G. Jagtap. 1982: Application of remote sensing techniques to study the distribution of mangroves along the estuaries of Goa. pp. 51–67. *In:* B. Gopal, R.E. Turner, R.G. Wetzel and D.F. Whigham, *Wetlands Ecology and Management.* Lucknow Publishing House, Lucknow, India.

Urbanič, G., J.A. Waringer and W. Graf. 2003: The larva and distribution of *Psychomyia klapaleki* Malicky, 1995 (Trichoptera: Psychomyiidae). *Lauterbornia* 46: 135–140.

Usinger, R.L. 1956: *Aquatic Insects of California.* University of California Press, Berkeley and Los Angeles. 508 pp.

van Achterberg, C. 1984: Essay on the phylogeny of Braconidae (Hymenoptera: Ichneumonoidea). *Entomologisk Tidskrift* 105: 41–58.99.

van der Goot, V.S. 1981: *De Zweefvliegen van Noordwest-Europa en Europees Rusland, in het bijzonder van de Benelux.* Koninklijke Nederlandse Natuurhistorische Vereniging, Amsterdam 496 pp.

Vilarino, A. and A.R. Calor. 2015: New species of Polycentropodidae (Trichoptera: Annulipalpia) from Northeastern Region, Brazil. *Zootaxa* 4007(1): 113–120.

Volonterio, O., R. Ponce de León, P. Convey and E. Krzeminska. 2013: First record of Trichoceridae (Diptera) in the maritime Antarctic. *Polar Ecology* 36: 1125–1131.

Vontas, J., E. Kioulos, N. Pavlidi, E. Morou, A. della Torre and H. Ranson. 2012: Insecticide resistance in the major dengue vectors *Aedes albopictus* and *Aedes aegypti. Pesticide Biochemistry and Physiology* 104(2): 126–131.

Vythalingam, I., W.M. NoorAzian, T.C. Huat, A.I. Jiram, Y.M. Yusri, A.H. Azahari, I. NorParina, A. NoorRain and S. LokmanHakim. 2008: *Plasmodium knowlesi* in humans, macaques, and mosquitoes in peninsula Malaysia. *Parasites & Vectors* 1(26): 10.1186/1756-3305/1-26.

Wagner, R. 1983: Aquatischen Empididen (Diptera) aus hessischen Mittelgebirgen und angrenzenden Gebieten. *Beiträgen zur Naturkunde Osthessen* 19: 135–146.

Wallace, J.B. and D. Malas. 1976: The fine structure of capture nets of larval Philopotamidae (Trichoptera), with special emphasis on *Dolophilodes distinctus. Canadian Journal of Zoology* 54(10): 1788–1802.

Wallace, I.D., B. Wallace and G.N. Philipson. 2003: *Keys to the case-bearing caddis larvae of Britain and Ireland.* Freshwater Biological Association.

Wesenberg-Lund, C. 1943: *Biologie der Süsswasserinsekten.* Verlag J. Springer, Berlin and Vienna. 682 pp.

West, R.C. 1956: Mangrove swamps of the Pacific coast of Colombia. *Annals of the Association of American Geographers* 46(1): 98–121.

Wheeler, W.M. 1942: Studies of Neotropical antplants and their ants. *Bulletin of the Museum of Comparative Zoology, Harvard* 90: 1–262 + pl. 1–56.

Wiggins, G.B. 2015: *Caddisflies: The Underwater Architects.* University of Toronto Press. 292 pp.

Williams, D.D. 2006: The Biology of Temporary Waters. Oxford University Press, Oxford. 347 pp.

Williams, W.D. 1998: Salinity as a determinant of the structure of biological communities in the salt lakes. *Hydrobiologia* 381: 191–201.

Winterbourne, M.J. and N.H. Anderson. 1980: The life history of *Philanisus plebeius* Walker (Trichoptera: Chathamiidae), a caddisfly whose eggs were found in a starfish. *Ecological Entomology* 5(3): 293–304.

Wise, K.A.J. 1965: An annotated list of the aquatic and semi-aquatic insects of New Zealand. *Pacific Insects* 7(2): 191–216.

Yankoviak, S.P. 1999: Effects of leaf litter species on macroinvertebrate community properties on mosquito yield in Neotropical tree hole microcosms. *Oecologia* 120: 147–155.

Younes, A.A., E.-S. Hanaa, G. Fathia and M. Marwa. 2016: *Sphaerodema urinator* Duforas (Hemiptera: Belostomatidae) as a predator of *Fasciola* intermediate Host, *Lymnaea natalensis* Krauss. *Egyptian Journal of Biological Pest Control* 26(2): 191–196.

Young, J.H. and D. Merritt. 2003: The ultrastructure and function of the silk-producing basitarsus in the Hilarini (Diptera: Empididae). *Arthropod Structure and Development* 32: 157–165.

Zaragoza, C.S. 1974: Coleópteros de algunas bromelias epífitas y doce neuvos registros de especies para la fauna Mexicana. *Anales del Instituto de Biología, Universidad Nacional Autónoma de México, Serie Zoología* 45: 111–118.

Zhao, W., M.-P. Zheng, X.Z. Hsu, X.-F. Liu, G.I. Guo and X.H. He. 2005: Biological and ecological features in saline lakes of northern Tibet, China. *Hydrobiologia* 541(1): 189–203.

Zillikens, A., I. de Souza Gorayeb, J. Steiner and C.B. Marcondes. 2005: Aquatic larvae and pupae of *Fidena (Laphriomyia) rufopilosa* (Ricardo) (Diptera: Tabanidae) developing in bromeliad phytotelmata in the Atlantic Forest of southern Brazil. *Journal of the Kansas Entomological Society* 78(4): 381–386.

Index

A

Abedus 101
aboriginis, Aedes 171
Acanthus 212
Acari 295
Acrididae 72, 133
Acrostichum 212
acuminata, Paulinia 69, 85, 88
acuta, Physa 249
Adephaga 10, 14–16, 78, 104, 120
Aechmea 156, 292, 293, 294. 295
Aedes 137, 138, 139, 154, 156, 171, 258, 259, 261, 262, 263, 264, 267, 268, 269, 270, 280
Aegialomyia 217
Aegicerus 212
aegypti, Aedes 137–139, 258, 259, 262, 263, 264, 267, 268, 269, 270, 280
Aelosomatidae 295
aemula, Palpada 195
aequinoctialis, Lemna 23, 87, 89
aerenchyme 68, 191
Aeschnidae 154
africana, Azolla 88
agallocha, Excoecaria 212
Agaonidae 245
Agrionidae 120
Agriotrypidae 243
Agromyzidae 148
agrorum, Palpada 195
Agyrtidae 215
alaskaensis, Culiseta 169
alatavicus, Culicoides 169
alba, Avicennia 212
alba, Nymphaea 56, 74, 86, 87, 89
alba, *Sonneratia* 212
albiguttalis, *Sameodes* 116
albimanus, Anopheles 273
albitarsus, Anopheles 273, 274, 283
albopictus, Aedes 156, 262, 264, 270
Alfavirus 264
aloides, Stratiotes 74. 88, 116
alpestre, Prosimulium 172
amazonensis, Anagrus 246
amazonica, Victoria 56
amazonum, Nymphaea 87, 89

ambulans, Badisis 165
americana, Periplaneta 129
amplexicaulis, Hymenachne 118
Ampulariidae 321
Anabaena 87
Anagrus 246
Anatopynia 154
angularis, Culicoides 154
angustifolia, Typha 191
Anisoptera 9, 21, 78, 120, 284
annulata, Trichocera 174, 175
annulirostris, Culex 280
Anodocheilus 84
Anomalopsychidae 112
Anopheles 137–139, 154, 257, 258, 272, 273, 274, 280, 283
Anophelinae 272
Anopleura 223
antarctica, Nothotrichocera 176
Antipodoeciidae 212
antipodum, Paracladura 176
Anurida 214
Anurophorus 145, 178
Apataniidae 112
Aphelinidae 245
Aphelocheiridae 9
Apiaciae 155
apiculata, Rhizophora 212
Apocrita 78
appasionata, Papaipema 166
Apsilops 247
Apterygota 2, 3, 13, 145
aquasalis, Anopheles 280
aquatica, Ipomoea 76
aquatica, Podura 4, 7, 60, 61, 62, 74
aquaticus, Sminthurides 62, 73, 74
aquaticum, Cornops 72, 73, 133
arabiensis, Anopheles 280
Araceae 105, 116
arbustorum, Eristalis 192
arctica, Tetracanthella 214
arcticum, Simulium 172
arcticus, Onychiurus 214
Arctiidae 117
Arctiinae 72, 117

argentina, Hydrometra 57
Argizala 132
arrhiza, Wolffia 55
Artemia 223, 225
Arthropleona 4, 9, 62, 178
Arthropoda 2, 3
asiaticus, Mosillus 216, 217
Asterinidae 218
atriceps, Culex 154
atrata, Microvelia 154
Atriplectididae 112
atroparvus, Anopheles 273
aucklandica, Nothotrichocera 176
aurantiaca, Paracles 118
aureum, Acrostichum 212
aureum, Simulium 172
auricularia, Lymnaea 249, 250
auriculata, Salvinia 69, 85, 87, 88
australis, Phragmites 68, 191
austrina, Cirrula 228
Austroniidae 245
Avicennia 212
Azolla 55, 74, 87, 88, 118
azollae, Anabaena 87
azollae, Paracles 118
azurea, Eichhornia 72, 133

B

Bacteria 295
Badisis 165
Baetidae 181
baffinense, Simulium 172
bancroftii, Wucheria 279, 280
Barbarochthonidae 112
batatas, Ipomoea 76
Bellamya 250
bellator, Anopheles 280
Belostoma 101
Belostomatidae 9, 17, 18, 22, 26, 91, 100, 102,
        103, 249, 321
Belostomatinae 101
Bembidion 65
Benacus 101
Beraeidae 112
berteroniana, Catopsis 150
bicalcarata, Nepenthes 161
bidentatus, Mosillus 216, 217
bigoti, Polybiomyia 195
Billbergia 150
bimaculatus, Stenus 65
bitaeniorhynchus, Culex 267, 280
Blatella 129
Blepharoceridae 120, 125, 169
bonariensis, *Nausigaster* 195
boops, Stenus 65
Brachycentridae 111
Brachycera 10, 69, 96, 97, 196
Brachyopini 102
Braconidae 10, 78, 105, 240, 241, 243, 245
bracteatus, Mosillus 216, 217
brevipenne, Cornops 133

brevistyla, Rhizophora 212
bridgesi, Pomacea 249
Bromeliaceae 147, 149, 157, 295
bromeliadicola, Mongoma 156
bromeliads 136, 146, 148, 150, 153, 155, 156,
        158, 292, 293, 294, 295, 324
Bromeliagrion 155, 156
bromeliarum, Limosina 156
bromeliicola, Erythrodiplax 156
bruesi, Ephydra 288
Brugia 279
Bruguiera 212
brunipennis, Erpelus 148

C

Cabomba 118
Cabombaceae 118
caerulea, Nymphaea 56
Calamoceratidae 112
calida, Scatella, 287
californica, Darlingtonia 162
Calliaspis 295
Calocidae 112
Calopterygidae 25, 120, 121
campestris, Rhingia 192
Camponotus 293
canadensis, Elodea 117, 118
candidoscutellum, Aedes 154
Canthydrus 85
Carabidae 15, 22, 65, 147
caseolaris, Sonneratia 212
castanea, Erythrodiplax 281, 282, 284
Cataclysta 23, 58, 74, 97, 115, 193
cataphylla, Aedes 171
Catopsis 151
cavicola, Metriocnemus 166
Cecidomyiidae 68, 295
Cepa 195
Cephalotus 165
Ceratophyllum 117
Ceratopogonidae 11, 20, 21, 125, 148, 151, 154,
        164, 165, 169, 170, 295
Ceriops 212
Chagas disease 258
Chalcididae 243, 245
Chalcidoidea 10, 245, 246, 247, 319
Chaoboridae 21, 25, 66, 96, 156
Chaoborus 98
Chathamiidae 112, 218
chikungunya fever 259, 264, 265
chiopterus, Culicoides 169
Chironomidae 20, 88, 93, 95–97, 112–114, 125,
        140, 148, 154, 155, 161, 164, 166, 180, 182,
        183, 185, 197, 199, 204, 219, 224, 225, 229,
        284, 286, 290, 295, 316
Chironominae 95, 114, 107, 183
Chironomus 289
Chloroperlidae 120
Chromatium 189, 192, 193
Chrysogaster 102, 192
Chrysomelidae 15, 295

Chrysopidae 77
cicindeloides, Stenus 65
cinereus, Aedes 171
cingulata, Nothotrichocera 176
Cirrula 228
Cladocera 88
Climacia 235, 238
Cloeon 95, 181
Clunio 184, 219, 227
Cnephia, 172
cobrerae, Eryngium 155
Coccinellidae 82
Cochliostema 156
Coelopa 216
Coelopidae 216
Coenagrionidae 121, 156, 295
Coleoptera 8, 10, 14, 18, 22, 56, 64, 73, 76, 78,
    82, 85, 98, 104, 120, 147, 156, 212, 213,
    215, 229, 295, 310
collaris, Tropisternus 84, 85
Collembola 2–9, 26, 56, 57, 58, 60, 61, 73, 74,
    89, 144–148, 178, 193, 199, 213, 214, 309,
    311, 315
Colocassia 76
communis, Aedes 171, 172,
conjugata, Rhizophora 212
Conocarpus 212
Conoesucidae 112
convexiusculus, Gyraulis 250, 251
cooki, Aedes 280
Coquillettidia 280
Corbicula 250
corbis, Simulium 172
cordata, Pontederia 133
Corethrella 154, 156
coriacea, Dermochelys 130
Corixidae 9, 16–18, 63, 91, 98, 229
corniculata, Aegiceras 212
Cornops 72, 73, 133
Corydalidae 9, 14, 96, 120, 121
corymbosum, Vaccinium 143
cottoni, Pontomyia 204,
coustani, Anopheles 272
cracens, Anopheles 273
Crambidae 23, 115, 116, 193, 247
crassipes, Eichhornia 70–72, 81, 89, 116, 118,
    133
crispus, Potamogeton 117
Crustacea 3, 223, 225
Culex 139, 154, 171, 179, 186, 261, 266, 267,
    280, 283
Culicidae 11, 20, 21, 66, 88, 91, 96, 97, 154, 156,
    164, 165, 166, 169, 170, 185, 196, 229, 257,
    259, 277, 278, 295
cucullata, Salvinia 87, 88
Culicoides 154, 169, 278
Culiseta 171
Curculionidae 15, 22, 58
Cyanobactera 87, 288
Cyclorrhapha 10, 191, 194, 196, 198, 227
Cyperaceae 86, 191
Cyperus 68, 305

D

darlingi, Anopheles 273, 280
Darlingtonia 162, 166
Dasyhelea 154
DDT 262, 264, 268, 269, 270
deaneorum, Anopheles 273
debilis, Hydrocanthus 85
decemarticulatus, Helodon 172
decorum, Simulium 172
demersum, Ceratophyllum 117
dengue fever 259, 262, 263
densa, Egeria 117
Dermochelys 130
Derris 212
destillatoria, Nepenthes 166
deyrolli, Kirkaldyia 101
diantaeus, Aedes 171
Diapriidae 245
dichopticus, Gymnopais 172
Dictyoptera 212
dicum, Prosimulium 172
dichlamydea, Aechmea 156
didactylus, Scapteriscus 129, 131
Dilaridae 77
Dimecoenia 217, 218, 227, 228
Dionaea 143
Dioscorea 76
Diplonychus 101
Diplura 3, 14
Dipseudopsidae 111
Diptera 8, 10, 14, 17, 20–22, 24, 36, 56, 65, 69,
    73, 76, 82, 88, 96, 97, 106, 107, 112, 120,
    144, 147, 148, 155, 162, 163–165, 170, 180,
    182, 185, 189, 191, 194, 196, 197, 205, 212,
    213, 215, 219, 223, 225, 229, 233, 295, 307,
    310, 311, 314, 322, 323
dipterum, Cloeon 95, 181
dirus, Anopheles 273
Dixella 156
Dixidae 21, 66, 96, 97, 148, 156, 196, 197
Dolichopodidae 58, 65, 229, 289, 317
domestica, Seira 6
downsi, Corethrella 156
Dryopidae 8, 10, 15, 19, 64, 96, 120
dufouri , Dasyhelea 154
Dytiscidae 10, 14, 15, 16, 78, 84, 91, 100, 104,
    105, 120, 122, 215, 229, 295

E

Echinochloa 117
Echinostoma 250, 251
Ecnomidae 111
Ectoprocta 236
edwardsii, Metriocnemus 166
Egeria 117
Eichhornia 70–72, 81, 89, 116, 118, 133
Elasmidae 245
elephantiasis 280
Elmidae 15, 19, 64, 120
Elodea 117, 118
Elophila 74

Embioptera 212
emergens, Cnephia, 172
Empididae 21, 22, 58, 59, 65, 96, 97, 165, 196, 321
Empidoidea 321
Encyrtidae 245
Endothenia 166
Entomology 50, 115, 298, 321
Entomobrya 146
Ephemeroptera 8, 13, 20, 95, 96, 180, 181, 185
Ephydra 288
Ephydridae 22, 59, 65, 96, 114, 165, 196, 197, 216, 218, 224, 230, 233, 286, 288–290, 317, 318
epineuston 303
Erebidae 72, 117
erectus, Conocarpus 212
eremites, Cnephia, 172
Ericaceae 143
Eristalinae 18, 26, 101, 189, 193, 194, 195
Eristalini 102
Eristalinus 189, 192, 195
Eristalis 189, 192
Erpelus 148
Eryngium 155
Erythrodiplax 156, 281, 282, 284
esculenta, Colocassia 76
Eucharitidae 245
Eulophidae 245
Eupelmidae 245
Eurotomidae 245
Excoecaria 212
excrucians, Aedes 171
Exyra 166

F

falciparum, Plasmodium 270, 271, 272, 273, 275
farauti, Anopheles 273, 280
fasciata, Palpada 195
fax, Exyra 166
femoratus, Camponotus 293
fernandezianum, Bromeliagrion 156
Fidena 156
Filaria 279
Filariasis 279
filiculoides, Azolla 87
fimetaria, Folsomia 62
fitchi, Aedes 171
flava, Sarracenia 166
flavescens, Aedes 171
Flaviviridae 262, 265
Flavivirus 262, 265
flavivirus 261
flexuosa, Utricularia 87, 88
florea, Myathropa 154, 192
fluviatilis, Mayaca 118
fluviatilis, Tonina 118
foliosa, Utricularia 87, 88
follicularis, Cephalotus 165
freeborni, Anopheles 273
frenatum. Cornops 133
fulvum, Prosimulium 172

funestus, Anopheles 272, 280
Fungi 295
fuscata, Macaca 287
fuscifemur, Dimecoenia 227
fuscipennis, Culicoides 169

G

gambiae, Anopheles 272, 273, 280
Gelastocoridae 9, 17, 18
geniculatus, Aedes 154
germanica, Blatella 129
germanus, Halobates 200
Gerridae 9, 16, 24, 40, 58, 63, 64, 96, 200, 202, 210, 316
Gerris 63, 99, 201
Gerromorpha 9, 16
gibba, Lemna 74
gigantea, Cirrula 228
glaberrima, Oryza 77
Glossosomatidae 111
Glyceria 65, 175, 191, 193
goeldii, Pachycondyla 293
Goeridae 111
Gomphidae 121
gouldingi, Simulium 172
grammicus, Suphisellus 85
grandiglumis, Oryza 118
griseola, Hydrellia 65
grisescens, Culicoides 169
griseus, Benacus 101
groenlandicus, Onychiurus 214
Gryllidae 132
Gryllotalpidae 129
gutae, Thalassomyia 219
Gymnopais 172
gymnorhiza, Bruguiera 212
Gynacantha 154
Gyrinidae 10, 15, 16, 18, 58, 59, 60, 64, 78
Gyrinus 58
Gyraulis 250, 251

H

hackeri, Anopheles 273
Haemagogus 261
Haliplidae 10, 15, 16, 78, 91, 100
Halobates 200, 201, 202, 203, 210, 212, 219
hancocki, Anopheles 272
hebesana. Endothenia 166
Hebridae 9, 16, 24, 58, 63
Helaeomyia 230, 231, 233
Heliamphora 161, 162
Helicophidae 111
Helicopsychidae 111
Helodidae 15, 16, 19, 96
Helodon, 172
Helophilus 192
Heloridae 245
helveticus, Culicoides 169
Hematobatidae 9
Hemerobiidae 77

Hemimetabola 9
Hemiptera 9, 14, 16, 17, 24, 57, 58, 62, 78, 85, 98, 100, 229, 246, 295, 299, 321
hemorrhagic fever 267
Hermatobatridae 16
Heteroceridae 15
heterophylla, Derris 212
Heteroptera 9, 14, 17, 56, 57, 58, 62, 63, 78, 100, 103, 202, 212
Hexapoda 2, 3, 24–26, 180, 295, 298
hiemalis, Isotoma 178
hiemalis, Trichocera 175
Hilara 197
Holometabola 9
holopticus, Gymnopais 172
Homoptera 14, 76, 82
horticola, Eristalis 192
Horvathinia 101
Horvathiniinae 101
humilis, Parydra 217
hunteri, Simulium 172
hybridus, Helophilus 192
Hydraenidae 10, 120
Hydrellia 65
Hydrilla 117
Hydrobiosidae 111
Hydrocanthus 85
Hydrocharitaceae 118
Hydrocleys 72
Hydrocyrtus 101
Hydrometra 57
Hydrometridae 9, 16, 24, 57, 57, 58, 63
Hydrophilidae 10, 11, 15, 16, 84, 91, 104, 155, 215, 229, 295
Hydropsychidae 111
Hydroptilidae 111
Hydrosalpingidae 112
Hydroscaphidae 15
Hydrotrephidae 9
Hygraula 117
Hymenachne 118
Hymenoptera 10, 13, 22, 66, 78, 94, 96, 105, 212, 213, 239, 240, 242, 243, 245, 248, 319, 320, 322
Hypogastrura 178, 214
Hypogastruridae 178

I

Ichneumonidae 78, 240, 241, 243, 425
Ichneumonoidea 10, 22, 96, 105 240, 241, 244, 247, 319
ilocanum, Echinostoma 250
impatiens, Azolla 88
Impatiens, Culiseta 171
impiger, Aedes 171
incana, Nymphaea 86
infuscata, Corethrella 154, 156
Insecta 2, 3, 8, 9, 26, 28, 62, 89, 180, 199, 213, 310
insidens, Culiseta 171
interpunctus, Helophilus 192
intricarius, Eristalis 192

intrudens, Aedes 171
Ipomoea 76
Isoptera 212
Isotoma 61, 74, 178
Isotomidae 61, 62
Isotomurus 61, 62, 74

J

Japanese encephalitis 259, 266
japanicus, Apsilops 247
japonica, Stratiomyia 286
javanica, Vivipara 250
juxtamansonia, Coquillettidia 280

K

Kandelia 212
Kirkaldyia 101
klagesi, Paracles 118
knabi, Metriocnemus 166
knowlesi, Plasmodium 271
kochi, Aedes 280
Kokiriidae 111
koliensis, Anopheles 280

L

laboulbeni, Palustra 118
laboulbeni, Paracles 118
Laccophilidae 10, 104
Laccophilinae 10, 104
Lactuca 118
lacustris, Gerris 99
laevigatus, Hydrocanthus 85
laevigatus, Megadytes 84
Lagerocyphon 117
Laguncularia 212
Lampyridae 15, 23, 80, 83, 249, 250
lanceolata, Pontederia 133
laricis, Anurophorus 145, 178
laselva, Erythrodiplax 156
latifolia, Typha 191
latifrons, Stenus 65
latipes, Simulium 172
leii, Luceola 250
Leishmania 277, 278
Lemna 23, 55, 58, 74, 87, 88, 89, 97, 115, 193
Lemnaceae 88
lemnae, Tanysphaerus 58, 74
lemnata, Cataclysta 23, 74, 97, 115, 193
Lentibulariacea 118, 143
Lepidoptera 10, 74, 91, 114, 115, 116, 166, 212, 246, 296, 307
Lepidostomatidae 111
Leptagrion 155, 156
Leptopodomorpha 9
Leptoceridae 111
Lethocerinae 101
Lethocerus 101
leucocephalus, Aedes 264
Leucospidae 245
Libellulidae 121, 156

Limnephilidae 110, 111
Limnichidae 15
Limnocentropodidae 112
Limnocharitaceae 72
Limnogeton 101
Limnogonus 201
Limoniidae 22, 96, 97, 156, 196, 197, 229, 295
Limosina 156
lindoensis, Corbicula 250
lindoensis, Echinostoma 250, 251
linearis, Xantholinus 65
lineatus, Helophilus 192
longicauda, Quichuana 195
loutoni, Paravelia 154
Luciola 250, 251
lutea, Nuphar 147
Lutzomyia 278
luteola, Lymnaea 249
Lycosidae 56
Lymnaea 249, 250

M

Macaca 287
macquarti, Chrysogaster 192
macrocarpon, Vaccinium 143
Macroveliidae 9, 16, 58, 63
maculatus, Anopheles 273
maculipennis, Trichocera 174
major, Lagerocyphon 117
major, Trichocera 175
malaria 257, 258, 260, 270, 271, 272, 273, 274,
    275, 276
malaria quartana 270, 271
malaria quotidiana 271
malaria tertiana 270, 271, 276
malaria tropica 270, 271
malariae, Plasmodium 270, 271, 275
malayanum, Echinostoma 250, 251
malayanus, Nepenthosyrphus 166
malayi, Brugia 179
malmgreni, Sminthurides 63
malyschevi, Simulium 172
manicata, Hilara 197
Manomachidae 245
Mansonia 261, 280
marajoara, Anopheles 273
margarita, Cepa 195
mariae, Toxorhynchites 156
Marilia 72, 73, 134
marina, Avicennia 212
Marine Biology 48
marinus, Clunio 184, 219, 227
maritima, Anurida 214
maritima, Xenylla 145, 178
martensi, Aechmea 292, 293, 294, 295
maxi, Culex 283
maxima, Glyceria 65, 175, 191, 193
Mayaca 118
Mayacaceae 118
Mecistogaster 155, 156

Mecoptera 13
Megadytes 84
Megaloptera 9, 13, 14, 20, 96, 103, 120, 121
Melanogaster 102
melas, Anopheles 280
Melosira 118
membranalis, Gynacantha 154
meningial encephalitis 267
meridionale, Simulium 172
Meromacris 195
merus, Anopheles 280
Mesoveliidae 9, 16, 58, 63
Metacnephia, 172
meticulosa, Neoascia 192
Metriocnemus 154, 166
mexicana, Azolla 87
Meyenia 236
micans, Halobates 200
Micropyzidae 165
Microvelia 154, 156
milesia, Meromacris 195
minor, Lemna 23, 74, 87, 89, 97, 115 193
minus, Cnephia 172
minutus, Suphus 85
mitchellii, Wyeomyia 151
modesta, Mecistogaster 156
Molannidae 112
mollis, Culex 154
Mongoma 156
morsitans, Culiseta 171
Mosillus 216, 217
Motschulskium 215
moucheti, Anopheles 272
mucronata, Rhizophora 212
multiplicalis, Samea 116
murmanum, Simulium 172
Muscidae 194, 196, 198
muscipula, Dionaea 143
mutata, Cnephia, 172
Myathropa 154, 192
Mycetophilidae 148
myersi, Paravelia 154
Mymaridae 245, 246
Mymarommatidae 245
Mymarommatoidea 245
Myriophyllum 117

N

Naididae 295
nallyi, Aechmea 292
natans, Pontomyia 182, 204, 205, 219
natans, Salvinia 69
natans, Trapa 76, 81
Naucoridae 9, 16, 17, 91, 103, 104
Nausigaster 195
nearcticus, Aedes 171
Nelumbo 30
Nematocera 10, 58, 96, 107, 112, 148, 155, 163
    196, 278, 279
Nematoda 278, 279, 280

nemorum, Eristalis 192
Neoascia 192, 194
Neocurtilla 130
Neoshoenobia 247
Nepenthaceae 161, 162, 165, 166
Nepenthes 160, 161, 162, 166
Nepenthosyrphus 166
Nepidae 9, 17, 18, 22, 91, 101, 103
Nepomorpha 9, 16, 17, 18
Neuroptera 9, 14, 20, 36, 56, 77, 96, 103, 120,
    121, 212, 235, 237, 319
Neuropterida 235
nigricans, Aedes 171
nigricoxum, Simulium 172
nili, Anopheles 272
nitens, Hygraula 117
nitida, Avicennia 212
nivalis, Entomobrya 146
Noctuidae 166
Noteridae 10, 14, 85, 104
Noterinae 10, 14, 85, 91, 100, 104
Notonectidae 9, 16, 17, 18, 91, 99, 100, 229
Nothotrichocera 176
nucifera, Nelumbo 30
Nuphar 147
Nymphaea 56, 74, 86, 87, 89
Nymphaeaceae 118
nymphaeata, *Elophila* 74
nymphoides, Hydrocleys 72

O

obsoletus, Culicoides 169
oceana, Pontomyia 204
ochropus, Parydra 217
Ochteridae 9, 17
Odonata 9, 13, 20, 21, 36, 56, 78, 103, 120, 121,
    151, 212, 246, 295
Odontoceridae 112
Oeconescidae 111
Oedogonium 118
officinalis, Avicennia 212
olivacea, Utricularia 118
Onchocerca 278, 279
Onchocerciasis 279
Onchocercidae 278
Oncorhynchus 124
Onychiurus 214
onychodacylum, Prosimulium 172
Opiinae 250
Ormyridae 245
orsinum, Prosimulium 172
Orthocladiinae 114
Orthonevra 102
Orthoptera 9, 72, 76, 128, 129, 133, 134, 212,
    308, 309
Oryza 75, 77, 118
Osmylidae 9, 14, 235
oudemansi, Nepenthosyrphus 166
ovale, Plasmodium 270, 271, 273, 276
ovatus, Canthydrus 85
oxydentalis, Anopheles 171

P

pachypus, Meromacris 195
pacifica, Pontomyia 204
pallidoventer, Trichoprosopon 154
Palpada 195
Palustra 118
palustris, Isotomurus 61, 62, 74
palustris, Paracles 72, 118
Papaipema 166
papyrus, Cyperus 68, 305
Paracladura 176
Paracles 72, 117, 118
Paracoenia 288
paraguayense, Cornops 133
Paraneonotus 131
Paraphrynoveliidae 9, 16
Parapoynx 74, 88, 116, 117
parasitoid 10, 13, 94, 105, 239–249, 320
Paravelia 154, 156
parviflora, Bruguiera 212
Parydra 217
Paspalum 118
Patiriella 218
paucicostata, Lemna 87, 89
Paulinia 69, 85, 88
Pauliniidae 69, 72, 85
Peliciera 212
Pelecinidae 245
pelocoroides, Horvathinia 101
pendulus, Helophilus 192
pennipes, Anatopynia 154
Peradeniidae 245
Perilampidae 245
Perilestidae 25
Periplaneta 129
Perlidae 120
Perlodidae 120
pertinax, Eristalis 192
petrolei, Helaeomyia 230–233
petrolei, Psilopa 230, 233
Petrothrincidae 112
pharaoensis, Anopheles 272
pheromones 185
Philanisus 218
Philopotamidae 111
Philorheithridae 112
Phlebotoma 278
Phlebovirus 267
Photinus 80, 81, 249, 250
Photuris 80, 81, 83
phragmicola, Stenodiplosis 68
Phragmites 68, 191
Phryganeidae 111
Phryganopsychidae 111
Physa 249
Phytotelm 136, 141, 143, 146, 147, 149, 150, 151,
    153, 156, 158, 160, 259, 294, 295, 310, 312
pictus, Thinopinus 215
Pinguicula 143
pinnata, Azolla 74, 87, 88
pionips, Aedes 171

pipiens, Culex 139, 179, 186, 266, 267, 280
pipiens, Syritta 192
Pirata 56
piraticus, Pirata 56
Pisauridae 56
Pistia 105, 116
Pisuliidae 112
Plannipennia 9, 96, 235
Plasmodium 137, 139, 258, 270, 271, 272, 273, 274, 275, 276. 277, 280, 283
plastron 19, 201, 306
plebeius, Philanisus 218
Plecoptera 9, 13, 20, 25, 36, 96
Plectotarsidae 111, 120
Pleidae 9, 17, 63
pleuralis, Helodon, 172
plumbeus, Anopheles 154
podagrica, Neoascia 192, 194
Podura 4, 7, 60, 61, 62, 74
Poduridae 60
Poecilia 232
Poissonia 101
Polybiomyia 195
Polycentropodidae 111
polynesiensis, Aedes 154, 280
Polyphaga 8, 10, 14, 15, 16, 104, 120
polyrrhiza, Spirodela 55, 74, 89
Pomacea 249
Pontederia 72, 133
Pontomyia 182, 204, 205, 219
Potamocoridae 8
Potamogeton 117
Prionocyphon 154
Proctotrupidae 243, 245
Prosimulium 172
Protoneuridae 25, 120
Protosialis 103
Protozoa 295
Protura 3
psammophila, Aegialomyia 217
psammophila, Stenotabanus 217
Psephenidae 15
pseudoscutellaris, Aedes 280
Pseudostigmatidae 156
pseudoterrestrium 68, 69, 73, 80, 82, 133, 304, 305
pseudotitillans, Mansonia 280
Psilopa 230
Psilopidae 233
Psilopinae 233
Psorophora 281, 282, 283
Psychidae 166
Psychodidae 21, 96, 140, 196, 278
Psychomyiidae 111
Pterygota 13
Pteromalidae 245
Ptiliidae 215
Ptilocoloepidae 111
pullatus, Aedes 169
punctatus, Culicoides 169
punctodes, Aedes 171
punctor, Aedes 171
punctulatus, Anopheles 280

purificata, Bellamya 250
purpurea, Sarracenia 166
pusilla, Eniocyla 109
pusillus, Toxorhynchites 154
Pyralidae 116
Pyraloidea 115
pyramidalis, Billbergia 150
Pyraustinae 116

Q

quadrimacula Anopheles 273
quadrioculatus, Folsomia 214
Quichuana 195
quinquefasciatus, Culex 280

R

racemosa, Laguncularia 212
rafflesiana, Nepenthes 160
rajah, Culex 166
rajah, Nepenthes 166
rajah, Toxorhynchites 166
Ranatridae 17, 18, 102, 103
regelationis, Trichocera 175
regularis, Patiriella 218
remipes, Marilia 72, 73, 133
repens, Paspalum 118
Research Opportunities 24, 299, 311, 322
reticulata, Poecilia 232
Rhagovelia 24, 59, 120, 121
Rhaphidia 9
rheedii, Kandelia 212
Rhingia 192
Rhizophora 212
rhizophora, *Peliciera* 212
Rhyacophilidae 109, 111
ridingsii, Exyra 166
Rift Valley fever 259, 267
ringueleti, Sminthurides 5
river blindness 278, 279
Roproniidae 245
Rossianidae 112
Rotifera 295
Rotoitidae 245
rotumae, Aedes 280
rubra, Calliaspis 295
rubtzovi, Simulium 172
rudipellis, Vivipara javanica 250
rufipes, Suphisellus 85
rufopilosa, Fidena 156
rufulus, Suphisellus 85
rusticum, Sphaerodema 249

S

Sabethes 261
saileri, Metacnephia 172
Saldidae 9
salina, Artemia 223
salinarius, Chironomus 224
salinarius, Tendipes 224
Salmo 124

saltator, Trichocera 175
Salvinia 69, 85, 87, 88, 104, 105
Samea 116
Sameodes 116
samoensis, Rhizophora 212
sarae, Anodocheilus 84
sarasinorum, Anisus (Gyraulus) 250
Sarcophagidae 194
Sargossum 200
Sarracenia 161, 162, 166
Sarraceniaceae 161, 162, 165, 166
sativa, Oryza 75
sativa, Lactuca 118
Scapteriscus 129, 131
scapularis, Aedes 280
Scatella 287, 289
schedocyclius, Trichoprosopon 154
Schistosoma 320
Sciomyzidae 10, 22, 248, 249
Scirtidae 154, 295
Seira 6
semicrocea, Exyra 166
semipunctatum, Bembidion 65
sepulcralis, Eristalinus 192
sericeus, Halobates 200
Sericostomatidae 112
serricornis, Prionocyphon 154, 166
setiger, Tropisternus 155
Sialidae 9, 14, 96, 103
Sialis 103
Signiphoridae 245
silvestrii, Anodocheilus 84
Simuliidae 11, 21, 66, 96, 108, 112, 113, 120, 169,
    170, 171, 279, 306, 307
Simulium 11, 172, 307
sinensis, Tendora 129
sinuaticolle, Motschulskium 215
siqueirai, Leptagrion 156
Sisyborina 235
Sisyra 235–238
Sisyrella 235
Sisyridae 9, 14, 235, 238
Sisyrina 235
Smithuridae 6, 62
Smithurides 5, 62, 73, 74
smithi, Wyeomyia 166
sobrinus, Halobates 200
socialis, Hypogastrura 178
socius, Hydrocanthus 85
solstitialis, Toxorhynchites 156
sommermani, Cnephia 172
Sonneratia 212
sonorensis, Culicoides 278
spectabilis, Echinochloa 117
Sphaeroceridae 22, 58, 65, 97, 148, 156, 165, 196,
    229, 317
Sphaerodema 101, 249
Sphagnaceae 143
Sphagnum 143, 144
spinosa, Dimecoenia 217, 227
Spirodela 55, 74, 89
splendens, Halobates 200
Spongilla 236, 319

Spongillidae 235
squamosus, Anopheles 272
stagnalis, Lymnaea 250
stagnalis,Scatella 287
Staphylinidae 22, 65, 215
Stratiomyia 286
stratiotes, Pistia 105, 116
stegmaieri, Mosillus 216
Stenodiplosis 68
Stenopelmatidae 128, 131
Stenopsychidae 111
Stenotabanus 217
Stenus 65
stephensi, Anopheles 273
stratiotata, Parapoynx 74, 88, 116, 117
Stratiotes 74, 88, 116
subpictus, Anopheles 273
subsaltans, Mosillus 216
substriatus, Gyrinus 58
sulfur bacteria 315
Suphisellus 85
Suphus 85
Symphypleona 5, 58, 62
Synechococcus 288
Syritta 192
Syrphidae 18, 82, 102 154, 165, 166, 189, 191,
    192, 193, 194, 195, 295
Syrphinae 189

T

Tabanidae 10, 22, 196, 217, 295, 317
tagal, Ceriops 212
Tanaostigmatidae 245
Tanysphyrus 58, 74
Tasimiidae 112
tasmanica, Nothotrichocera 176
teatacealis, Neoshoenobia 247
Tendipes 224
Tendora 129
tentans, Chironomus 289
tenur, Neoascia 192
terebrella, Nothotrichocera 176
territans, Culex 171
Tetracampidae 245
Tetracanthella 214
Tetrigidae 88
Thalassomyia 219
theobaldi, Toxorhynchites 154, 156
thermarum, Scatella 289
Thinopinus 215
Thremmatidae 112
Thysanura 3
tibialis, Mosillus 216
tigrinus, Nepenthes 166
Tillandsia 149
Tipulidae 22, 96, 97, 196, 197
Tipuloidea 18
tlicifolius, Acanthus 212
Tonina 118
tonnoiri, Nothotrichocera 176
Tortricidae 166

Torymidae 245
Toxorhynchites 154, 156, 166
transfugus, Helophilus 192
Trapa 76, 81
travisi, Prosimulium 172
Trichocera 174, 177
Trichoceridae 173, 176, 314
Trichocerinae 176
Trichogrammatidae 245
Trichoprosopon 154
Trichoptera 10, 13, 20, 56, 96, 103, 106, 108,
    110–112, 115, 117, 166, 218, 306, 307
triphyllum, Myriophyllum 117
triseriatus, Aedes 154
trisulca, Lemna 88
tritaeniorhynchus, Culex 267
trivittatus, Helophilus 192
trivittata, Photuris 80
Tropisternus 84, 155
tuberosum, Simulium 172
tullbergi, Hypogastrura 214
turbida, Paracoenia 288
turionifera, Lemna 55, 89
Typha 191
Typhaceae 86

U

Uenoidae 112
uniformis, Mansonia 280
usneoides, Tillandsia 149
Utricularia 87, 88, 118, 143

V

vanduzeei, Coelopa 216
Vaccinium 143
vanduzeei, Wyeomyia 151, 156
Vanhorniidae 245
varipes, Psorophora 281, 282, 283
Veliidae 9, 16, 24, 59, 63, 96, 120, 154, 295
venustum, Simulium 172

Vertigopus 178
verticillata, Hydrilla 117
Victoria 56
viduata, Chrysogaster 192
vigilax, Aedes 280
violacea, Isotoma 178
viridis, Isotoma 62, 74
vittatum, Simulium 172
vivax, Plasmodium 270, 271, 273, 274
Vivipara 250
Viviparidae 250
Viviparus 250
volvulus, Onchocerca 278, 279
vriesianun, Leptagrion 156
vulgaris, Pinguicula 143
vulgaris, Utricularia 87, 88

W

Weberiella 101
wellcomei, Anopheles 280
West Nile virus, 259, 266
westerlundi, Vertigopus 178
Wolffia 55
Wucheria bancrofti 279, 280
Wyeomyia 151, 156, 166
wygodzinskyi, Dixella 156

X

Xantholinus 65
Xenylla 145, 178
Xiphocentronidae 111

Y

yellow fever 258, 259, 260, 261, 262, 263

Z

zika virus 259, 265
Zygoptera 9, 25, 78, 120, 121, 151

9 780367 781163